Topics in
Phosphate Chemistry

Topics in
Phosphate Chemistry

M-T Averbuch-Pouchot
A Durif

Lab.de Cristallographie–CNRS, Grenoble

World Scientific
Singapore • New Jersey • London • Hong Kong

Published by

World Scientific Publishing Co. Pte. Ltd.

P O Box 128, Farrer Road, Singapore 912805

USA office: Suite 1B, 1060 Main Street, River Edge, NJ 07661

UK office: 57 Shelton Street, Covent Garden, London WC2H 9HE

British Library Cataloguing-in-Publication Data
A catalogue record for this book is available from the British Library.

ISBN 981-02-2634-9

This book is printed on acid-free paper.

Printed in Singapore by Uto-Print

PREFACE

During the past century, much attention was devoted to the chemistry of phosphates. An important number of new materials were clearly characterized and the foundations of their present classification elaborated in their main lines. Some pioneer investigations of this period are still valuable and useful documents. Rather neglected in the first half of this century, phosphate chemistry developed rapidly during the past thirty years. This development can be explained by several factors. The progresses of chromatographic analysis and chiefly the improvement of X-ray diffraction structure analysis were fundamental tools to clarify, for instance, some problems of polymerization unsolved since the last century. But, this renewal of interest is also due to the applications of phosphate materials themselves. If their traditional use as fertilizers, detergents and food adjuvants induced a huge number of investigations, many other new domains were greatly improved thanks to new phosphate materials. Let us simply remember the great amount of investigations dealing with the ferroelectric properties of KDP and similar phosphates, the discovery of the first stoichiometric laser materials in the ultraphosphate family, the progresses made by non-linear optics since the discovery of KTP, the important development of biomaterials and phosphate glasses, the renew of interest for quartz-like piezoelectric compounds as $GaPO_4$, the very recent discovery of new efficient molecular sieves in the field of alumino- and gallo-phosphates... In the next future, one can expect the industrial development of asbestos ersatz, whose production is optimized for now more than ten years and the substitution of some traditionnal water soluble fertilizers by less polluting, slowly degradable phosphate materials.

Moreover, during these past three decades, fundamental progresses were performed in the knowledge of oligophosphates and large ring cyclophosphates. In this specific last domain, one can simply report that during the past ten years were optimized processes authorizing the convenient syntheses of cyclohexa-, cycloocta-, cyclodeca- and cyclododecaphosphates. The burgeoning of investigations in phosphate chemistry is clearly demonstrated by the following example. Among the twenty-six articles published in the February 1991 issue of "Journal of Solid State Chemistry", seven were reporting characterizations of new inorganic phosphates. This profusion of literature has sometimes negative consequences. The great majority of these investigations are of very good quality, but so many new non-conventional compounds were described that some anarchy appeared in the nomenclature, already very confusing since the past century. Most of the articles reporting the characterization of new phosphates include accurate determinations of the atomic arrangements, but only a few lines to describe more or less reproducible chemical preparations and very often nothing concerning the basic properties of the described compounds. The present system of transmission of the scientific information has certainly a part of responsability for this situation.

This rapid expansion would by itself justify a survey of the present state of the art, even if only to explain or repeat what is present classification and suggest the improvements necessary to clarify it. But, is it possible without reporting the present state of phosphate chemistry? So, a good part of this book will be devoted to illustrate this important aspect, with the help of some selected representative examples for this book cannot be an encyclopedia. Time for encyclopedia terminated when large computerized data banks appeared.

Several well documented reviews dealing with the development of phosphate or phosphorus chemistry or the state of the art in these matters have been published during the past forty years by Corbridge [1], Topley [2], Van Wazer [3], Thilo [4], Kalliney [5], Emsley and Hall [6] and Griffith *et al.* [7]. Unfortunately, these reviews were all published before the huge expansion of X-ray structural analysis and so, cannot report many fundamental results obtained through this technique. Several recent surveys dealing with more restricted topics of phosphate chemistry must also be reported. Crystal chemistry of oligophosphates [8], cyclohexaphosphates [9], ultraphosphates [10] and cyclophosphates [11] have been published by the authors, a general review of condensed phosphates by Durif [12] and a detailed survey of calcium phosphates by Elliot [13]. So, the chapter devoted to the "Present State of Phosphate Chemistry" will include only abridged reports for these kinds of phosphates, except when fundamental new results appeared since these surveys were published.

But, how explain what is phosphate chemistry without some words about phosphorus itself, its oxides and the corresponding acids? A good opportunity to discover that in spite of a gigantic amount of investigations accumulated during more than a century, the knowledge of these basic materials with some rare exceptions did not benefit of the major improvements of some modern techniques. Our knowledge of phosphorus for instance is still fragmentary and deceiving mainly in its structural aspects.

Along this survey, we also discovered the importance of the networks built by acidic phosphoric anions. An important chapter is devoted to their review.

The geometry of the main building unit of all phosphoric anions, the PO_4 tetrahedron was thoroughfully analyzed when it appears as an isolated entity in the case of monophosphates, but, up to now no survey of the main features of this entity in condensed anions exists. A wide number of accurate determinations of atomic arrangements of condensed phosphates is today performed, so, we did an attempt to examine this type of anions in a final chapter.

In many of its aspects, this book can appear as non satisfactory. The reader must not forget it has been written by two crystallographers, whose the natural tendency is to emphasize on the structural aspect of chemistry. This inclination may be sometimes benefic, since, for instance, the first chapters provide the opportunity to find along a few pages the present state of the structural knowledge of some basic materials as elemental phosphorus, phosphorus pentoxide and phosphoric acids whose data were, up to now, dispersed in literature and sometimes uncompletely or unadequately reported in textbooks.

Many things remain to be analyzed in such a wide field. May be the scope for a new book.

References

1. D. E. C. Corbridge, *The Structural Chemistry of Phosphorus,* (Elsevier Scientific Publishing Company, Amsterdam, 1966).

2. B. Topley, *Quart. Rew.* **3**, (1949), 345–368.

3. J. R. Van Wazer, *Phosphorus and its Compounds*, Vol. 1, (Interscience, New York, 1966).

4. E. Thilo, *Adv. Inorg. Chem. Radiochem.* **4**, (1962), 1–75.

5. S. Y. Kalliney, Cyclophosphates in *Topics in Phosphorus Chemistry*, Vol. 7, ed. E. J. Griffith and M. Grayson (Interscience, New York, 1972), p. 255–309.

6. J. Emsley and D. Hall, *The Chemistry of Phosphorus,* (Harper and Row Publishers, 1976).

7. *Environmental Phosphorus Handbook*, ed. E. J. Griffith, A. Beeton, J. M. Spencer, and D. T. Mitchell (Wiley and Sons, 1973).

8 M. T. Averbuch-Pouchot and A. Durif, Crystal Chemistry of Oligophosphates, in *Annu. Rev. Mater. Sci.*, **21**, (1991), 65–92.

9. M. T. Averbuch-Pouchot and A. Durif, Crystal Chemistry of Cyclohexaphosphates, *Eur. J. Solid State Inorg. Chem.*, **8**, (1991), 9–22.

10. M. T. Averbuch-Pouchot and A. Durif, Crystal chemistry of ultraphos-

phates, *Z. Kristallogr.*, **201**, (1992), 69–92.

11.   M. T. Averbuch-Pouchot and A. Durif, Crystal chemistry of cyclophos-
      phates, in *Stereochemistry of organometallic and inorganic Compounds*,
      Vol. 5, ed. P. Zanello (Elsevier, 1994), p. 1–160.

12.   A. Durif, *Crystal Chemistry of Condensed Phosphates,* (Plenum Press, to
      be published).

13.   J. C. Elliot, *Structure and Chemistry of the Apatites and Other Calcium
      Orthophosphates,* (Elsevier, 1994).

CONTENTS

Contents xv

CHAPTER 1

Elemental Phosphorus

1.1. INTRODUCTION

Phosphorus, the common component of all the compounds we shall describe in this book, merits well a short introductory chapter at the beginning of this survey so much the more that the complex behavior of this element is probably unique.

It is generally accepted that elemental phosphorus was for the first time isolated by Brand in 1669. However, Van Wazer [1] referring to a French book of the past century [2] attributes this discovery to Alchid El Bechil, an Arabian alchemist of the 12th century; Kabbaj [3] also reports the same possible origin for this discovery. We evidently have more details on the Brand's work. A good history of this discovery is reported by Lemoine [4].

It is now well established, for more than eighty years, that three forms of phosphorus exist, white, black, and red phosphorus. Behind this very simplified scheme, reality is much more complicated. In fact, each of these three forms has itself several crystalline modifications and in many cases the mechanisms of transformation are complex and sometimes not reproducible.

A huge amount of data have been accumulated on the three forms of phosphorus since the Brand's discovery. Nevertheless, we must confess that the knowledge of this element is today, in many of its aspects still confuse and fragmentary. For instance, data on the atomic arrangements, corresponding to these multiple crystalline forms are relatively rare. In 1965, Brown and Rundquist [5] when publishing a refinement of the orthorhombic form of black phosphorus could write: "Elementary phosphorus occurs in a number of modifications, but so far only the crystal structure of the black orthorhombic modification has been determined". Along the greatest part of this section, we summarize and gather well-known data that one can find in any valuable text book. So, we restrict the list of references to articles dealing with the structural aspects of phosphorus chemistry, generally not clearly or incompletely reported in chemistry books.

1.2. PREPARATION AND USES

The preparation of elemental phosphorus is possible by reduction of any kind of compounds containing phosphorus. In the early time of phosphorus chemistry, this production was conducted by reduction of a mixture of dessicated urines, sand and charcoal. Now, the method is not fundamentally different, but the starting materials are various phosphate rocks reduced by a mixture of sand and coke in an electric furnace. Being given the diversity of compostions of phosphate rocks, it is not possible to summarize this process by a general equation. We try below to imagine what is probably such a mechanism when using a pure hydroxyapatite:

$$Ca_5[OH](PO_4)_3 + 5SiO_2 + 7C \longrightarrow 5CaSiO_3 + 7CO + 1/2H_2O + 3P$$

Things are in fact much more complicated for most of the phosphate rocks used in this industry are fluor-apatites and a great amount of gaseous fluorinated residuals have to be carefully eliminated. During this operation, the temperature of the electric furnace is kept close to 1673–1773 K as to obtain the melting of the slag ($CaSiO_3$). The vapors of phosphorus so obtained are condensed and stocked under water. The operation is costly, requiring more than 10,000 KWh per ton of phosphorus.

At the laboratory scale, one can produce elemental phosphorus by reduction of other types of phosphates. For instance, lead or bismuth phosphates can be reduced by hydrogen at temperatures lower than 850 K.

For a long time, the main use of phosphorus was the production of matches, but this activity decreases rapidly. In France, for instance, the last phosphorus factory closed in 1966. In fact, little elemental phosphorus is required as such for industrial purposes. Most of the production is converted into other materials as PCl_3, P_4S_{10}, $POCl_3$ and P_4O_{10}.

1.3. THE VARIOUS FORMS

1.3.1. White Phosphorus

This form discovered by Brand, appears as a waxy translucid matter burning spontaneously when in contact with the oxygen of air. Soluble in several classical solvents as benzene and carbon disulphide, white phosphorus does not react with water and can be easily stocked, melted or transported when protected by a layer of water. This form is very poisonous, the lethal dose being as low as 50 mg.

The very low values measured for its melting (317.1 K) and boiling temperatures (553 K) as well as its solublity in organic solvents suggest that this form corresponds to a molecular compound.

White phosphorus is the most reactive of the three phosphorus forms and reacts with most of the elements.

It oxidizes spontaneously and vigorously when in contact with oxygen, and also reacts with carbon dioxide. In both cases the final product is P_4O_{10}:

$$P_4 + 5O_2 \longrightarrow P_4O_{10}$$

$$P_4 + 10CO_2 \longrightarrow P_4O_{10} + 10CO$$

The reaction with oxygen has been extensively investigated for in some conditions, reduced pressure of oxygen for instance, it is associated with a greenish glow. In spite of many investigations, this phenomenon which appears as complex, is not yet clearly understood.

White phosphorus has at least two crystalline forms. In normal conditions of temperature and pressure, the stable form is usually called α. At 195.2 K, this variety transforms reversibly into the β-form. This second form having a high birefringence, the transformation can be easily observed with a polarizing microscope. Since the discovery of X-ray diffraction, many unit cells and possible atomic arrangements were proposed for these two forms. A good number of these measurements based on powder data are probably incorrect. Today, it seems well established that the room temperature form (α) is body-centred cubic with a unit cell of 18.61 Å and $I\bar{4}3m$ as possible space group. According to Corbridge [6], the atomic arrangement is probably a stacking of fifty eight P_4 units arranged inside the unit cell as manganese atoms in the α-form of this metal. The β-form is said to be hexagonal, but no structural data exist.

1.3.2. Red Phosphorus

About red phosphorus, discovered in 1847 by Von Schrotter, the present state of our knowledge is also fragmentary and still confuse. Several processes can be used to transform white phosphorus into red phosphorus. On can obtain red phosphorus by heating the white form at 525–550 K in a sealed vessel. This transformation is accelerated if traces of iodine are added to white phosphorus. To the difference with the white form, red phosphorus does not ignite when in contact with oxygen and cannot be dissolved in the classical solvents of white phosphorus. Its melting temperature, 863–873 K, is considerably higher than that of the white form. All these basic properties suggest that this variety is not a molecular compound. The structural investigations confirm this hypothesis. When melted, red phosphorus leads to an amorphous mass becoming polycrystalline when slowly reheated. During this recrystallization, several forms occur. Chemical literature reports at least five forms of red phosphorus, but their preparations appear more or less reproducible. The atomic arrangement of one of these forms, often said Hittorf's form or "violet form", has been accurately determined. This form of phosphorus is monoclinic, $P2/c$, with $Z = 84$ and the following unit-cell dimensions:

$$a = 9.27, \quad b = 9.17, \quad c = 22.61 \text{ Å}, \quad \beta = 106.1°$$

The crystal structure determination has been performed by Thurn and Krebs [7]. The crystal, used by these authors, was obtained by slow cooling of a melt containing 1 g of white phosphorus dissolved in 30 g of lead. This mixture was first kept for a short time at 903 K and then slowly cooled (10 K/day) down to 793 K. Crystals appear as platelets after elimination of the lead flux. From this structure determination, the calculated density is 2.361.

The atomic arrangement of the Hittorf's form is rather complicated. The phosphorus atoms assemble as to build tubes of pentagonal cross-section.

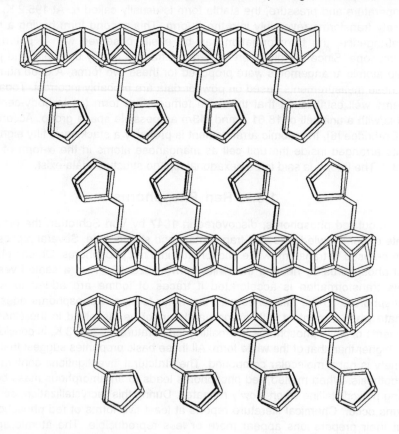

Figure 1.3.1. Projection, along the [1$\bar{1}$0] direction, of the Hittorf form of red phosphorus. Only the P–P bonds are represented by double lines. The **c** axis is vertical.

These tubes are themselves arranged in layers perpendicular to the **c** direction and organized around the planes $z = (2n+1)/8$. There are two types of layers. Those located around the planes $z = 1/8$ and $7/8$ are built by tubes parallel to the [110] direction, while the layers built around the planes $z = 3/8$ et $5/8$ are made by the same type of tubes, but parallel to the [1$\overline{1}$0] direction. Figure 1.3.1, a projection along the [1$\overline{1}$0] direction, explains well this situation. Layers located around the planes $z = 1/8$ and $3/8$ are connected by P–P bonds as well as those around planes $z = 5/8$ and $7/8$, but no bonding exists between the layers located around planes $z = 3/8$ and $5/8$. Details of the linkage of phosphorus atoms in a pentagonal channel is illustrated by Figure 1.3.2. Each phosphorus atom involved in this arrangement is connected to three neighbors with P–P distances spreading between 2.178 to 2.299 Å. The P–P average distance is 2.219 Å and the P–P–P average angle is 101°.

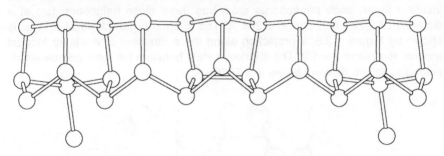

Figure 1.3.2. Details of the linkage of phosphorus atoms in a pentagonal channel.

1.3.3. Black Phosphorus

The black form, discovered by Bridgeman in 1914, can be considered as a high-pressure form of the white phosphorus. At room temperature, white phosphorus transforms into the black form under 35,000 atm. At 473 K, this same transformation ocurs with only 12,000 atm. Another way to produce black phosphorus is to heat a mixture of mercury and white phosphorus seeded with crystals of the black form.

Black phosphorus is not soluble in the classical solvents of the white form suggesting this form is not molecular, but a polymeric arrangement. In addition, its graphite-like aspect suggests a layer organization. It is the most stable of the three forms of phosphorus and is chemically the least reactive. For instance, unlike white phosphorus, it does not ignate spontaneously. It is also the densest. Under normal conditions of pressure, at 823 K, black phosphorus transforms irreversibly into red phosphorus.

According to various authors, this form of phosphorus has two or three crystalline forms. In fact, only one of them, an orthorhombic one, is structurally well-established. This form of black phosphorus can be prepared as single crystals by dissolving phosphorus in liquid bismuth. Crystals so obtained appear as needles. Its crystal structure was first investigated by Hultgren *et al.* [8], Gingrich and Hultgren [9–10] and Hultgren and Warren [11–12] and later on refined by Brown and Rundquist [5]. Its orthorhombic unit cell, with:

$$a = 3.314, \quad b = 4.376, \quad c = 10.477 \text{ Å}$$

contains eight phosphorus atoms and the space group is *Bmab*.

Phosphorus atoms are located in layers perpendicular to the **c** direction. These layers are organized around the planes $z = 0$ and 1/2. The two projections reported below, Figure 1.3.3 along the **a** direction and Figure 1.3.4 along the **b** direction, show clearly the layer organization, but do not exhibit that, inside a layer, each phosphorus atom has three close neighbors, two at a distance of 2.224 and one at 2.244 Å. The internal organization of a layer is shown by Figure 1.3.5, a projection along the **c** direction of the layer located around the plane $z = 1/2$. The shorter contact between two phosphorus atoms belonging to two different layers is 3.592(2) Å.

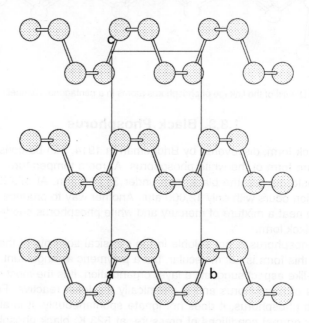

Figure 1.3.3. Projection, along the **a** direction, of the orthorhombic form of black phosphorus.

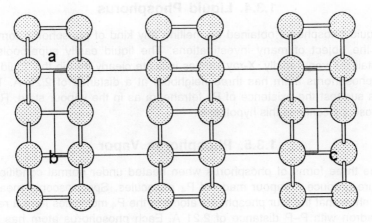

Figure 1.3.4. Projection, along the **b** direction, of the orthorhombic form of black phosphorus.

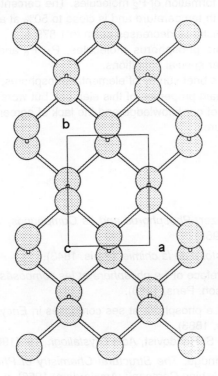

Figure1.3.5. Projection, along **c**, of the phosphorus layer located around the plane $z = 1/2$.

1.3.4. Liquid Phosphorus

Liquid phosphorus obtained by melting any kind of phosphorus form has been the object of many investigations. The liquid easily supercools and recrystallizes very rapidly. X-ray studies indicate clearly that in the liquid state each phosphorus atom has three neighbors at a distance of 2.25 Å. These results suggest the existence of P_4 tetrahedra as in the vapour state. Raman spectroscopy confirms this hypothesis.

1.3.5. Phosphorus Vapor

The three forms of phosphorus when heated under normal conditions of pressure produce a vapour made of P_4 molecules. Spectroscopic measurements show that the four phosphorus atoms in the P_4 molecules form a regular tetrahedron with P–P distance of 2.21 Å. Each phosphorus atom has three neighbors and the P–P–P angles are evidently 60°. Above 1073 K, a dissociation occurs with the formation of P_2 molecules. The percentage of diatomic molecules increases with temperature and is close to 50% at about 2000 K. In this molecule, the P–P distance decreases down to 1.875 Å.

Other polyatomic phosphorus molecules, P_3, P_6 and P_8, seem to have been observed under special conditions.

At the end of this brief survey of elemental phosphorus, we tried to summarize in a table the main properties of this element, but were discouraged by the fragmentary aspect of our knowledge and the lack of agreement between some data found in literature.

References

1.   J. R. Van Wazer, *Phosphorus and its Compounds*, Vol. 1, (Interscience Publishers, 1966).

2.   F. Hoefer, *Histoire de la chimie*, (Paris, 1843).

3.   M. Kabbaj, Preface of *Le phosphore et les composés phosphorés*, ed. R. Dumon (Masson, Paris, 1980).

4.   G. Lemoine, Le phosphore et ses composés in *Encyclopédie Chimique*, (Dunod, Paris, 1883).

5.   A. Brown and S. Rundqvist, *Acta Crystallogr.*, **19**, (1965), 684–685.

6.   D. E. C. Corbridge, *The Structural Chemistry of Phosphorus*, (Elsevier Scientific Publishing Company, Amsterdam, 1966), p. 13–24.

7. H. Thurn and H. Krebs, *Acta Crystallogr.*, **B25**, (1969), 125–135.

8. R. Hultgren, N. S. Gingrich, and B. E. Warren, *J. Chem. Phys.*, **3**, (1935), 351–355.

9. N. S. Gingrich and R. Hultgren, *Phys. Rev.*, **47**, (1935), 808.

10. N. S. Gingrich and R. Hultgren, *Bull. Am. Phys. Soc.*, **10**, (1935), 30.

11. R. Hultgren and B. E. Warren, *Phys. Rev.*, **47**, (1935), 808.

12. R. Hultgren and B. E. Warren, *Bull. Am. Phys. Soc.*, **10**, (1935), 30.

CHAPTER 2

Phosphorus Oxides

2.1. INTRODUCTION

A series of phosphorus oxides, now well-known, range between the compositions P_4O_{10} and P_4O_6. All correspond to adamantine-like molecules. The evolution from P_4O_{10} to P_4O_6 can be schematized by a progressive removal of the four unshared oxygen atoms of P_4O_{10}. So, one obtains P_4O_{10}, P_4O_9, P_4O_8, P_4O_7 and P_4O_6, successively. Figure 2.1.1 represents this set of molecules. Evidently in this set of oxides, the valency of phosphorus varies from V in P_4O_{10} to III in P_4O_6, the intermediate oxides corresponding to a mixture of valencies.

This book being devoted to phosphate chemistry, we shall limit in this section our description to the pentoxide P_4O_{10}. In another section dealing with some substituted phosphates (thiocyclophosphates), we shall bring attention with more details to some $P_4O_{10-x}S_x$ sulfur-substituted compounds used as starting materials for the preparation of these thiocyclophosphates.

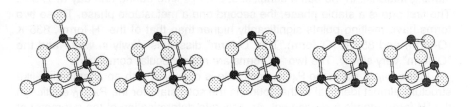

Figure 2.1.1. The five phosphorus oxydes: P_4O_{10}, P_4O_9, P_4O_8, P_4O_7 and P_4O_6.

2.2. PREPARATION

P_4O_{10} is normally obtained by burning white phosphorus in a dry oxidizing atmosphere. P_4O_{10} is the main product of this combustion. An almost theoretical yield can be obtained if the reaction is performed in an excess of oxygen.

The vapours of P_4O_{10} are condensed in a cold tank and a further purification can be made by sublimation in an atmosphere of oxygen. The form of P_4O_{10} produced by this reaction is that usually called "H-form".

2.3. PROPERTIES

Two other forms of P_4O_{10} are presently well-known. They have rather different properties which can be easily interpreted by the examination of their atomic arrangements.

Among these three forms, the first one known as the *"H-form"*, is a molecular compound. It is very hygroscopic, colorless, and melts at about 700 K only when rapidly heated, for it sublimes from 635 K. Its density is 2.30. This form of P_4O_{10} transforms rapidly into monophosphoric acid when in contact with water at room temperature or above:

$$P_4O_{10} + 6H_2O \longrightarrow 4H_3PO_4$$

but, under controlled conditions, cyclotetraphosphoric acid can be obtained:

$$P_4O_{10} + 2H_2O \xrightarrow{\ T < 280\,K\ } H_4P_4O_{12}$$

Two other forms commonly called *"O- and O'-forms"* can be prepared from the "H-form". The "O-form" can be obtained by heating the "H-form" for about two hours at 673 K in a closed vessel, the "O'-form" by heating the same starting material, in the same conditions, but this time during one day at 723 K. The first one is a stable phase, the second one a metastable phase. These two forms have melting points significantly higher than that of the "H-form", 835 K (O-form) and 853 K (O'-form). The "O-form" dissolves slowly in water and the "O'-form" very slowly. The two "O" forms are not molecular compounds.

In the *vapor state*, the P_4O_{10} molecule is observed with an adamantine-like structure almost identical to that found in the solid state for the P_4O_{10} entities in the "H form" atomic arrangement. An accurate determination of the geometry of this molecule was performed by Beagley *et al.* [1]. Various glassy forms were also reported.

2.4. USES OF PHOSPHORUS OXIDE

The main use of P_4O_{10}, in its "H-form", is the "dry process" production of phosphoric acid we describe in the next chapter, but some other applications are to be reported.

Its great avidity for water makes of this compound one of the best dessicant. This property is widely used in chemistry laboratories for the production of electric bulbs, and in the synthesis of acrylic resins. Some processes of organic polymerizations and the hardening of asphalts use also this property.

P_4O_{10} is also used for its catalytic properties in petroleum industry and as a fundamental starting material for the production of pesticides, organic phosphates, dyes, enamels and drugs.

2.5. ATOMIC ARRANGEMENTS

The atomic arrangements of the three forms of P_4O_{10} are today well-established. The structure of the "H-form" is a simple stacking of P_4O_{10} molecules. The "O-form" or "stable form" is a three dimensional network, while the third one, "O'-form" or "metastable form" is a layer structure. A common feature of these three forms is that all the PO_4 tetrahedra, involved in these arrangements, share three of their oxygen atoms with three different adjacent tetrahedra.

2.5.1. The "H-form"

This form is rhombohedral, $R3c$, with a bimolecular unit cell:

$$a = 7.43 \text{ Å} \qquad \alpha = 87.0°$$

In the hexagonal setting, Z becomes 6 and the unit-cell dimensions are:

$$a = 10.31 \qquad c = 13.3 \text{ Å}$$

The atomic arrangement was first determined by de Decker and MacGillavry [2] and later refined by Cruikshank [3]. A perspective view of an isolated P_4O_{10} entity is given in Figure 2.5.1. Such a groupement has a ternary symmetry. One of the four phosphorus atoms and its unshared oxygen atom are both located on the ternary axis. Inside such a group, the P–O distances corresponding to the P–O–P bonds range between 1.53 and 1.60 Å, while these

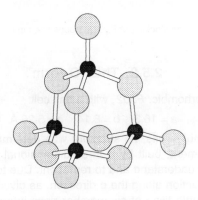

Figure 2.5.1. Perspective view of an isolated P_4O_{10} group.

distances are 1.41 and 1.51 Å for the P–O bonds involving the unshared oxygen atoms. The two P–P distances, 2,79 and 2,80 Å, are very similar, so that the tetrahedron built by the four phosphorus atoms is almost regular. Figure 2.5.2 represents the stacking of the P_4O_{10} groups in projection along the **a** axis of the rhombohedral unit cell.

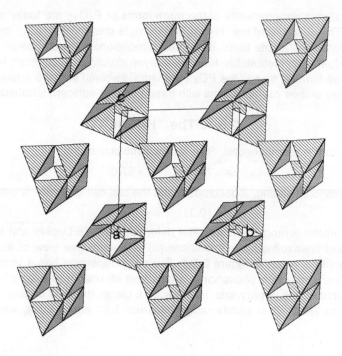

Figure 2.5.2. Projection of the stacking of P_4O_{10} groups seen in projection along the **a** direction of the rhombohedral cell.

2.5.2. The "O-form"

This form is orthorhombic, *Fdd*2, with a unit cell:

$$a = 16.3, \quad b = 8.14, \quad c = 5.26 \text{ Å}$$

containing eight formula units. Its structure was determined by de Decker [4]. This atomic arrangement, built by a three-dimensional network of PO_4 tetrahedra, is not easy to understand and to represent. Due to the high symmetry of the structure, its projection along the **c** direction, as given in Figure 2.5.3, could appear as a very simple tiling of six-member rings interconnected by common oxygen atoms. In fact, each pseudo-ring is the projection of a spiral of corner-

Figure 2.5.3. Projection, along the **c** direction, of the atomic arrangement in the "O-form" of P_4O_{10}.

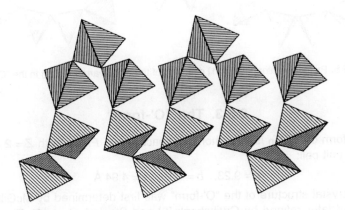

Figure 2.5.4. Projection, along the **b** direction, of an isolated spiral.

sharing tetrahedra, spreading along the **c** axis, with a period of six tetrahedra. An isolated spiral, seen in projection along the **b** direction, is given in Figure 2.5.4. So, the best way to understand and describe this arrangement is to consider it as built by a stacking of these spirals interconnected along the **a** and **b** directions by sharing oxygen atoms. Another surprising aspect of this complicated arrangement is provided by its projection along the [011] direction reported in Figure 2.5.5. This time, it appears as a compact tiling of ten-member rings sharing some oxygen atoms. Inside this arrangement, built by only one crystallographically independent tetrahedron, the three P–O distances involved in P–O–P bonds are 1.61, 1.61 and 1.62 Å, while the unshared oxygen atom corresponds to a P–O distance of 1.39 Å. The two P–P distances are 2.79 and 2.92 Å.

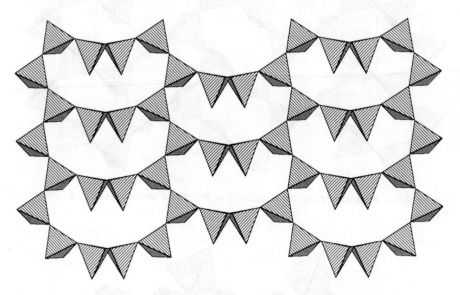

Figure 2.5.5. Projection, along the [011] direction, of the atomic arrangement in the "O-form" of P_4O_{10}.

2.5.3. The "O'-form"

This form of P_4O_{10} is orthorhombic, space group *Pnam*, with $Z = 2$ and the following unit cell:

$$a = 9.23, \quad b = 7.18, \quad c = 4.94 \text{ Å}$$

The crystal structure of the "O'-form" was first determined by McGillavry *et al.* [5] and later refined by Cruikshank [6] and Stachel *et al.* [7]. The atomic arrangement is a typical layer organization. Figure 2.5.6, a projection along the **c** direction, shows clearly how thick layers perpendicular to the **a** direction alternate along this direction. In fact, wo such layers, organized around the planes x = 1/4 and 3/4, cross the unit cell. The internal organization of such a layer is explained by Figure 2.5.7 which reports the projection along the **a** direction of an isolated layer. On can describe such an arrangement as a tiling of rings made by six PO_4 groups or consider it as built by a succession of infinite $(PO_3)_n$ chains, spreading parallel to the **b** direction, and interconnected

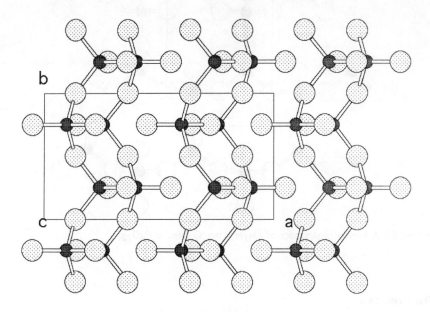

Figure 2.5.6. Projection, along the **c** direction, of the "O'-form" of P_4O_{10}.

in the **c** direction by some common oxygen atoms. The two independent phosphorus atoms are located on the mirror plane, y = 1/4, and so all the PO_4 tetrahedra involved in this arrangement adopt this symmetry. As in all the other forms of P_4O_{10}, each PO_4 tetrahedron shares three oxygen atoms with three different adjacent neighbors. Inside the layer, the interatomic distances and bond angles are not fundamentally different of the ones observed in various condensed phosphoric anions. Here, the P–O distances involved in the P–O–P bonds spread between 1.565 and 1.574 Å and those corresponding to the unshared oxygen atoms are 1.432 and 1.451 Å. The two P–P distances are 2.952 and 2.983 Å, significantly longer than in the other forms.

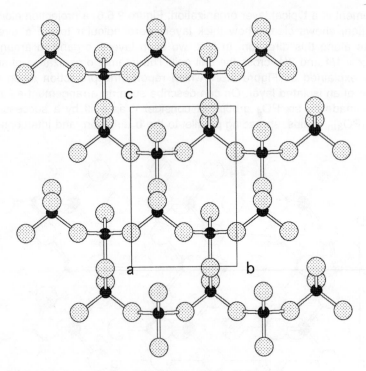

Figure 2.5.7. Internal organization of a layer in the "O'-form" of P_4O_{10}.

References

1.   B. Beagley, D. W. J. Cruickshank, and T. G. Hewitt, *Trans. Faraday Soc.*, **63**, (1967), 836–845.

2.   H. C. J. de Decker and C. M. MacGillavry, *Rec. Trav. Chim. Pays-Bas*, **60**, (1941), 153–175.

3.   D. W. J. Cruickshank, *Acta Crystallogr.*, **17**, (1964), 677–679.

4.   H. C. J. de Decker, *Rec. Trav. Chim. Pays-Bas*, **60**, (1941), 413–427.

5.   C. M. MacGillavry, H. C. J. de Decker, and L. M. Nijland, *Nature*, **164**, (1949), 448–450.

6.   D. W. J. Cruickshank, *Acta Crystallogr.*, **17**, (1964), 679–680.

7.   D. Stachel, I. Svoboda, and H. Fuess, *Z. Kristallogr.*, (1994), in press.

CHAPTER 3

Phosphoric Acids

3.1. INTRODUCTION

Many phosphoric acids could be described in this section, but this book being essentially devoted to the crystal chemistry of phosphates, we deliberately limit our survey to the basic one, H_3PO_4, its hemihydrate and to phosphorous acid which corresponds to the simplest substitution in the PO_4 tetrahedron:

$$[PO_4]^{3-} \longrightarrow [PO_3H]^{2-}$$

The numerous salts of this latter acid, the phosphites, are, in our opinion, belonging to the family of monophosphates. Hypophosphorous acid, $H(PO_2H_2)$, corresponding to the substitution of two oxygen atoms by two hydrogen atoms in the anionic entity also exists, but very few of its derivatives have been studied.

A good part of our survey of phosphate chemistry will concern condensed phosphates, salts deriving from the various polymeric forms of H_3PO_4, but in this field, to the exception of diphosphoric acid, $H_4P_2O_7$, the corresponding acids have not been isolated and crystallized. Nevertheless, these condensed phosphoric acids can be prepared in aqueous solutions by action of ion-exchange resins on the corresponding alkali salts or in the case of cyclotetraphosphoric acid by low-temperature hydrolysis of P_4O_{10}:

$$P_4O_{10} + 2H_2O \xrightarrow{\text{T<280 K}} H_4P_4O_{12}$$

3.2. MONOPHOSPHORIC ACID

3.2.1. Preparation and General Properties

Monophosphoric acid was first reported, as early as 1680, by Boyle who noticed the acidic properties of the liquid obtained by action of water on the products of combustion of phosphorus. Various preparations of phosphoric acid are known, but today two types of processes are dominating the industrial production. They correspond to very different paths.

For the first one, commonly called the *wet process*, the starting material is a calcium salt. Calcined bones, used for a long time, are now replaced by phosphate rocks. In this process, H_3PO_4 is obtained by action of H_2SO_4 on the calcium salt. Schematically, with a fluor-apatite, the reaction could be written:

$$Ca_5(PO_4)_3F + 5H_2SO_4 + 10H_2O \longrightarrow 3H_3PO_4 + HF + 5CaSO_4.2H_2O$$

The sparingly soluble calcium sulphate is removed by filtration. The acid produced by this way is far from pure. A large number of processes have been described to eliminate the residual impurities.

The second process, called the *dry process*, based on the hydration of phosphoric anhydride, is more costly and only used to obtain a pure phos-

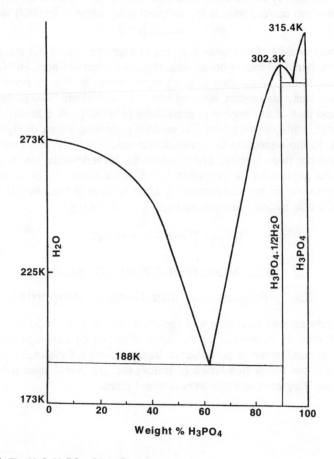

Figure 3.2.1. The H_2O–H_3PO_4 phase diagram.

phoric acid necessary for laboratory experiments, manufacture of food products and drugs, for instance. P_4O_{10} obtained, as mentioned in Chapter 2, by the oxidation of white phosphorus is hydrated immediately after its formation :

$$P_4O_{10} + 6H_2O \longrightarrow 4H_3PO_4$$

A treatment with H_2S is sometimes necessary to remove arsenic contamination by precipitation of arsenic sulfide. The phosphoric acid obtained by this process is also known as the "thermal acid".

The greatest part of phosphoric acid, 75–80%, is produced by the wet process. During certain periods, the production by the dry process increased for economical reasons governed by the ratio between the cost of H_2SO_4 and the cost of electricity.

In crystalline state, phosphoric acid exists as H_3PO_4 and as an hemihydrate, $H_3PO_4.1/2H_2O$. Another hydrate, $10H_3PO_4.H_2O$, was also reported, but its existence was not confirmed. This hydrate does not appears in the H_2O–H_3PO_4 phase-equilibrium diagram which has been carefully investigated. This diagram is presented in Figure 3.2.1. Monophosphoric acid is a colorless solid melting at 315.3 K. A deuterated derivative can be prepared by action of P_2O_5 on D_2O or hydrolysis of $POCl_3$ by D_2O. Deuterophosphoric acid, D_3PO_4, has properties slightly different of those of normal phosphoric acid. This compound was used for the preparation of various deutereted KDP-like compounds for neutron-diffraction experiments.

3.2.2. Uses of Phosphoric Acid

Phosphoric acid is used largely for the production of phosphate fertilizers as well as of various derivatives with specific uses, detergents, fire retardant, tooth paste, human and animal foods, water softening. Other applications are in the beer and cocacola industries and as an additive to some animal foods. We shall examine along our survey the different uses of phosphates.

The free acid itself has many industrial applications. An important one is its use for metal treatment against corrosion (phosphatizing). The chemical polishing of aluminum and the electropolishing of stainless steel articles are also to be mentioned.

3.2.3. Atomic Arrangements of Phosphoric Acids

3.2.3.1. H_3PO_4

The crystal structure of H_3PO_4 was the object of several investigations by Furberg [1–2] and Smith *et al.* [3]. It seems now well established that only one crystalline form exists. H_3PO_4 is monoclinic, $P2_1/c$, with $Z = 4$ and the following

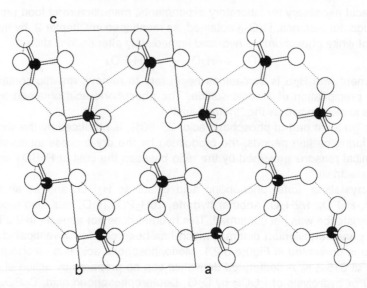

Figure 3.2.2. Projection of the structure of H_3PO_4 along the **b** direction. H–bonds are figured by dotted lines. This view induces the feeling that some H–bonds connect two O atoms of the same phosphoric tetrahedron, but they are, in fact, established between two superimposed tetrahedra.

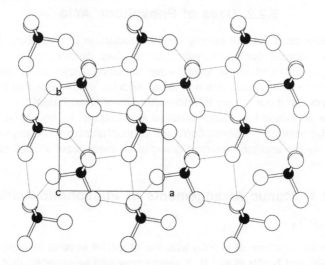

Figure 3.2.3. Projection of a layer of H_3PO_4 along the **c** direction. Hydrogen bonds are figured by dotted lines.

unit-cell dimensions:

$$a = 5.80, \quad b = 4.85, \quad c = 11.62 \text{ Å}, \quad \beta = 95.20° \text{ (3)}$$

Inside the PO_4 tetrahedron, one observes one P–O distance of 1.52 Å and three longer ones (1.57 Å) corresponding to the P–OH bonds. The H_3PO_4 entities are interconnected by two types of hydrogen bonds, two strong ones (O–O = 2.53 Å) and a weak one (O–O = 2.84 Å). As shown by Figure 3.2.2, a projection along the **b** direction, the atomic arrangement is a layer organization. Layers spread in the (a, b) planes perpendicular to the **c** direction. Two layers, approximately centred at planes $z = 1/4$ and $3/4$, cross the unit cell. No hydrogen bond interconnects these layers. Hydrogen atoms were not located by the authors, thus hydrogen bonds are deduced from considerations based on the P–O and O–O distances observed in the arrangement. Figure 3.2.3 reports, in projection along the **c** direction, the internal organization of an isolated layer.

3.2.3.2. $H_3PO_4.1/2H_2O$

$H_3PO_4.1/2H_2O$ is monoclinic, $P2_1/a$, with $Z = 8$ and the following unit-cell dimensions:

$$a = 7.92(1), \quad b = 12.99(2), \quad c = 7.47(1) \text{ Å}, \quad \beta = 109.9(1)°$$

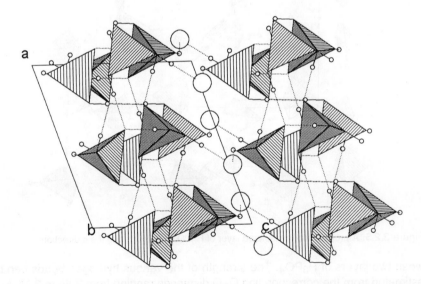

Figure 3.2.4. A projection of the structure of $H_3PO_4.1/2H_2O$ along the **b** direction. Large empty circles represent water molecules and small ones the hydrogen atoms. The PO_4 tetrahedra are given in polyhedral representation. Hydrogen atoms of the water molecules have been omitted.

Its crystal structure was determined almost simultaneously by Mighell *et al.* [4], Mootz *et al.* [5] and Mootz and Goldmann [6]. This atomic arrangement can be described as that of H_3PO_4 by a layered organization. Thick layers built by the H_3PO_4 molecules alternate with sheets of water molecules perpendicular to the **c** direction. This stacking is reported by Figure 3.2.4, a projection along the **b** direction. The organization inside a layer of H_3PO_4 molecules is given by Figure 3.2.5, showing how each H_3PO_4 entity is connected by hydrogen bonds to its four adjacent neighbors. The apparently non connected hydrogen atoms correspond in fact to the hydrogen bonds connecting this layer to the adjacent sheets of water molecules. As in the anhydrous form, no H–bond exists bet-

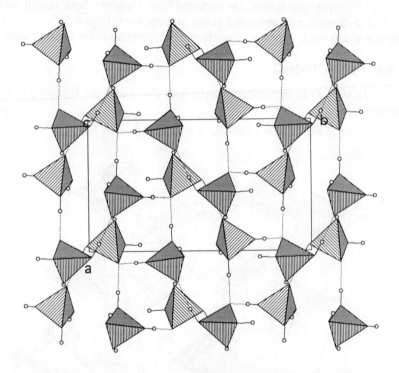

Figure 3.2.5. A projection of an isolated layer of $H_3PO_4.1/2H_2O$ along the **c** direction.

ween two layers of H_3PO_4. The strength of the various hydrogen bonds can be estimated from the corresponding O–O distances ranging from 2.55 to 2.71 Å.

The corresponding arsenic acid hemihydrate, $H_3AsO_4.1/2H_2O$, is isotypic. Its crystal structure was described by Worzala [7].

3.3. PHOSPHOROUS ACID

3.3.1. Chemical Preparation and Properties

The hydrolysis of PCl_3 remains the only convenient process of preparation for this acid:

$$PCl_3 + 3H_2O \longrightarrow H_2(PO_3H) + 3HCl$$

Phosphorous acid is a colorless deliquescent solid, melting at 343.1 K. Its density is 1.65. This acid, as well as its salts, usually named phosphites, are strong reducing agents. Chromic and sulphuric acids, for instance, can be reduced according to the following schemes:

$$H_2(HPO_3) + H_2SO_4 \longrightarrow H_3PO_4 + SO_2 + H_2O$$

$$3H_2(HPO_3) + 2H_2CrO_4 \longrightarrow 3H_3PO_4 + Cr_2O_3 + 2H_2O$$

The possible existence of phosphorous in two tautomeric forms:

$$P(OH)_3 \longleftrightarrow H_2(PO_3H)$$

was a matter of discussion for some time. The atomic arrangements of nume-

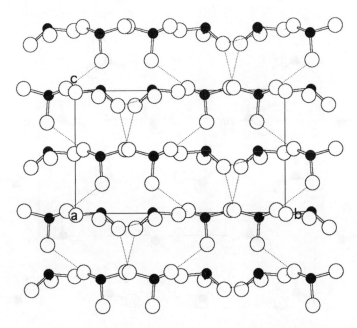

Figure 3.3.1. Projection of the structure of phosphorous acid along the **a** direction. Black circles figure the P atoms, empty ones the O atoms. Hydrogen bonds are represented by dotted lines.

rous phosphites as well as that of phosphorus acid itself show clearly the existence, in the solid state, of a PO₃H tetrahedral anionic group.

3.3.2. Atomic Arrangement of $H_2(PO_3H)$

Phosphorus acid is orthorhombic with a unit cell:

$$a = 7.27, \quad b = 12.06, \quad c = 6.85 \text{ Å}$$

containing eight formula units. The atomic arrangement was described by Furberg and Landmark [8], using space group $Pna2_1$. The structure is a three-dimensional network of PO₃H groups interconnected by hydrogen bonds. As shown in Figure 3.3.1, a projection along the **a** direction, the structure can be described by a set of PO₃H layers perpendicular to the **c** direction and located around the planes $z = 0$ and 1/2. These layers are interconnected by hydrogen bonds.

The details of the internal atomic arrangement inside a layer are given in Figure 3.3.2, which shows a projection of an isolated layer along the **c** direction. The main feature is the existence of wide ribbons of PO₃H entities running parallel tó the **a** direction. In the two crystallographically independent PO₃H

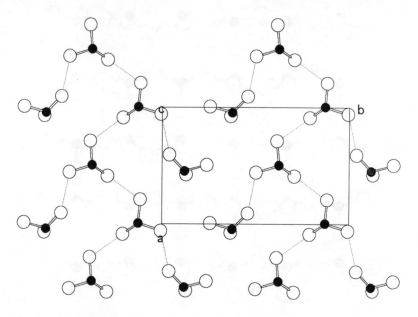

Figure 3.3.2. Projection, along the **c** direction, of the layer located around the plane $z = 1/2$. Conventions for the representation are similar to those used in Figure 3.3.1.

groups, the P–O distances corresponding to the P–OH bonds are between 1.526 and 1.552 Å, while the two P–O bond distances are 1.451 and 1.485 Å. The hydrogen atoms were not located by the authors, but the O–O distances corresponding the H–bonds indicate clearly that they are strong (between 2.58 and 2.63 Å).

3.4. OTHER PHOSPHORIC ACIDS

In many phosphoric derivatives, the anionic entities are condensed or substituted. In the field of *condensed phosphoric acids*, very little is known. The chemical literature reports many unsuccessful attempts made in the last century to isolate various "metaphosphoric acids". To date, using exchange resins, most of these condensed acids can be obtained in aqueous solutions. One of them, $H_4P_4O_{12}$, is commonly produced by the low-temperature hydrolysis of P_4O_{10}. With the exception of diphosphoric acid, they have never been crystallized. Even for this latter acid, the state of our knowledge is still confused and the reported data contradictory. For some authors, diphosphoric acid is a crystalline solid melting at 334 K, for others, it has two crystalline forms, one melting at 327.3, the second at 344.5 K. Thus, one must be circumspect about the two sets of unit-cell dimensions found in the literature for these two modifications:

$$\text{Modification I} \quad a = 13.69, \quad b = 20.08, \quad c = 6.49 \text{ Å}$$
$$\text{Modification II} \quad a = 11.05, \quad b = 19.21, \quad c = 10.40 \text{ Å}$$

Among the acids corresponding to *substituted monophosphates*, the well-known phosphorous acid that we described above can be easily crystallized. The two fluoromonophosphoric acids, H_2PO_3F and HPO_2F_2, unlike the condensed phosphoric acids are very stable in aqueous solutions, but, nevertheless, were never obtained as crystalline materials.

References

1.   S. Furberg , *Acta Chem. Scand.*, **9**, (1955), 1557–1566.

2.   S. Furberg , *Acta Chem. Scand.*, **8**, (1954), 532.

3.   J. P. Smith, W. E. Brown, and J. R. Lehr, *J. Am. Chem. Soc.*, **77**, (1955), 2728–2730.

4.   A. D. Mighell, J. P. Smith, and W. E. Brown, *Acta Crystallogr.*, **B25**, (1969), 776–781.

5.   D. Mootz, J. Goldmann, and H. Wunderlich, *Angew. Chem.*, Internat. Edit., **8**, (1969), 116.

6.   D. Mootz and J. Goldmann, *Z. anorg. allg. Chem.*, **368**, (1969), 231–242.
7.   H. Worzala, *Acta Crystallogr.*, **B24**, (1968), 987–991.
8.   S. Furberg and P. Landmark, *Acta Chem. Scand.*, **11**, (1957), 1505–1511.

CHAPTER 4

Definition, Classification and Nomenclature

4.1. DEFINITION

Strictly speaking, phosphates are salts with anionic entities built of one PO_4 tetrahedron or as a result of a condensation of several PO_4 groups sharing one, two or three of their oxygens. From this very simple geometric definition, one can develop an infinite number of geometries for condensed phosphoric anions, but we will limit our survey to the presently well-known examples of condensation. There are a few exceptions to this definition. Many compounds in which one or several oxygen atoms of the basic PO_4 building unit are substituted by other atoms as, H, F or S, must also be considered phosphates. For example, in the well-known class of *phosphites,* one oxygen atom of the PO_4 tetrahedron is substituted by one hydrogen atom, leading to a $[PO_3H]^{2-}$ anion. In the same way, *thiophosphates* correspond to PO_3S, PO_2S_2, POS_3 or PS_4 entities and *fluorophosphates* to PO_3F or PO_2F_2 groups. This kind of substitution has been observed mainly in isolated tetrahedra, but recently some substituted cycloanions such as $[P_3O_6S_3]^{3-}$ and $[P_4O_8S_4]^{4-}$ have been characterized. These types of anions are usually called *substituted anions.* In the final chapter devoted to a reflexion on possible improvements of the present nomenclature and classification, we will discuss the most appropriate location for the substituted phosphates in a coherent classification. One question about the totally substituted tetrahedra remains: are compounds containing PS_4 and PN_4 groups still to be considered as phosphates?

4.2. CLASSIFICATION AND PRESENT NOMENCLATURE

As early as 1940, a good idea of the classification of silicates could be obtained thanks to the numerous structural investigations of silicates performed by various schools of crystallographers. The second volume of "Strukturbericht", published in 1937 and summarizing the crystallographic data obtained between 1928 and 1932, reports almost one hundred structures of silicates, some of them very complex and classified according to a system not fundamentally different from their present classification. Considering with the very restricted structural knowledge of phosphates at that time, this abundance of results might appear surprising. The very confusing chemical literature dealing with condensed phosphates and the difficulty of their preparation, whereas numerous kinds of silicate crystals are generously provided by Nature, can explain this scarcity of investigations. Nevertheless, one must say that during this early period of structural crystallography, a number of monophosphates were investigated and that the structural evidence for the existence of two condensed phosphoric anions was given by Levi and Peyronnel (1935) for the P_2O_7 diphosphate group and by Pauling and Sherman (1937) for the P_4O_{12} ring. It is not until the early fifties that interest came for structural investigations of phosphates. Several facts, almost simultaneous, can explain this renew of interest. The possibilities offered by the major improvements made in chromatographic techniques to measure the degree of polymerization of the condensed phosphoric anions and the great development of X-ray structural analysis. The main impulse came from the Thilo's school in Berlin. Associating a deep knowledge of chemistry, a systematic use of the modern chromatographic techniques recently developed by Westman and Ebel and X-ray structural analysis, this group clarified rapidly the field of condensed phosphates and can be considered responsible for its modern evolution. In addition, the pioneering work of Griffith and Van Wazer in USA and the enormous amount of results produced in the former URSS by the Tananaev's school opened the way to a rather logical classification, certainly not perfect, probably temporary, but accepted today as satisfactory in spite of some minor deficiencies.

Like in all classes of compounds in which the anionic part can condense, the classification of phosphates must be supported by the knowledge of the various geometries of the anions. This necessity will bring us, at least in this section, to dicuss with a language closer to that of geometry than to that of chemistry.

Phosphate materials can be presently classified in four families:
- monophosphates
- condensed phosphates
- adducts
- heteropolyphosphates.

4.2.1. Monophosphates

The appellation of monophosphates is given to compounds whose anionic entity, $[PO_4]^{3-}$, is composed by an almost regular tetrahedral arrangement of four oxygen atoms centered by a phosphorus atom. The corresponding salts for a long time known as orthophosphates are today called monophosphates, but the former appellation is still frequently encountered in scientific literature.

Among the various categories of phosphates, monophosphates are the most numerous, not only because they were the first to be investigated, but also because they are the most stable and therefore the only phosphates to be found in Nature. Their stability will be explained when we will discuss the hydrolysis of the P–O–P bond in the section of the Chapter 5 devoted to long-chain polyphosphates.

A detailed numerical example reporting the main geometrical features of a $[PO_4]^{3-}$ anion will be given in the section of Chapter 5, dealing with the present state of monophosphate chemistry.

4.2.2. Condensed Phosphates

The term 'condensed phosphates' is applied to salts containing condensed phosphoric anions. A strict definition of this type of anion can be given by saying that any phosphoric anion including one or several P–O–P bonds is a condensed phosphoric anion. The formation of these bonds can be performed by many different processes we will examine along our survey of phosphate chemistry. Let us simply report the simplest of these processes, commonly used to produce the P_2O_7 diphosphate anion, built by two corner-sharing PO_4 tetrahedra:

$$2Na_2HPO_4 \xrightarrow{700-900\,K} Na_4P_2O_7 + H_2O$$

This reaction, performed as early as 1827 by Clark, is at the origin of the discovery of condensed phosphates and of their first classification, later elaborated by Graham in 1833.

At the present state of our knowledge, one can distinguish three types of condensations for the phosphoric anions that we will examine now.

4.2.2.1. Polyphosphates

The first type of condensation corresponds to a progressive linear linkage, by corner sharing, of PO_4 tetrahedra, this association leading to the formation of finite or infinite chains. The corresponding phosphates are generally known as *polyphosphates*. The general formula for such anions is given by:

$$[P_nO_{3n+1}]^{(n+2)-}$$

where n is the number of phosphorus atoms in the anionic entity. The first terms of this condensation, corresponding to small values of n, are commonly named *oligophosphates* and are today well characterized up to n = 5. In Figures 4.2.1 and 4.2.2, we show two examples of such anions.

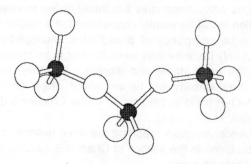

Figure 4.2.1. A triphosphate group as seen in $FeH_2P_3O_{10}$ by Averbuch-Pouchot and Guitel [1].

The terminology used for oligophosphates is now well established:

– for n = 2, the anion is $[P_2O_7]^{4-}$ and the corresponding phosphates for a long time called *pyrophosphates* are now designated by the appellation of *diphosphates*.

– for n = 3, the formula of the anion becomes $[P_3O_{10}]^{5-}$. The former appellation of *tripolyphosphates* for the corresponding salts is now replaced by *triphosphates*

– for n = 4, the anion is $[P_4O_{13}]^{6-}$ and the term *tetraphosphates* is now prefered to the former *tetrapolyphosphates*.

– and so on...

In these finite anions, the phosphorus atom belonging to a tetrahedron located at one extremity of the chain is called *terminal phosphorus*, while the one belonging to tetrahedra located inside the chain is called *internal phosphorus* .

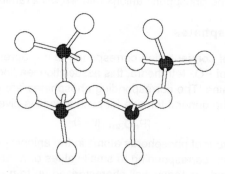

Figure 4.2.2. A tetraphosphate group as observed in $Cr_2P_4O_{13}$ by Lii *et al.* [2].

When n is becoming very large, the P/O ratio ⟶ 1/3 and the anion can be described as an infinite $(PO_3)_n$ chain. The corresponding phosphates are commonly called *long-chain polyphosphates* or more simply *polyphosphates*. Figure 4.2.3 gives the representation of such a chain.

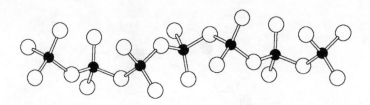

Figure 4.2.3. The infinite $(PO_3)_n$ chain observed in $CaH_2(PO_3)_4$ by Averbuch-Pouchot [3].

4.2.2.2. Cyclophosphates

The second type of condensation corresponds to the formation of cyclic anions still built by a set of corner-sharing PO_4 tetrahedra leading to the following anionic formula:

$$[P_nO_{3n}]^{n-}$$

These rings are now well characterized for n = 3, 4, 5, 6, 8, 9, 10 and 12. We represent, in Figures 4.2.4–4.2.6, some phosphoric ring anions. The corresponding salts were during a long time called *metaphosphates*. Today the use is to employ a more descriptive terminology and to say:

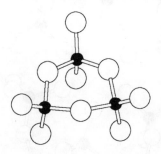

Figure 4.2.4. The P_3O_9 ring anion observed in $Na_3P_3O_9$ by Ondik [4].

   – *cyclotriphosphates* instead of the former *trimetaphosphates* for derivatives containing $[P_3O_9]^{3-}$ anions
   – *cyclotetraphosphates* replace *tetrametaphosphates* for compounds with

tetrameric $[P_4O_{12}]^{4-}$ anions
– and so on...

In Table 4.2.1, we sum up the present classification and nomenclature for the first two types of condensation.

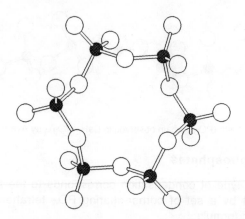

Figure 4.2.5. The P_6O_{18} ring anion observed in $Ag_6P_6O_{18}.H_2O$ by Averbuch-Pouchot [5].

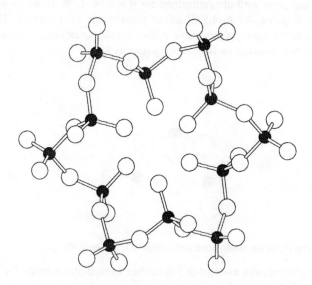

Figure 4.2.6. The $P_{12}O_{36}$ ring anion observed in $Cs_3V_3P_{12}O_{36}$ by Lavrov *et al.* [6].

Table 4.2.1
Summary of the present nomenclature for the first two types of condensed phosphates.

Polyphosphates		
n	Anions: $[P_nO_{3n+1}]^{(n+2)-}$	Names of the salts
2	$[P_2O_7]^{4-}$	Diphosphates
3	$[P_3O_{10}]^{5-}$	Triphosphates
4	$[P_4O_{13}]^{6-}$	Tetraphosphates
5	$[P_5O_{16}]^{7-}$	Pentaphosphates
∞	$[PO_3]∞$	Long-chain polyphosphates

Cyclophosphates		
n	Anions: $[P_nO_{3n}]^{n-}$	Names of the salts
3	$[P_3O_9]^{3-}$	Cyclotriphosphates
4	$[P_4O_{12}]^{4-}$	Cyclotetraphosphates
5	$[P_5O_{15}]^{5-}$	Cyclopentaphosphates
6	$[P_6O_{18}]^{6-}$	Cyclohexaphosphates
8	$[P_8O_{24}]^{8-}$	Cyclooctaphosphates
9	$[P_9O_{27}]^{9-}$	Cyclononaphosphates
10	$[P_{10}O_{30}]^{10-}$	Cyclodecaphosphates
12	$[P_{12}O_{36}]^{12-}$	Cyclododecaphosphates

n = number of phosphorus atoms in the anionic entity

4.2.2.3. Ultraphosphates

In the first two types of condensation, each PO_4 tetrahedron shares no more than two of its corners with the adjacent ones, whereas in this third type some PO_4 groups of the anionic entity share three of their corners with the adjacent ones. This type of condensation occurs in all condensed phosphates whose the anionic entities are richer in P_2O_5 than the richest members of the two classes of condensed anions we already described, long-chain anions and ring anions. From this definition, the formulation of the anion is given by:

$$n[PO_3]^- + mP_2O_5 \text{ or } [P_{(2m+n)}O_{(5m+3n)}]^{n-}$$

The corresponding salts are generally called *ultraphosphates*. This very general formulation corresponds to an infinite number of branches according to the values of m and n. To date, the only unambiguously characterized anions of this type are those corresponding to $m = 1$ and whose the general formula is given by:

$$[P_{(n+2)}O_{(3n+5)}]^{n-}$$

Such anions are now well-known for n = 2, 3, 4, 5 and 6.

The phosphorus atoms belonging to the PO_4 tetrahedra sharing three of their corners with the adjacent ones are named *ternary phosphorus* or very often *branching phosphorus*.

The ultraphosphate anions exist as finite groups like in $FeNa_3P_8O_{23}$ (Figure 4.2.7), as infinite ribbons like in $YCaP_7O_{20}$ (Figure 4.2.8), as layers or three dimensional networks. To the difference with the first two types of condensed anions, there is no correlation between the formula of an ultraphosphate anion and its geometry. For instance, a $[P_5O_{14}]^{3-}$ anion can appear as an infinite ribbon in $SmP_5O_{14}(I)$ or build an infinite three-dimensional network as in $HoP_5O_{14}(II)$.

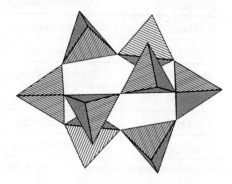

Figure 4.2.7. A discrete ultraphosphate anion, $[P_8O_{23}]^{6-}$, as observed in $FeNa_3P_8O_{23}$ [7].

No clear or coherent nomenclature exists for ultraphosphates. Most of time, these compounds are simply called "ultraphosphate of ...". This absence of nomenclature is at the origin of some very confusing mistakes. For instance, LnP_5O_{14} ultraphosphates have been designated by physicists according to the number of phosphorus atoms in the formula unit and since these investigations known as "pentaphosphates", an appellation already used in the classification of oligophosphates. We shall discuss of a possible coherent nomenclature in the last chapter.

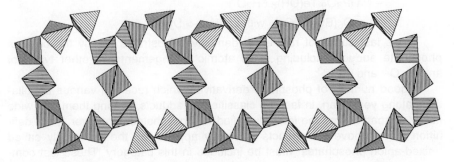

Figure 4.2.8. An infinite ultraphosphate-ribbon anion as observed in $YCaP_7O_{20}$ [8].

4.2.3. Other Types of Phosphates

Phosphoric anions, condensed or not, are also found in many other compounds which cannot find their place in the above classification. After a long and careful examination of these numerous compounds, we came to the conclusion that they could all be classified in two classes, *adducts and heteropolyphosphates*. A long part of the Chapter 5 will be devoted to these two kinds of derivatives and we suggest, in the last chapter which deals with possible improvements of phosphate nomenclature, a logical system for their appellation in order to clarify this still confusing aspect of phosphate chemistry.

4.2.3.1. Adducts

In our opinion, the proper definition of an adduct, in the domain of phosphate chemistry, can be formulated as follow. An adduct is a compound including in its atomic arrangement a phosphoric anion belonging to one of the categories we described above and in addition another anion or group of anions, phosphoric or not, coexisting as "independent units". By independent units we mean that, in these salts, the two anionic entities do not share any atom. In the present state of our knowledge, these additional anions are very various, O^{2-}, OH^-, F^-, Cl^-, Br^-, I^-, $[NO_3]^-$, $[BO_3]^{3-}$, $[CO_3]^{2-}$, $[SO_4]^{2-}$, $[CrO_4]^{2-}$, $[C_2O_4]^{2-}$, or any kind of phosphate or silicate anions. Some other entities as telluric acid or ethylenediamine can also form adducts with phosphates. We will review with more details these compounds in the Chapter 5, thus we simply report below a few examples we consider as well representative:

– Mg_2FPO_4 or $Mg_3(PO_4)_2.MgF_2$

– $Hg_4(NO_3)(PO_4).H_2O$ or $Hg_3PO_4.HgNO_3.H_2O$

– $MnNa_3(PO_4)(CO_3)$

– $Na_3P_3O_9.Te(OH)_6.6H_2O$

– $Ln_7O_6(BO_3)(PO_4)_2$ with Ln (La –> Dy)

These last series of rare-earth derivatives illustrates clearly the case of phosphate adducts including in its atomic arrangement two other types of anions, O^{2-} and $[BO_3]^{3-}$.

A good number of phosphate derivatives which received various appellations along years can, in fact, be classified as adducts. Among them, the wide family of apatites and the formerly called oxyphosphates. Moreover, if the definition given above for adducts is strictly applied, all the previously called "mixed-anion phosphates" must be included in this category. These last compounds correspond to salts including in their arrangements two types of phosphoric anions. We report briefly two examples of anion stackings in adducts.

We describe, first, $Na_3P_3O_9.Te(OH)_6.6H_2O$ [9]. In this $P6_3/m$ hexagonal compound whose unit-cell dimensions are:

$$a = 11.67(1), \quad c = 12.12(1) \text{ Å}$$

two types of anionic entities coexist as independent units, a cyclic P_3O_9 group

Figure 4.2.9. Projection of the anionic arrangement in $Na_3P_3O_9.Te(OH)_6.6H_2O$ along the **c** direction. The two entities are given in polyhedral representation. Sodium atoms and water molecules are not drawn.

and an octahedral $Te(OH)_6$ group. Figure 4.2.9, a projection along the **c** direction, represents the repartition of these two entities.

In the second example, $Hg_4(NO_3)(PO_4).H_2O$ [10], the stacking of the two anions is very simple. Both NO_3 groups and PO_4 tetrahedra are located in layers spreading around the planes $y = 1/4$ and $3/4$ of a $P2_1/n$ monoclinic cell:

$$a = 18.38(9), \quad b = 8.258(3), \quad c = 5.952(2) \text{ Å}, \quad \beta = 91.2(1)°$$

containing 4 formula units. In order to show the organization inside one layer, we give, in Figure 4.2.10, the projection, along the **b** direction, of such a slice.

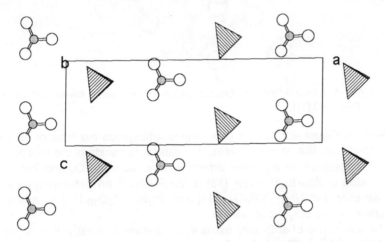

Figure 4.2.10. Projection, along the **b** direction, of a layer of NO_3 groups and PO_4 tetrahedra in the arrangement of $Hg_4(PO_4)(NO_3).H_2O$. The phosphoric anion is drawn in polyhedral representation, smaller circles are nitrogen atoms and larger ones oxygen atoms of the NO_3 groups.

4.2.3.2. Heteropolyphosphates

It seems necessary to give first a definition as clear as possible of what is usually meant by the term of heteropolyanion. In our opinion an heteropolyanion is:

i) a finite or infinite condensed anion

ii) built up by a set of XO_n and YO_n polyhedra sharing edges or corners and thus includes X–O–Y and possibly X–O–X and Y–O–Y bonds.

In the case of heteropolyphosphates, X or Y = P.

The boundaries of this last class of compounds are difficult to assign precisely and some ambigiuities will persist in spite of the clear definition we gave of them. It is rather evident that in the particular case of finite heteropolyanions as those we shall meet when describing phosphochromates or the well-known phosphotungstates and phosphomolybdates no major problems appear. We report in Figure 4.2.11 some examples of these heteropolyanions. The ribbon built by an association of P_2O_7 groups and SiO_4 tetrahedra in $Cd_2SiP_4O_{14}$ we

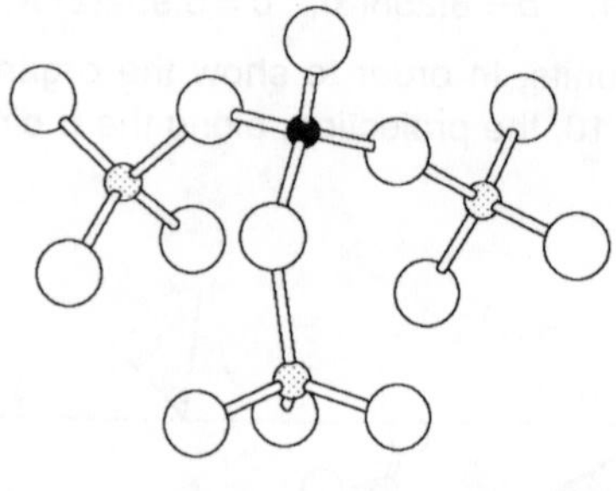

Figure 4.2.11. The finite $[PCr_3O_{13}]^{3-}$ heteropolyanion as observed by Averbuch-Pouchot *et al.* in $Na_3PCr_3O_{13}.3H_2O$ [11].

represent in Figure 4.2.12 is a good and unambiguous example of an infinite heteropolyanion. But, in many cases, it is difficult to decide of the proper appellation. For instance, in the three-dimensional network of PO_4 and ZnO_4 tetrahedra found in $Zn_3Rb_2(P_2O_7)_2$ [13] is $[Zn_3P_4O_{14}]^2$ an heteropolyanion and shall we consider $(NH_4)_2SiP_4O_{13}$ [14] and $Be_2NH_4P_3O_{10}$ [15] as heteropolyphosphates or as polyphosphates?

The same type of ambiguity exists in the silicate chemistry. It is sometimes not very easy to say clearly if a given compound is an aluminosilicate, a silicate of aluminium or even an aluminosilicate of aluminium and of...

In the absence of strict and clear rules based upon the nature of the X–O

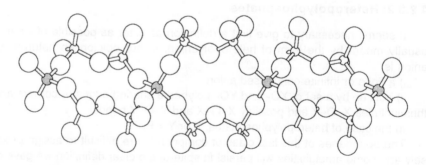

Figure 4.2.12. The infinite $[SiP_4O_{14}]^{4-}$ ribbon anion as obseved in $Cd_2SiP_4O_{14}$ (12).

and Y–O bonds, it seems difficult to imagine or elaborate a coherent and logical nomenclature. A simple, but perhaps very academic question remains: shall we consider, phosphochromates and phosphosilicates, by instance, as belonging to phosphate chemistry or to chromate or silicate chemistry?

References

1   M. T. Averbuch-Pouchot and J. C. Guitel, *Acta Crystallogr.*, **B33**, (1977), 1613–1615.

2   K. H. Lii, Y. B. Chen, C. C. Su, and S. L. Wang, *J. Solid State Chem.*, **82**, (1989), 156–160.

3   M. T. Averbuch-Pouchot, *Z. anorg. allg. Chem.*, in press.

4.   H. M. Ondik, *Acta Crystallogr.*, **18**, (1965), 226–232.

5.   M. T. Averbuch-Pouchot, *Z. Kristallogr.*, **189**, (1989), 17–23.

6.   A. V. Lavrov, V. P. Nikolaev, G. G. Sadikov, and M. Ya. Voitenko, *Dokl. Akad. Nauk SSSR*, **259**, (1981), 103–106.

7.   N. N. Chudinova, K. K. Palkina, N. B. Komarovskaya, S. I. Maksimova, and N. T. Chibiskova, *Dokl. Akad. Nauk SSSR*, **306**, (1989), 635–638.

8.   A. Hamadi and T. Jouini, *J. Solid State Chem.*, **111**, (1994), 443–446.

9.   N. Boudjada, M. T. Averbuch-Pouchot, and A. Durif, *Acta Crystallogr.*, **B37**, (1981), 645–647.

10.   A. Durif, I. Tordjman, R. Masse, and J. C. Guitel, *J. Solid State Chem.*, **24**, (1978), 101–105.

11.   M. T. Averbuch-Pouchot, A. Durif, and J. C. Guitel, *J. Solid State Chem.*, **33**, (1980), 325–333.

12.   M. Trojan, D. Brandova, J. Fabry, J. Hybler, K. Jurek, and V. Petricek, *Acta Crystallogr.*, **C43**, (1987), 2038–2040.

13.   M. T. Averbuch-Pouchot, *Z. Kristallogr.*, **171**, (1985), 113–119.

14.   A. Durif, M. T. Averbuch-Pouchot, and J. C. Guitel, *Acta Crystallogr.*, **B32**, (1976), 2957–2960.

15.   M. T. Averbuch-Pouchot, A. Durif, J. Coing-Boyat, and J. C. Guitel, *Acta Crystallogr.*, **B33**, (1977), 203–205.

CHAPTER 5

The Present State of Phosphate Chemistry

5.1. MONOPHOSPHATES

5.1.1. Introduction

Monophosphates are the most common and the simplest of all phosphates since their anionic entity is an isolated PO_4 tetrahedron. They also constitute the widest family within the chemistry of phosphate because they have been well-known for a very long time and are very stable. In addition, all natural phosphates are monophosphates.

This last assumption deserves a short parenthetical comment. In all the other families of phosphates, the phosphoric anion includes P–O–P bonds which are sensitive to hydrolysis phenomena. This type of bond more or less rapidly broken constitutes a weak point in the atomic arrangement. The final term of such a degradation is always a monophosphate or a mixture of monophosphates. The various parameters ruling the speed of these decondensation processes will be noted when we examine the general properties of polyphosphates. Let us simply add that all condensed inorganic anions have the same behavior, and in some cases like chromates, sulphates or arsenates, the degradation is very rapid and sometimes instant. Among compounds including condensed inorganic anions, phosphates are nevertheless the most stable after silicates, germanates, and borates. As we shall have the opportunity to discover in this section that in spite of the very simple geometry of their anion, the monophosphates have very intricate atomic arrangements which are often difficult to explain and represent in the solid state.

5.1.2. Monovalent-Cation Monophosphates

These salts correspond to various formulæ, $M_3^I PO_4$, $M_2^I HPO_4$ or $M^I H_2 PO_4$, but examples of monophosphates including PO_4, HPO_4 and H_2PO_4 groups simultaneously in their atomic arrangements are frequently observed.

5.1.2.1. General properties

Solubility – This property is very dependent on the nature of the associated cation and on the degree of acidity. Thus, Li_3PO_4, Ag_3PO_4, and Tl_3PO_4 are almost or completely insoluble, whereas all the other $M_3^IPO_4$ monophosphates are very soluble. All the $M_2^IHPO_4$ or $M^IH_2PO_4$ monophosphates are soluble.

Thermal behavior – The thermal behavior of monovalent-cation monophosphates is very dependent on the degree of acidity of the anions.

$M_3^IPO_4$ monophosphates are very stable, often with relatively high melting point, for instance, K_3PO_4 melts at 1613 K.

The thermal evolution of monohydrogenmonophosphates, $M_2^IHPO_4$, is different and through a dehydration-condensation reaction leads to diphosphates according to the following scheme:

$$2M_2^IHPO_4 \longrightarrow M_4^IP_2O_7 + H_2O$$

The temperatures of formation of the diphosphates are very dependent on the nature of the associated cations. In the case of sodium, the reaction occurs between 590 and 625 K.

The thermal behavior of dihydrogenmonophosphates, $M^IH_2PO_4$, is much more complex and was the object of many investigations. In most cases, the reaction is:

$$M^IH_2PO_4 \longrightarrow M^IPO_3 + H_2O$$

The final term is always a M^IPO_3 long-chain polyphosphate when the reaction is conducted up to the melting temperature of the corresponding polyphosphate or close to this temperature, but in many cases intermediate steps of condensation are observed. For example, in the case of sodium, a dihydrogendiphosphate, $Na_2H_2P_2O_7$, forms when the reaction is performed between 473 and 513 K.

$$2NaH_2PO_4 \longrightarrow Na_2H_2P_2O_7 + H_2O$$

With some $M^IH_2PO_4$ and in well optimized conditions, this dehydration-condensation phenomenon can exhibit very different aspects. We report below two examples which have very important applications for the chemical preparation of two fundamental starting materials used in the syntheses of cyclophosphates.

– Dehydration-cyclization of LiH_2PO_4. In a rather detailed study of the thermal behavior of LiH_2PO_4, Schülke and Kayser [1] investigated the various paths of its dehydration-condensation and observed that if pure LiH_2PO_4 is kept at 623 K for 30 min the reaction is:

$$LiH_2PO_4 \longrightarrow LiPO_3 + H_2O$$

with the formation of lithium long-chain polyphosphate.

If the same process is applied to the same starting material, but seeded with 10 to 20% of $Li_6P_6O_{18}$, the final product is pure lithium cyclohexaphosphate:

$$6LiH_2PO_4 \longrightarrow Li_6P_6O_{18} + 6H_2O$$

This discovery is the source of the large development of cyclohexaphosphate chemistry during the past ten years.

– Dehydration-cyclization of NaH_2PO_4. In the case of the sodium salt, the thermal dehydration-cyclization phenomena were carefully investigated by Thilo and Grunze [2]. These authors observed that, if NaH_2PO_4 is heated above 823 K for some hours, the insoluble sodium long-chain polyphosphate is obtained:

$$NaH_2PO_4 \longrightarrow NaPO_3 + H_2O$$

but, if the reaction is conducted at 823 K for at least 5 h, pure sodium cyclotriphosphate is obtained with an almost theoretical yield.

$$3NaH_2PO_4 \longrightarrow Na_3P_3O_9 + 3H_2O$$

This last process is now universally used for the production of $Na_3P_3O_9$, a fundamental material for the synthesis of cyclotriphosphates. More details on this synthesis will be discussed in the section devoted to the chemical preparations of cyclotriphosphates.

– Other attempts of dehydration-cyclization. Various investigations were performed by Thilo [3] and Verdier [4] in order to prepare alkali or ammonium cyclophosphates from the corresponding dihydrogenmonophosphates. Thilo was able to produce $Na_3P_3O_9$ and $K_3P_3O_9$ by heating NaH_2PO_4 or KH_2PO_4 for several days at 363 K in a 1:1 mixture of acetic acid and acetic anhydride. Later, Verdier confirmed these results for the potassium salt and proved that, in this case, condensation is made possible by using only acetic anhydride.

5.1.2.2. M_2O–H_3PO_4–H_2O systems

Most of the M_2O–H_3PO_4–H_2O systems were thoroughly investigated and the numerous monophosphates isolated are now well-known. However, some recent characterizations showed that this domain can still be enriched. For instance, in 1985, a new form of RbH_2PO_4 [5] was characterized and, in the same year, a new complex monophosphate, $Rb_5H_7(PO_4)_4$ [6], was discovered in the same system.

As an example of the richness of these systems, listed below are the sodium monophosphates characterized in the Na_2O–H_3PO_4–H_2O system:

– Na_3PO_4, $Na_3PO_4.1/2H_2O$, $Na_3PO_4.6H_2O$, $Na_3PO_4.8H_2O$, $Na_3PO_4.12H_2O$
– Na_2HPO_4, $Na_2HPO_4.2H_2O$, $Na_2HPO_4.7H_2O$, $Na_2HPO_4.8H_2O$, $Na_2HPO_4.12H_2O$
– NaH_2PO_4, $NaH_2PO_4.H_2O$, $NaH_2PO_4.2H_2O$

– Na$_2$HPO$_4$.NaH$_2$PO$_4$, Na$_2$HPO$_4$.2NaH$_2$PO$_4$.2H$_2$O
– NaH$_5$(PO$_4$)$_2$

Much work is still to be done for their structural investigations. It is only very recently that structures of some of them were performed: in 1983 for Na$_2$HPO$_4$ [7], Na$_3$PO$_4$.1/2H$_2$O [8], and Na$_3$PO$_4$.8H$_2$O [9] and in 1994 for K$_2$HPO$_4$ [10].

5.1.2.3. Some structural examples

Atomic arrangement of Na$_3$PO$_4$.12H$_2$O – This compound, not strictly stoichiometric, is known to retain a small amount of NaOH and its formula should be written as Na$_3$PO$_4$.(NaOH)$_x$.12H$_2$O, where x is within the range of 0.0–0.25. This compound is trigonal, with the $P\bar{3}c1$ space group, $Z = 4$ and the following unit-cell dimensions:

$$a = 11.890(6), \quad c = 12.671(7) \text{ Å}$$

An accurate determination of its atomic arrangement was performed by Tillmans and Baur [11]. It can be described as a succession of thick layers of edge- and corner-sharing Na(H$_2$O)$_6$ octahedra alternating perpendicular to the **c** axis and separated by a distance of $c/2$. The stacking of these layers creates two kinds of channels parallel to the **c** direction. The channels located along

Figure 5.1.1. Projection of a part of the atomic arrangement of Na$_3$PO$_4$.(NaOH)$_x$.12H$_2$O along the **c** direction. The octahedral void including the NaOH fragment is not shown.

the internal ternary axes accommodate the PO_4 tetrahedra as shown in Figure 5.1.1. The authors were able to localize the small amount of NaOH situated in an octahedral void at the origin of the unit cell, in the second type of channel. Figure 5.1.2 shows the arrangement of an octahedral layer around the origin. The three-dimensional cohesion of such a stacking is performed by hydrogen bonds only. The corresponding arsenate and vanadate are isotypic. Their structures are described in the same study.

In the field of phosphate chemistry, atomic arrangements in which the phosphoric anion does not share any oxygen atom with the associated cations are rare and only observed in highly hydrated salts like in the present compound. Some examples are provided by β-$Na_2HPO_4.12H_2O$ investigated by Catti *et al.* [12], $Na_2HPO_4.7H_2O$ by Baur and Khan [13] and $Cr_2P_6O_{18}.21H_2O$ by Bagieu-Beucher *et al.* [14].

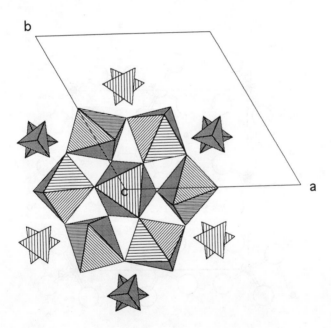

Figure 5.1.2. Projection of a part of the structure of $Na_3PO_4.(NaOH)_x.12H_2O$ along the **c** direction. The octahedron located around the origin of the unit cell contains the NaOH fragment.

Atomic arrangements in the KDP family – Among the numerous compounds of this family, the most famous is KH_2PO_4. Its ferroelectric, optical, and piezoelectric properties induced a good number of investigations and a large production of single crystals.

At room temperature, this salt is tetragonal, with the space group $I\bar{4}2d$, $Z = 4$ and the unit-cell parameters:

$$a = 7.448, \quad c = 6.977 \text{ Å}$$

Figure 5.1.3 shows the atomic arrangement projected along the **c** direction. The simple stacking which is difficult to represent can be understood as a set of alternating $-K-PO_4-K-$ arrays parallel to the **c** axis. These arrays are themselves interconnected by a network of strong and quasi-symmetric hydrogen bonds.

The ferroelectric transition occurs at 122 K and below this temperature the symmetry of the compound becomes orthorhombic with:

$$a = 10.53, \quad b = 10.44, \quad c = 6.90 \text{ Å}, \quad Fdd2, \quad Z = 8 \quad (116 \text{ K})$$

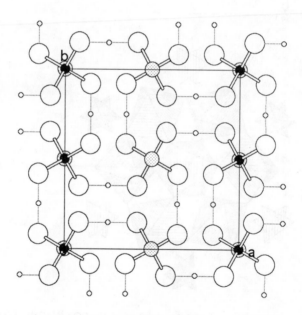

Figure 5.1.3. The room-temperature form of KH_2PO_4 is projected along the **c** direction. The largest open circles are oxygen atoms, black circles are phosphorus atoms, pale grey circles are potassium atoms and the smallest open circles are hydrogen atoms. The hydrogen bonds are represented by dotted lines.

This reversible transition induces small changes in the atomic arrangement. In order to compare these two forms, it is sometimes useful to describe the tetragonal body centered room-temperature form using a diagonal unit cell corresponding to a non conventional face-centered tetragonal unit cell:

$$a' = a\sqrt{2} = 10.48, \quad c' = c = 6.90 \text{ Å}$$

In such a description, the space group becomes $F\overline{4}d2$ and $Z = 8$.

The room-temperature form of the corresponding ammonium compound, $(NH_4)H_2PO_4$, is isotypic with the room-temperature form of KH_2PO_4 with:

$$a = 7.499, \quad c = 7.548 \text{ Å}$$

and also undergoes the same type of transformation, but at 143–153 K leading, this time, to an orthorhombic unit cell:

$$a = 7.49, \quad b = 7.51, \quad c = 7.48 \text{ Å}, \quad P2_12_12_1, \quad Z = 4 \quad (143 \text{ K})$$

As in the case of the potassium salt, this transition does not induce a great variation in the atomic arrangement.

Since the discovery, more than 70 years ago, of the interesting dielectric behavior of KH_2PO_4, thousands of studies dealing with the structural aspects of the problem or with the physical properties and their applications have been published. Also all the $M^IH_2PO_4$ monophosphates with more or less analogous arrangements and the corresponding deuterated compounds have been extensively investigated.

5.1.3. Divalent-Cation Monophosphates

$M_3^{II}(PO_4)_2$, $M^{II}HPO_4$ and $M^{II}(H_2PO_4)_2$ and their various hydrates are the simplest possible formulæ in this category of salts, but combinations of the various anionic entities in the same compound are also frequently found in monophosphates.

5.1.3.1. General properties

The *solubility* is evidently dependent mainly on the type of anion. All the $M_3^{II}(PO_4)_2$ are insoluble in water, whereas many $M^{II}HPO_4$ and $M^{II}(H_2PO_4)_2$ monophosphates are moderately soluble, often incongruently.

The *thermal behavior* is more complex to analyze. All the neutral monophosphates, $M_3^{II}(PO_4)_2$, are very stable, most of them melting congruently at rather high temperatures. Reported, in Table 5.1.1, are some of these melting points.

$M^{II}HPO_4$ monohydrogenmonophosphates are normally transformed by heating into diphosphates according to the following reaction:

$$2M^{II}HPO_4 \longrightarrow M_2^{II}P_2O_7 + H_2O$$

Table 5.1.1
Melting temperatures of some divalent-cation monophosphates.

Formula	mp (K)	Ref.	Formula	mp (K)	Ref.
$Co_3(PO_4)_2$	1433	[15]	$Mg_3(PO_4)_2$	1630	[19]
$Ni_3(PO_4)_2$	1623	[15]	$Ca_3(PO_4)_2$	2029–2073	[20–21]
$Zn_3(PO_4)_2$	1333	[16]	$Sr_3(PO_4)_2$	1343 (d)	[22]
$Cd_3(PO_4)_2$	1453	[17]	$Ba_3(PO_4)_2$	1878	[23]
$Pb_3(PO_4)_2$	1288	[18]			

d: decomposition

The temperatures of these transformations are strongly dependent on the nature of the associated cation. Reported below are some examples:

$$2CaHPO_4.2H_2O \xrightarrow[-4H_2O]{393\,K} 2CaHPO_4 \xrightarrow[-H_2O]{723\,K} \gamma\text{-}Ca_2P_2O_7 \xrightarrow{1023\,K} \beta\text{-}Ca_2P_2O_7 \quad [24]$$

$$2ZnHPO_4.H_2O \xrightarrow{343-423\,K} 2ZnHPO_4 \xrightarrow{538\,K} Zn_2P_2O_7 \quad [25]$$

$$2MnHPO_4.3H_2O \xrightarrow{93-413\,K} 2MnHPO_4.H_2O \xrightarrow{428-603\,K} 2MnHPO_4 \xrightarrow{603-813\,K} Mn_2P_2O_7 \quad [26]$$

These temperatures of dehydration-condensation are greatly influenced by the experimental conditions. The effect of water vapor seems to be an important factor as well as the conditions of heating, dynamic or quasi-isothermal. In addition, one observes the existence of an amorphous phase in a large range of temperature before the crystallization of the diphosphate in many cases.

For dihydrogenmonophosphates, $M^{II}(H_2PO_4)_2$, the thermal evolution can correspond to various types of dehydration-condensation and is also dependent on the nature of the cation. In general, the reaction can always be schematized by:

$$M^{II}(H_2PO_4)_2 \longrightarrow M^{II}(PO_3)_2 + 2H_2O$$

But, the formula $M^{II}(PO_3)_2$ can, in fact, correspond to very different types of condensed phosphates. Thus, for M^{II} = Co, Mg, Cu, Mn, Ni, Zn, Fe, the above reaction must be written:

$$2M^{II}(H_2PO_4)_2 \longrightarrow M_2^{II}P_4O_{12} + 4H_2O$$

as in these cases the condensation produces cyclotetraphosphates $M_2^{II}P_4O_{12}$. On the contrary, for M^{II} = Be, Ca, Sr, Ba, Pb, Cd and Hg, one must write:

$$M^{II}(H_2PO_4)_2 \longrightarrow M^{II}(PO_3)_2 + 2H_2O$$

the final term of the condensation is a long-chain polyphosphate.

In some chemical literature, it has been asserted that cyclotriphosphates, $M_3^{II}(P_3O_9)_2$, can be prepared through the above reaction, but such assumptions have not yet been confirmed.

Here also, as in the case of monovalent cations, one must not forget the existence of the intermediate steps leading to the formation of dihydrogendiphosphates. Reported below are some examples.

$$Sr(H_2PO_4)_2 \xrightarrow{463-483\,K} SrH_2P_2O_7 + H_2O \quad [27-28]$$

$$Mg(H_2PO_4)_2.2H_2O \xrightarrow{368-403\,K} Mg(H_2PO_4)_2 + 2H_2O \xrightarrow{488-523\,K} MgH_2P_2O_7 \quad [29]$$

The corresponding manganese diphosphate, $MnH_2P_2O_7$, can be obtained from $Mn(H_2PO_4)_2.2H_2O$ through an almost identical scheme [30] as well as $CdH_2P_2O_7$ [31]:

$$Cd(H_2PO_4)_2.2H_2O \xrightarrow{373\,K} Cd(H_2PO_4)_2 \xrightarrow{453\,K} CdH_2P_2O_7 \xrightarrow{473\,K} Cd_2P_4O_{12}$$

5.1.3.2. Some structures of divalent-cation monophosphates

We cannot give here a complete survey of what has been established in this very rich category of salts which have been the subject of hundreds of investigations since the last century. Our first idea was to illustrate this section with a review of calcium monophosphates whose importance is fundamental for both biology and industry. Recently, Elliot [32] performed a detailed study of these salts. To avoid duplication, we will illustrate this section with a survey of zinc monophosphates. A good number of them have been thoroughly studied and the corresponding collection of data dealing with these salts is, in our opinion, sufficient to give a good general idea of what is chemistry and crystal chemistry of divalent-cation monophosphates.

The $ZnO-P_2O_5-H_2O$ system or parts of this system have been investigated by several authors: Katnack and Hummel [16], Goloshchapov and Filatova [33], Salmon and Terrey [34], Komrska and Satava [35], and Cudennec *et al.* [36]. Among the various compounds characterized during these studies, nine were the object of crystallographic investigations. Reported first, in Table 5.1.2, are their main crystallographic features, and in the following pages their atomic arrangements are described when these are known.

Some other zinc monophosphates are also reported in the literature:

– $Zn_3(PO_4)_2.2H_2O$ by Fred *et al.* [48], Thilo and Schulz [49] and Saison [50]

– $Zn_3(PO_4)_2.3H_2O$ by Salmon and Terrey [34] and Komrska and Satava [35]. Recently, Cudennec *et al.* [36] confirmed the existence of these two hydrates, but could not obtain single crystals.

Table 5.1.2
Main crystallographic data of zinc monophosphates.

Formula	a α	b β	c (Å) γ (°)	S. G.	Z	Ref.
α-Zn$_3$(PO$_4$)$_2$	8.14(2)	5.63(1) 105.13(8)	15.04(4)	$C2/c$	4	[37]
β-Zn$_3$(PO$_4$)$_2$	9.393(3)	9.170(6) 125.73(10)	8.686(3)	$P2_1/c$	4	[38]
γ-Zn$_3$(PO$_4$)$_2$	5.074(8)	8.469(1) 120.85(2)	8.766(15)	$P2_1/c$	2	[39]
Zn$_3$(PO$_4$)$_2$.H$_2$O	8.696(6)	4.891(2) 94.94(3)	16.695(9)	$P2_1/c$	4	[40]
Zn$_3$(PO$_4$)$_2$.4H$_2$O (a)	10.597(3)	18.318(8)	5.031(1)	$Pnma$	4	[41–43]
Zn$_3$(PO$_4$)$_2$.4H$_2$O (b)	5.773(6) 93.40(16)	7.546(9) 91.10(16)	5.276(5) 91.33(16)	$P\bar{1}$	1	[44]
Zn(H$_3$O)PO$_4$	10.689(2)	10.689(2)	8.708(1)	$P6_3$	8	[45]
ZnHPO$_4$.H$_2$O	6.429(13) 74.11(10)	7.726(11) 81.35(9)	12.285(16) 80.99(9)	$P\bar{1}$	2	[46]
Zn(H$_2$PO$_4$)$_2$.2H$_2$O	7.266(2)	9.901(5) 94.76(1)	5.332(1)	$P2_1/n$	2	[47]

a:hopeite
b:parahopeite

– Zn$_3$(PO$_4$)$_2$.5H$_2$O and Zn$_3$(PO$_4$)$_2$.6H$_2$O were reported in the literature of the last century, but their existence has not yet been confirmed.

– Zn$_3$(PO$_4$)$_2$.8H$_2$O was prepared by de Schulten [51]. No crystal data could be found for this salt, but one can suspect its isomorphy with koettigite, Zn$_3$(AsO$_4$)$_2$.8H$_2$O, the corresponding arsenate.

One may be surprised by the absence in Table 5.1.2 of some well-known zinc salts like: Zn$_2$(OH)PO$_4$ (tarbutite), Zn$_2$(OH)PO$_4$.3/2H$_2$O (spencerite), and Zn$_7$(OH)$_2$(PO$_4$)$_4$.7H$_2$O (hibbenite). These compounds, presently known as hydroxyphosphates, must in fact be classified among addition compounds and denominated as *hydroxide-monophosphates*. This assumption is justified in the Chapter 7.

Another salt has been reported several times during the various investigations of the ZnO–P$_2$O$_5$–H$_2$O system, namely Zn(H$_2$PO$_4$)$_2$.2H$_3$PO$_4$. This derivative is also evidently to be classified as an adduct. There are valid reasons to suspect its isotypy or at least strong structural analogies with Cd(H$_2$AsO$_4$)$_2$.2H$_3$AsO$_4$ and Co(H$_2$PO$_4$)$_2$.2H$_3$PO$_4$, two isomorphous salts described respectively by Boudjada *et al.* [52] and Herak *et al.* [53].

$Zn_3(PO_4)_2$ – A phase diagram has been reported by Katnack and Hummel [16] for the ZnO–P_2O_5 system. From this investigation, $Zn_3(PO_4)_2$ appears as a dimorphous congruent melting compound (mp = 1333 K). The transition temperature between the low-temperature α-form and the high-temperature β-form is given as 1215 K. This transformation is reversible, but slow. This last feature can be easily explained by comparing the two atomic arrangements. When a part of the zinc is substituted by Mn, Mg or Cd, a third form called γ appears. This last form which is isotypic with $Mg_3(PO_4)_2$, is in fact a solid solution whose composition range varies with the nature of the foreign cation.

Described above are the atomic arrangements of these three forms, whose main crystallographic features can be found in Table 5.1.2.

– α-$Zn_3(PO_4)_2$. The crystal structure of this form, determined by Calvo [37], is a three-dimensional network of ZnO_4 and PO_4 tetrahedra. Two crystallographically independent zinc atoms coexist inside the arrangement. Zn(1) is located on a twofold axis and Zn(2) in a general position. The Zn(1)O_4 tetrahedron shares its corners with the four adjacent PO_4 tetrahedra and two Zn(2)O_4 tetrahedra, whereas the Zn(2)O_4 tetrahedra is assembled in pairs by sharing edge to build Zn_2O_6 units. These pairs have two common edges with adjacent PO_4 groups and share corners with two other phosphoric groups and two Zn(1)O_4 tetrahedra. Figure 5.1.4 represents the atomic arrangement projected along the **b** direction. This direction of projection exhibits clearly the layer arrangement of this structure. Layers of ZnO_4 tetrahedra parallel to the (101) plane alternate with parallel layers containing the phosphoric entities. The Zn–O distances, in the Zn(1)O_4 tetrahedron, are 1.93 and 2.03 Å, whereas they vary from 1.86 to 2.03 Å in the Zn(2)O_4 tetrahedron. The P–O distances are observed between 1.50 and 1.58 Å, with an average value of 1.53 Å.

– β-$Zn_3(PO_4)_2$. The crystal structure determination of the β-form was performed by Calvo [38]. In this atomic arrangement, there are three crystallographically independent zinc atoms, all in general position. Zn(1) has a fourfold tetrahedral oxygen coordination, whereas Zn(2) and Zn(3) have a fivefold. As in the α-form, the Zn(1)O_4 tetrahedra is assembled in pairs by sharing a common edge. These Zn(1)$_2O_6$ units are centrosymmetric and are all located in a layer centered by the plane $x = 0$. The Zn(3)O_5 polyhedra form a thick layer spreading around the plane $x = 1/2$. Inside this layer, these entities do not share any edge or corner. The Zn(2)O_5 polyhedra located approximately halfway between the Zn(1) and Zn(3) layers are spread around the $x = \pm 0.30$ planes. The location of the PO_4 tetrahedra around the planes $x = \pm 0.14$ and $x = \pm 0.36$ confirms the layer arrangement of this structure. Figure 5.1.5 shows this

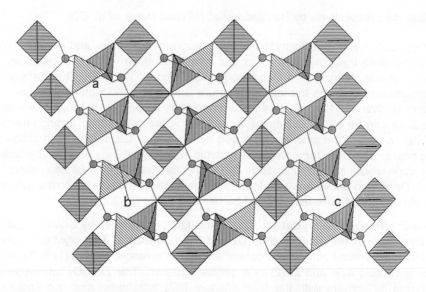

Figure 5.1.4. The α-Zn$_3$(PO$_4$)$_2$ atomic arrangement is projected along the **b** direction. ZnO$_4$ groups are given by polyhedral representation, phosphorus atoms are represented by grey circles and P–O bonds by solid lines.

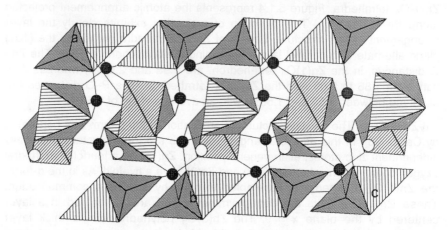

Figure 5.1.5. The atomic arrangement of β-Zn$_3$(PO$_4$)$_2$ projected along the **b** direction. Zn(1)O$_4$ and Zn(3)O$_5$ polyhedra are given by polyhedral representation, whereas Zn(2) atoms are shown by open circles. Phosphorus atoms are represented by grey circles and P–O bonds by solid lines.

arrangement in projection along the **b** direction. Each PO_4 group shares two oxygen atoms with two different $Zn(1)O_6$ clusters, one with the $Zn(2)O_5$ polyhedron and the last one with the $Zn(3)O_5$ polyhedron.

The Zn–O distances vary between 1.886 and 2.105 Å in the Zn(1) coordination, 1.994 and 2.276 Å in the $Zn(2)O_5$ polyhedron and 1.924 and 2.174 Å around Zn(3), whereas the P–O distances in the two independent PO_4 tetrahedra vary from 1.502 to 1.558 Å. The existence of a Zn(1)–O distance of 2.549 Å and of a Zn(2)–O distance of 2.509 Å was a matter for discussion. From bond-strength calculations using the Brown and Shannon's formula [54], Nord and Kirkegaard [55] concluded that a fivefold oxygen coordination for Zn(1) and a sixfold for Zn(2) are to be preferred.

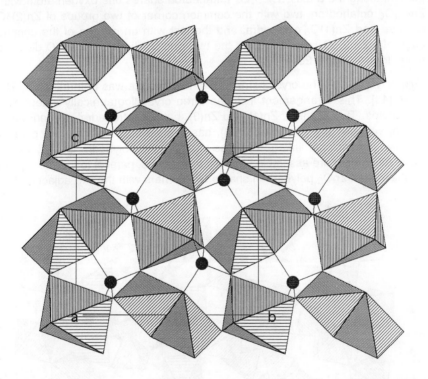

Figure 5.1.6. The structure of γ-$Zn_3(PO_4)_2$ projected along the **a** axis. The Zn polyhedra are shown by polyhedral representation, whereas the P atoms are represented by grey circles and P–O bonds by solid lines. This projection gives the impression that each PO_4 tetrahedron has a common edge with one of its adjacent ZnO_6 octahedron. Each PO_4 is, in fact, connected with two different octahedra, superimposed in projection and belonging to two successive layers of Zn polyhedra.

– γ-$Zn_3(PO_4)_2$. In this last form of $Zn_3(PO_4)_2$, whose crystal structure determination was also performed by Calvo [39], two crystallographically independent zinc atoms coexist. The first one, Zn(1), located on an inversion center has a sixfold oxygen coordination, whereas the second one, Zn(2), in general position has a fourfold. As in the two other forms, the ZnO_4 tetrahedra are assembled in pairs to build centrosymmetric Zn_2O_6 entities. As shown in Figure 5.1.6, each ZnO_6 octahedron shares four of its corners with four adjacent Zn_2O_6 groups and in the same way each Zn_2O_6 group shares four oxygen atoms with four ZnO_6 octahedra. This linkage of polyhedra builds a two-dimensional network around the plane $x = 0$. The successive layers of this type do not share any oxygen atoms. The phosphoric entities perform the cohesion of this arrangement along the **a** axis. The PO_4 tetrahedron shares one oxygen atom with a $Zn(1)O_6$ octahedron, two with the common corner of two groups of $Zn(2)_2O_6$ clusters and $Zn(1)O_6$ octahedra, and the last with an oxygen of the common edge of a $Zn(2)_2O_6$ cluster. Inside the two zinc polyhedra, the Zn–O distances range from 1.95 to 2.20 Å.

$Zn_3(PO_4)_2.H_2O$ – The crystal structure of this hydrate was determined by Riou *et al.* [40]. This arrangement includes three crystallographically independent zinc atoms. Two of them, Zn(1) and Zn(3), have a tetrahedral oxygen coordination, whereas the third one, Zn(2), has fivefold building a square pyramid. The two-dimensional linkage of these various zinc polyhedra is easy to describe. All of them are assembled to build thick layers, parallel to the (201) plane, in which each ZnO_5 polyhedron shares two corners with the two adjacent ZnO_5

Figure 5.1.7. Projection of a layer of zinc polyhedra approximately along the [201] direction in $Zn_3(PO_4)_2.H_2O$.

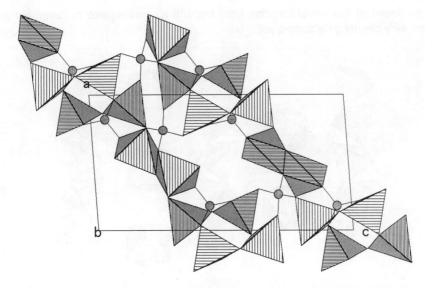

Figure 5.1.8. Projection, along the **b** axis, of a part of the atomic arrangement of $Zn_3(PO_4)_2.H_2O$. Zn polyhedra are given by polyhedral representation, whereas P atoms are shown by grey circles and P–O bonds by thin lines. This direction of projection corresponding to the axis of the $(ZnO_5)_n$ chain gives the impression of a common edge between two adjacent ZnO_5 groups.

groups to build a chain spreading along the **b** direction. The ZnO_4 groups form a centrosymmetric ring of four corner-sharing tetrahedra. Each Zn_4O_{12} ring shares two oxygen atoms with two adjacent $(ZnO_5)_n$ chains. The internal arrangement of an isolated slab of these zinc polyhedra is represented in Figure 5.1.7. The Zn–O distances, in the Zn polyhedra, vary between 1.886 and 2.102 Å. The water molecule is part of the coordination polyhedron of the pentacoordinated zinc atoms. The two hydrogen bonds correspond to O(W)–O distances of 2.765 and 2.800 Å.

Figure 5.1.8 shows how the two successive layers of Zn polyhedra are interconnected by the PO_4 tetrahedra to build a thick slab. These slabs do not share any oxygen atom, the three-dimensional cohesion of this network is being performed by the set of hydrogen bonds.

$Zn_3(PO_4)_2.4H_2O$: hopeite and parahopeite – Hopeite and parahopeite were the subject of many investigations since their discovery in Zambia by Spencer [56] in 1908. Various studies by Hill and Jones [41], Liebau [42], Gamidov *et al.* [43], Chao [44] and Whittaker [57] were devoted to the determinations of their atomic arrangements. The syntheses and thermal behavior were studied by de Schulten [51] and Hill and Jones [41]. In spite of these various studies, the

conditions of the transformation from hopeite to parahopeite or reversely are not very clearly established yet.

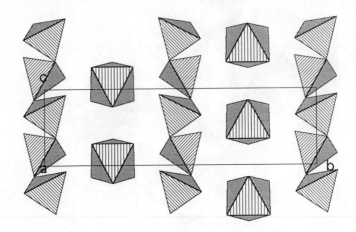

Figure 5.1.9. Projection, along the **a** axis, of an isolated slab of Zn polyhedra in hopeite.

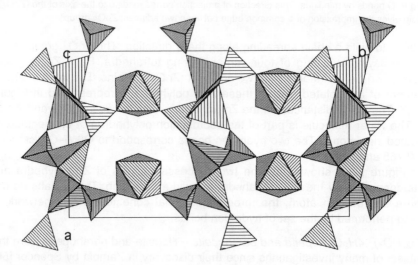

Figure 5.1.10. Projection, along the **c** axis, of the atomic arrangement of hopeite showing how the layers of zinc polyhedra are interconnected by the PO$_4$ groups, represented by the smaller tetrahedra. This direction of projection that corresponds to the axis of the (ZnO$_4$)$_n$ chain can give the impression of a common edge between two adjacent ZnO$_4$ groups.

– Hopeite. The atomic arrangement of hopeite, as mentioned above, was determined several times. For the present description, we use the data reported by Hill and Jones [41]. In this structure, one third of the zinc atoms have a sixfold coordination built by two oxygen atoms and four water molecules, two third are in tetrahedral coordination. All the zinc polyhedra are located in thick slabs perpendicular to the **a** direction and centered by the planes $x = 1/4$ and $3/4$. Inside such a layer, they are arranged differently. ZnO_6 polyhedra are independent, whereas ZnO_4 tetrahedra are assembled by corner sharing to build an infinite chain spreading along the **c** direction. As to depict the internal arrangement of such a layer, Figure 5.1.9 shows one of them projected along the **a** axis. As shown in Figure 5.1.10, these layers of zinc polyhedra are inter-connected along the **a** direction by the PO_4 tetrahedra. This projection also

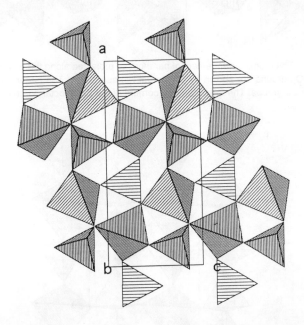

Figure 5.1.11. The internal arrangement of a layer of PO_4 and ZnO_4 tetrahedra in hopeite as seen projected along the **b** direction.

shows that both ZnO_4 and PO_4 tetrahedra are located in thick layers centered around the planes $z = 0$ and $1/2$. The internal arrangement of such a layer is given in Figure 5.1.11. Inside this sheet, each ZnO_4 tetrahedron of the $(ZnO_4)_n$ chain shares its four corners with four PO_4 groups and with its two adjacent ZnO_4 tetrahedra, whereas each PO_4 shares only three corners with the

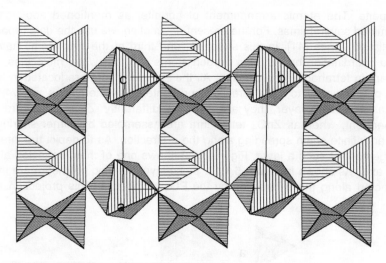

Figure 5.1.12. The atomic arrangement of parahopeite projected along the **c** direction. The PO_4, ZnO_4 and ZnO_6 groups are shown by polyhedral representation. The larger tetrahedra correspond to the ZnO_4 groups. The four unshared corners of the ZnO_6 octahedron correspond to the water molecules.

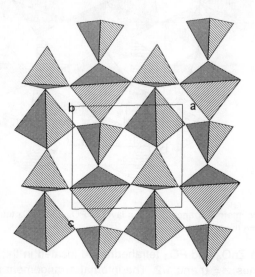

Figure 5.1.13. The internal arrangement of a layer of PO_4 and ZnO_4 tetrahedra as seen in projection along the **b** direction.

adjacent ZnO_4, the fourth one being shared with a ZnO_6 group to perform the cohesion along the **b** direction.

– Parahopeite. The atomic arrangement of parahopeite was determined by Chao [44]. This structure is a three-dimensional staking of PO_4, ZnO_4 tetrahedra and ZnO_6 octahedra. The centrosymmetric ZnO_6 group is located around the inversion center at $(0,0,0)$ and as in the corresponding polyhedra in hopeite is surrounded by two oxygen atoms and four water molecules. The ZnO_4 and PO_4 groups are assembled to build a compact layer of corner-sharing tetrahedra centered around the plane $y = 1/2$. Unlike hopeite, the ZnO_4 tetrahedra in this structure do not form chains. Figure 5.1.12, a projection along the **c** direction, shows this layer arrangement, whereas Figure 5.1.13, the projection along the **b** axis of an isolated layer of PO_4 and ZnO_4 tetrahedra, depicts their linkage. In this slab, each PO_4 group shares its four corners with four adjacent ZnO_4 groups and vice versa. As shown in Figure 5.1.12, the cohesion between these slabs is performed by the ZnO_6 octahedra.

ZnHPO₄.H₂O – The structure of this salt was determined by Riou *et al.* [46]. Three crystallographically independent zinc atoms coexist in the atomic arran-

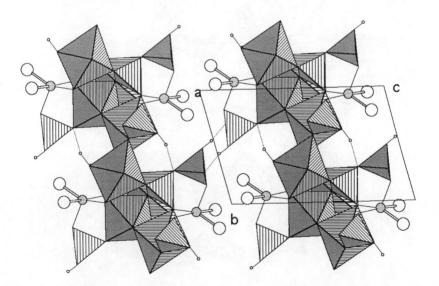

Figure 5.1.14. Projection along the **a** direction of the atomic arrangement of $ZnHPO_4.H_2O$. The ZnO_6 groups and the PO_4 tetrahedra are given in polyhedral representation. The Zn(3) atoms are shown by grey circles and the water molecules of the ZnO_5 polyhedra by open circles. Hydrogen atoms of the phosphoric groups are represented by small open circles.

gement. Two of them have a sixfold coordination, built by six oxygen atoms for Zn(1) and five oxygen atoms and one water molecule for the Zn(2). The Zn(3) atom has a fivefold coordination of three oxygen atoms and two water molecules. As shown in Figure 5.1.14, the atomic arrangement can be described by a succession of complex bidimensional stackings of polyhedra parallel to the (a, b) plane and interconnected along the **c** axis by hydrogen bonds. The central part of such a layer is made by a slab of edge-sharing ZnO_6 groups. This slab is lined by the HPO_4 and the $ZnO_3(H_2O)_2$ groups. The cohesion between the successive layers along the **c** direction is only performed by the hydrogen bonds. Among the various zinc polyhedra the Zn–O distances vary from 1.981 to 2.383 Å, whereas the Zn–(H_2O) distances vary between 2.011 and 2.202 Å. The P–OH distances in the two independent HPO_4 groups are 1.571 and 1.584 Å and the hydrogen bonds correspond to O–O or O(W)–O distances varying from 2.538 to 2.992 Å.

Zn(H₃O)PO₄ – This salt, whose atomic arrangement was determined by Sandormirskii *et al.* [45], has a global formula similar to that of the acidic mono-

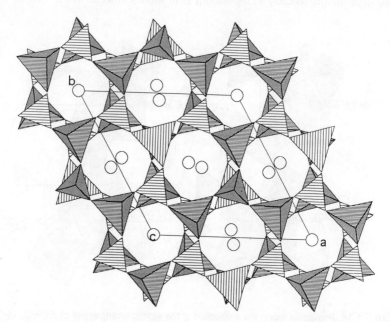

Figure 5.1.15. The atomic arrangement of $Zn(H_3O)PO_4$ projected along the **c** direction. The PO_4 and ZnO_4 tetrahedra are polyhedral representation. The larger tetrahedra correspond to the ZnO_4 groups. White open circles represent the $[H_3O]^+$ entities.

phosphate described above. Its formulation as $Zn(H_3O)PO_4$ suggested by the authors, following the localization of a part of the hydrogen atoms through low-temperature experiments was for a time questionable so much the more that no detail was given concerning the chemical preparation. However, a careful examination of the stacking of the heavy atoms, Zn, P, O, seems to exclude the possibility of the existence of an acidic HPO_4 group and comfort the assumption of the authors. Later, Cudennec *et al.* [36] suggested that this salt could, in fact, be $ZnNH_4PO_4$.

The atomic arrangement is a typical tridymite-like structure, built by a three-dimensional stacking of six-member rings of alternating ZnO_4 and PO_4 tetrahedra. As shown in Figure 5.1.15, this tetrahedron staking creates large channels parallel to the **c** axis, in which are located the $[H_3O]^+$ entities. In the two independent ZnO_4 tetrahedra, the Zn–O distances range from 1.895 to 1.959 Å, whereas the P–O distances in the two PO_4 tetrahedra vary between 1.509 and 1.546 Å. The hydrogen bonds connecting the $[H_3O]^+$ entities are not very strong since they correspond to an average of 2.84 Å for the O–O distances.

$Zn(H_2PO_4)_2.2H_2O$ – As early as 1879, Demel [58] mentioned the existence of this salt. It is later identified by Averbuch-Pouchot [47] as an isotype of the corresponding cadmium salt, whose atomic arrangement was previously

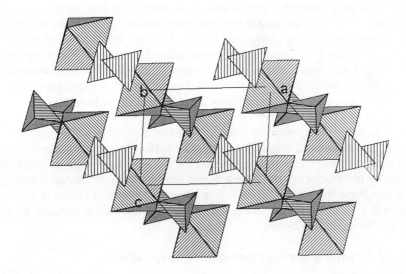

Figure 5.1.16. Projection of the atomic arrangement of $Zn(H_2PO_4)_2.2H_2O$ along the **b** direction. The PO_4 tetrahedra and ZnO_6 octahedra are polyhedral representation and the hydrogen atoms are removed.

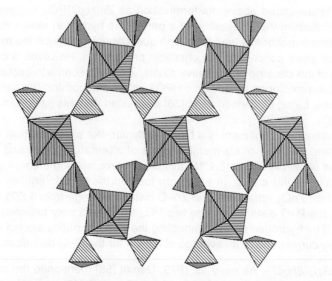

Figure 5.1.17. Projection along the [10$\bar{1}$] direction of an isolated layer of ZnO$_6$ and PO$_4$ polyhedra in Zn(H$_2$PO$_4$)$_2$.2H$_2$O. Drawing conventions are identical to those used in Figure 5.1.16.

described by Averbuch-Pouchot and co-workers [59–60]. Since the crystal structure of the zinc compound has not been refined, we examine and discuss this arrangement by using the data obtained for the isotypic cadmium salt. As shown in Figure 5.1.16, this is a layer structure. Layers, built by centrosymmetric ZnO$_4$(H$_2$O)$_2$ octahedra and PO$_4$ tetrahedra connected by corner-sharing, alternate parallel to the (101) plane. In Figure 5.1.17, a projection along the [10$\bar{1}$] direction describes the internal arrangement of an isolated layer. Inside this sheet, a ZnO$_6$ octahedron shares its four oxygen atoms with the four adjacent PO$_4$ tetrahedra, whereas each tetrahedron shares two of its corners with two adjacent ZnO$_6$ groups. For the ZnO$_6$ octahedra, the unshared corners correspond to the water molecules and for the PO$_4$ tetrahedra to the oxygen atoms of the P–OH bonds. These layers do not share any oxygen atom and thus the three-dimensional cohesion is uniquely performed by the H–bonds. In Figure 5.1.18, a projection along the **c** direction shows the intricate arrangement of the hydrogen bonds.

5.1.3.3. Crystal growth of zinc monophosphates

At the end of the section devoted to monophosphates, we will report the most general procedures used for their preparations. Nevertheless, to complete

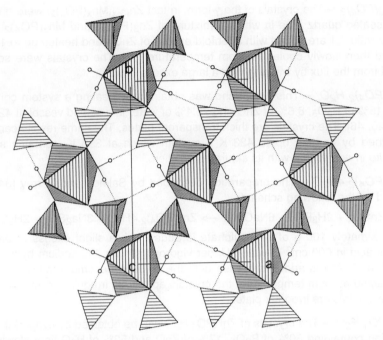

Figure 5.1.18. Projection, along the **c** direction, of the Zn(H₂PO₄)₂.2H₂O atomic arrangement. PO₄ tetrahedra and ZnO₆ octahedra are polyhedral representation, hydrogen atoms represented by small open circles and hydrogen bonds by dotted and solid lines.

this detailed survey of zinc monophosphates, it would be interesting to describe briefly the chemical preparations corresponding to the above described compounds. Most of the preparations that are described below were used or optimized by crystallographers during these investigations. Thus, this short parenthesis will illustrate some processes commonly used when suitable single crystals are needed for structural investigations.

α-Zn₃(PO₄)₂ – The crystals were grown by using a flux of ZnCl₂ [37]. A mixture of ZnCl₂ and Zn₃(PO₄)₂ with a tenfold molar excess of the chloride was placed in a vycor tube closed with quartz wool. The zinc monophosphate is completely dissolved at 923 K. Crystals are then grown by slow cooling of the melt down to room temperature. They are isolated from the flux by the addition of water and they appear as needles elongated along the **b** direction.

β-Zn₃(PO₄)₂ – The crystals were obtained by slow cooling a melt of Zn₃(PO₄)₂ down to its melting temperature and quenching it to room temperature [38].

γ-$Zn_3(PO_4)_2$ – The crystals of the γ-form, in fact $Zn_{(3-x)}Mn_x(PO_4)_2$, were grown in a sealed quartz tube in which a mixture of $Zn_3(PO_4)_2$ and $Mn_3(PO_4)_2$ in a molar ratio 7:1 are added with a tenfold excess of $ZnCl_2$ and heated up to 1173 K and then slowly cooled to room temperature [39]. The crystals were separated from the flux by washing with a large excess of water.

$Zn_3(PO_4)_2.H_2O$ – This compound was obtained by heating a system containing 15% of P_2O_5, 8.6% of ZnO and 76.4% of water in a sealed vessel at 423 K [36, 40, 46]. The crystals are thick transparent plates. The same results can be obtained by heating up to 423 K a system which at 393 K, contains solid hopeite in equilibrium with its saturated solution.

$Zn_3(PO_4)_2.4H_2O$ – The preparation described by Salmon and Terrey [34] is based on the following scheme:

$$3ZnSO_4 + 2H_3PO_4 + 6NaOH \longrightarrow Zn_3(PO_4)_2.4H_2O + 3Na_2SO_4 + 2H_2O$$

Approximately 100 g of zinc sulphate is added with a slight excess of phosphoric acid in 600 cm^3 of water. Under vigorous stirring, the sodium hydroxide solution (1–2N) is added dropwise until the solution becomes neutral. When conducted at room temperature, the crystals appear as irregular prisms. At 353 K, the crystals are irregular plates.

$ZnHPO_4.H_2O$ – The crystals of $ZnHPO_4.H_2O$ can be obtained by evaporating a solution containing 30% of P_2O_5, 17% of ZnO and 53% of H_2O very slowly at 353 K [36, 40, 46]. The crystals appear as transparent truncated parallelepiped needles.

$Zn(H_3O)PO_4$ – The authors did not report any details about the production of this compound which is simply said to have been prepared by hydrothermal synthesis [45].

$Zn(H_2PO_4)_2.2H_2O$ – This salt is prepared by slowly adding zinc carbonate to phosphoric acid till a precipitation occurs at room temperature [47]. A further evaporation, at the same temperature, leads to the formation of large single crystals.

5.1.3.4. For fun: a last example of divalent-cation monophosphate

Among divalent-cation monophosphates, we sometimes met unexpected compounds. We cannot resist the pleasure of describing one of them: a rare example of a divalent rhodium salt, $Rh(H_2PO_4)_2.H_2O$. This compound, studied by Dikareva *et al.* [61], is monoclinic with:

$$a = 7.285, \ b = 13.564, \ c = 7.225, \ \beta = 93.59, \ Z = 4$$

It was described in space group *C2* although the centrosymmetric group *C2/m* seems the most probable based on the data reported by the authors. The basic building unit is a centrosymmetric arrangement made by two central rhodium atoms separated by 2.487 Å. This pair of atoms is surrounded by four $[H_2PO_4]^-$ groups and two water molecules. Each phosphoric group shares two O atoms with the Rh pair. Each Rh atom is thus surrounded by a distorted octahedron built by a Rh atom, four O atoms from the phosphoric groups and one water molecule. The Rh–O distances vary from 2.02 to 2.08 Å and the Rh–H_2O distance is 2.292 Å. These entities are interconnected by a three-dimensional network of H–bonds. A perspective view of this building unit is shown in Figure 5.1.19.

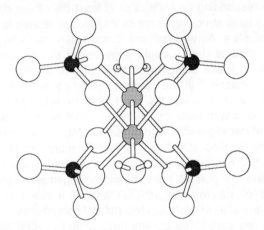

Figure 5.1.19. A perspective view of the $[Rh_2(H_2PO_4)_4(H_2O)_2]$ unit. The grey circles are the Rh atoms, the black ones are P atoms, the large open ones are O atoms and water molecules. H atoms of the phosphoric groups are omitted, those of the water molecule are small open circles.

5.1.4. Divalent-Monovalent Cation Monophosphates

A great variety of stoichiometries exist for the divalent-monovalent cation monophosphates. For instance, the existence of many compounds as different as $Cu_3Na_5(PO_4)_2(HPO_4)$, $Mg_2Ca_4Na_4(PO_4)_{18}$ and $CaMn_7Na_2(PO_4)_6$ is well established to date, but this class of compounds is mainly dominated by the $M^{II}M^IPO_4$ monophosphate family that we have selected as examples in this section. The choice of this category of compounds provides the opportunity to survey a good number of well-known types of structures. Due to the interesting ferroic properties of some of them, they were thoroughly investigated. A few of them like $BeNaPO_4$ (beryllonite), $FeLiPO_4$ (triphylite), $MnLiPO_4$ (lithiophilite), $MnNaPO_4$ (natrophilite) are natural compounds.

As can be inferred from their chemical formulæ, $M^{II}M^{I}PO_4$ monophosphates are not water soluble. Many of them are polymorphic. For instance, $ZnTlPO_4$ undergoes two transformations at 536 and 723 K, before melting congruently at 1363 K. $ZnCsPO_4$ also has three crystalline forms with transition temperatures at 523 and 583 K, before decomposing at 1713 K.

At a structural point of view, most $M^{II}M^{I}PO_4$ monophosphates crystallize with atomic arrangements that are more or less similar to well established structure types. In the first step, a good part of these compounds can roughly be classified into three categories:

– when one of the associated cations adopts a tetrahedral oxygen coordination, the corresponding derivatives or at least their three-dimensional tetrahedron networks have structures more or less closely related to that of the *high-tridymite* form of silica. As we shall see below when discussing some of these arrangements, none of them adopts a geometry strictly identical to that of tridymite. Some of these compounds have atomic arrangements which are similar to those observed in the room-temperature form of NH_4LiSO_4 or in $BeNaPO_4$; other ones, although their unit cells are related to that of tridymite, cannot be compared with certainty to known structure types due to the lack of enough structural investigations. To date, Li, Be, Mg, Co, Zn, and Mn have been observed in tetrahedral coordination in $M^{II}M^{I}PO_4$ monophosphates.

– when the two associated cations are in octahedral oxygen coordination, the corresponding compounds have atomic arrangements identical or closely related to that of *olivine* (Mg_2SiO_4) and in a few cases to *maricite* $[(Fe_{0.9}Mn_{0.06}Mg_{0.03}Ca_{0.01})NaPO_4]$, two natural compounds.

– when the two associated cations have higher oxygen coordinations, the structural arrangement of the corresponding salts is related in many cases to the orthorhombic room-temperature form of K_2SO_4 (*arcanite*) or to structure types as *aphtitalite* ($NaKSO_4$) or *glaserite* $[NaK_3(SO_4)_2]$ both of which are very similar to that of the high-temperature form of K_2SO_4.

As in all family of compounds, some of them cannot be classified as in the case of $NiKPO_4$ and $MgKPO_4$ in which the divalent cations have a pentahedral oxygen coordination and of some copper derivatives.

5.1.4.1. Tridymite-like compounds

As we already said, a good number of $M^{II}M^{I}PO_4$ monophosphates have atomic arrangements more or less closely related to that of high-tridymite. There are three forms of tridymite: high-, middle- and low-tridymite, but only the high-form arrangement has been clearly determined. High-tridymite is hexagonal, $P6_3/mmc$, with $Z = 4$ and the following unit-cell dimensions:

$$a = 5.04, \quad c = 8.24 \text{ Å}$$

The structure, like that of the other classical polymorphs of silica, is based on a three-dimensional linkage of corner-sharing SiO_4 tetrahedra. In order to facilitate further comparisons with the various types of tridymite-like $M^{II}M^{I}PO_4$ monophosphates, we describe this arrangement by considering the planes of SiO_4 tetrahedra alternating perpendicular to the **c** direction. In Figure 5.1.20, a projection of high-tridymite along the **c** direction shows clearly the internal arrangement of such a plane. SiO_4 tetrahedra are assembled to build six-member rings sharing oxygen atoms. These rings are linked together to create large empty channels around the 6_3 helical axes. In a given ring, the tetrahedron apices are alternatively directed up and down, so that each tetrahedron pointing upward in a layer is surrounded by three tetrahedra pointing downward and vice versa. Figure 5.1.21 shows how the successive planes of rings are assembled. All the oxygen atoms, shared between adjacent planes, are located in the mirror planes, $z = 1/4$ and $3/4$, so that each plane can be considered to be reproduced from the adjacent one by a mirror operation. The two projections of this arrangement show the existence of large empty channels along the three directions of the unit cell.

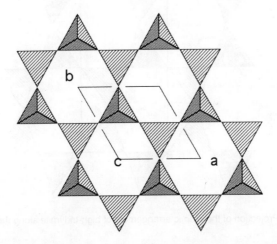

Figure 5.1.20. Projection of the atomic arrangement of high-tridymite along the **c** direction.

The substitution scheme, leading to the formation of silica-like $M^{II}M^{I}PO_4$ monophosphates, is evidently:

$$2Si \longrightarrow P + M^{II}$$

building the $M^{II}PO_4$ tetrahedral networks, the second associated cation being located in the above mentioned channels.

The tetrahedral framework of these compounds never adopts an arrangement and a symmetry strictly analogous to that of high-tridymite. Most of them
have orthorhombic or monoclinic unit cells related to the hexagonal unit cell of
high-tridymite by the relationship:

$$a = a_t, \quad b = a_t\sqrt{3}, \quad c = c_t$$

These distortions, induced both by the mixed composition of the tetrahedron
framework and by the nature of the associated cations located inside the
channels, create some changes in the geometry of the networks of tetrahedra,
so that tridymite-like $M^{II}M^{I}PO_4$ monophosphates must be sorted in three
different categories that we examine below. The distinction is based on the
geometry of the tetrahedron networks and not on the unit-cell values.

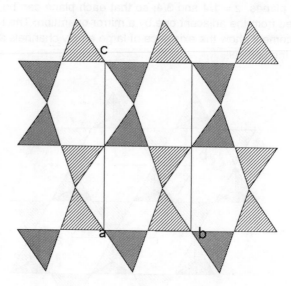

Figure 5.1.21. Projection of the atomic arrangement of high-tridymite along the **a** direction.

NH_4LiSO_4-like compounds – Some salts have an anionic framework with a
geometry identical to that observed in the room-temperature form of NH_4LiSO_4.
We report their main crystal data in Table 5.1.3 and explain their structural
features using the sulphate structure.

NH_4LiSO_4 is an orthorhombic compound [62], with the space group $Pnb2_1$,
$Z = 4$ and the following unit-cell dimensions:

$$a = 8.786, \quad b = 9.140, \quad c = 5.280 \text{ Å}$$

The relationship between this unit cell and that of tridymite is:

$$a = c_t, \quad b = a_t\sqrt{3}, \quad c = a_t$$

Figure 5.1.22, a projection of this arrangement along the **a** direction shows the arrangement of one layer of tetrahedra. Its general features are roughly similar to that of tridymite, but the six-member rings are built by alternating LiO_4 and SO_4 tetrahedra and, in a given ring, three adjacent tetrahedra have their apices upward and the other three downward.

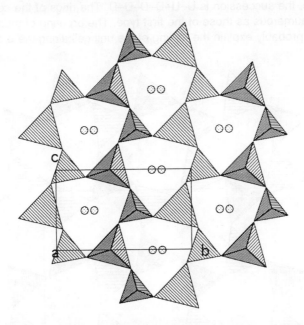

Figure 5.1.22. Projection of the atomic arrangement of NH_4LiSO_4 along the **a** direction. Open circles represent the ammonium groups. The larger tetrahedra correspond to the LiO_4 entities.

Beryllonite-like compounds – Some tridymite-like $M^{II}M^{I}PO_4$ monophosphates adopt the structure of the beryllonite, $BeNaPO_4$. We report their unit-cell dimensions in Table 5.1.4.

As can be seen in this table, the relationship between the unit cells of tridymite and that of beryllonite is given by:

$$a = a_t\sqrt{3}, \quad b = c_t, \quad c = 3a_t$$

An exception is $ZnNa_{0.79}K_{0.21}PO_4$ [78], prepared by hydrothermal synthesis and whose crystallographic data are:

$a = 17.609(5)$, $c = 8.112(2)$ Å, $Z = 24$, and space group: $P6_3$

As in all tridymite-like compounds, the tetrahedral framework of the beryllonite arrangement is built by a succession of planes of corner-sharing tetrahedra, in this case perpendicular to the **b** direction. As usual, in such planes, the tetrahedra are assembled to form interconnected six-member rings, but in beryllonite two different ring geometries coexist inside a plane. As shown in Figure 5.1.23, one category of rings is similar to those observed in tridymite itself with alternating up (U) and down (D) vertices of the tetrahedra, whereas, in the second type, the succession is U–U–D–D–U–D. The rings of the second type are twice as numerous as those of the first type. The ordering of these two kinds of rings can probably explain the tripling of the unit cell along the **c** direction.

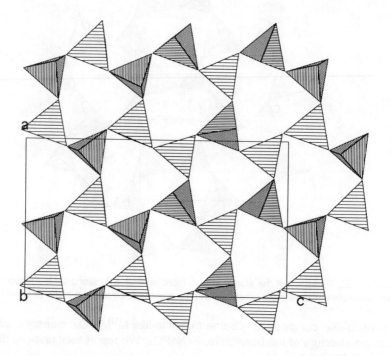

Figure 5.1.23. Projection, along the **b** direction, of a plane of tetrahedra in the beryllonite structure. The sodium atoms are not represented.

Possible tridymite-like $M^{II}M^{I}PO_4$ monophosphates – A good number of other $M^{II}M^{I}PO_4$ monophosphates, with unit-cell dimensions diversely but closely related to those of trydimite, crystallize probably with tetrahedral frameworks more

Table 5.1.3

Main crystal data for the $M^{II}M^{I}PO_4$ monophosphates, crystallizing with a NH_4LiSO_4-type tetrahedron framework.

Formula		a α	b β	c (Å) $\gamma(°)$	S. G.	Z	Ref.
$BeKPO_4$	(a)	8.506(4)	4.937(4)	8.344(5)	$Pc2_1n$	4	[63–64]
$BeCsPO_4$	(a)	8.713(4)	8.836(5)	5.147(4)	$Pnam$	4	[64]
$BeTlPO_4$	(b)	9.298(1)	8.101(5)	4.843(1)	$Pn2_1a$	4	[65]
$CoRbPO_4$	(c)	8.862(1)	5.431(1) 90.18(1)	8.997(1)	$P2_1$	4	[66]
$MgCsPO_4$		8.935(2)	5.526(1)	9.646(2)	$Pnma$	4	[66]
$ZnCsPO_4$	(a)	9.191(5)	5.493(5)	9.406(5)	$Pnma$	4	[67]
$ZnCsPO_4$	(d)	9.246(9)	5.468(6)	9.359(7)	$Pn2_1a$	4	[67]
$ZnCsPO_4$	(c)	18.33(8) $= 2c_t$	5.45(4) 90.14(8)	9.25(6)	$P2_1/a$	8	[67]
$ZnTlPO_4$		8.828(2)	5.462(1) 90.61(2)	8.729(1)	$P2_1$	4	[65, 68–69]
$ZnRbPO_4$		8.855(3)	5.408(1) 90.30(3)	8.956(4)	$P2_1$	4	[68, 70]
$MnCsPO_4$	(b)	9.575(2)	9.128(3)	5.595(1)	$Pbn2_1$	4	[71]

a: high-temperature form
b: hydrothermal synthesis
c: room-temperature form
d: intermediate-temperature form

Table 5.1.4

Unit-cell dimensions of $M^{II}M^{I}PO_4$ monophosphates isotypic of beryllonite. The common space group is $P2_1/n$ and $Z = 12$.

Formula	a α	b β	c (Å) $\gamma(°)$	Ref.
$BeNaPO_4$	8.178	7.818 90.00	14.114	[72–74]
$ZnNaPO_4$	8.667(1)	8.114(1) 90.116(7)	15.262(2)	[73, 75–77]
$ZnAgPO_4$	8.762	8.125 90.25	15.46	[73, 75]

or less similar to that of this compound, but because of a lack of structural characterizations, they cannot be classified in one of the types of structure that we have examined above. This lack of structural informations can be partly explained by the fact that many crystals of this series are often twinned. Table 5.1.5 contains some crystallographic data for these compounds.

Table 5.1.5
Crystallographic data for various $M^{II}M^{I}PO_4$ monophosphates crystallizing with a tridymite-like tetrahedral framework.

Formula	a α	b β	c (Å) γ (°)	S. G.	Z	Ref.
BeKPO$_4$ (a)	8.52	8.34 90.0	4.94	$P2_1/n$	4	[79]
BeRbPO$_4$ (a)	8.63	8.58 90.0	5.01	$P2_1/n$	4	[79]
BeRbPO$_4$ (b)	8.636(3)	5.012(2)	8.587(3)	$Pc2_1n$ or $Pcmn$	4	[64]
ZnNaPO$_4$ (b)	8.735	8.735	8.060	$P6_3$?	6	[75]
ZnKPO$_4$	18.14(1)	18.14(1)	8.504(5)	$P6_3$	24	[80–81]
ZnNH$_4$PO$_4$	8.960	5.438 90.35	8.781	$P2_1/n$?	4	[82]
ZnNH$_4$PO$_4$	10.67	10.67	8.69	$P6_3$?	8	[82]
MgRbPO$_4$	9.303(3)	5.338(2)	8.829(2)	$Pc2_1n$	4	[65, 66]
MgTlPO$_4$	8.383(1)	5.336(1) 91.91(1)	9.538(1)	$P2_1/m$ or $P2_1$	4	[65]
CoKPO$_4$ (b)	8.944(2)	5.297(1)	8.551(2)	$P2_12_12_1$	4	[66]
CoCsPO$_4$	18.44(1)	5.476(2) 90.40(1)	9.294(2)	Pa	8	[66]
CoRbPO$_4$ (b)	9.006(2)	5.419(1)	8.843(1)	$P2_12_12_1$	4	[66]
CoTlPO$_4$	8.965(3)	5.495(2) 90.53(1)	8.775(3)	$P2_1/m$ or $P2_1$	4	[66]

a: room-temperature form
b: high-temperature form

5.1.4.2. Olivine and maricite-like compounds

Olivine, Mg_2SiO_4, is a well-known silicate structural type, whose atomic arrangement was studied as early as 1926 by Bragg and Brown [83]. The magnesium atoms are located on two different crystallographic sites. One is on an inversion center, the other in a mirror plane. In the case of the $M^{II}M^{I}PO_4$ monophosphates crystallizing with the olivine arrangement, the two cations are

Table 5.1.6

Main crystallographic data for the olivine-like $M^{II}M^{I}PO_4$ monophosphates. Various crystallographic settings were used by the authors.

| Formula | a | b | c (Å) | S. G. | Z | Ref. |
	α	β	γ (°)			
NiLiPO$_4$	10.03	5.85	4.68	Pnma	4	[84]
CoLiPO$_4$	10.200(1)	5.920(1)	4.690(2)	Pnma	4	[84–85]
FeLiPO$_4$	4.67	10.34	6.00	Pbnm	4	[86]
MgLiPO$_4$	10.147	5.909	4.692	Pnma	4	[84, 87]
MnLiPO$_4$	6.10(2)	10.46(3)	4.744(10)	Pmnb	4	[88–90]
MnNaPO$_4$	10.523(5)	4.987(2)	6.312(3)	Pnam	4	[91–92]
CdLiPO$_4$	10.713(4)	4.801(1)	6.284(3)	Pna2$_1$	4	[70, 93]
CdNaPO$_4$	10.832	6.494	5.055	Pnma	4	[94–95]

ordered on these two sites and it is now well established that the monovalent cations have a preference for the centrosymmetric one with the exception of the few maricite-like derivatives in which the reverse is observed. Thus, the two

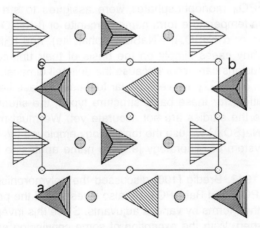

Figure 5.1.24. Projection, along the **c** direction, of the MnLiPO$_4$ structure. PO$_4$ groups are shown in polyhedral representation, larger circles represent manganese atoms, smaller ones are the lithium atoms.

structure types are closely related. Table 5.1.6 contains the main crystallographic data for the $M^{II}M^{I}PO_4$ monophosphates belonging to the olivine-structure type and Table 5.1.7 those of the maricite-like derivatives.

The structure of $MnLiPO_4$ as determined by Geller and Durand [88] is an example of olivine arrangement. The Li and Mn atoms are ordered. The LiO_6 groups are centrosymmetric and the MnO_6 octahedra have a mirror symmetry and both have distorted octahedral coordinations. By sharing edges, the LiO_6 polyhedra are assembled to build chains parallel to the **a** direction in the planes $y = 0$ and $1/2$, whereas the MnO_6 octahedra build layers of corner-sharing octahedra centered by planes $y = 1/4$ and $3/4$. Figure 5.1.24 gives a projection of this arrangement along the **c** direction.

Table 5.1.7

Main crystallographic data for the maricite-like $M^{II}M^{I}PO_4$ monophosphates. The common space group is *Pmnb* and $Z = 4$.

Formula	a	b	c (Å)	Ref.
$MnNaPO_4$ (a)	6.904(1)	9.088(1)	5.113(1)	[96]
$FeNaPO_4$	6.861	8.987	5.045	[97]

a: hydrothermal synthesis

5.1.4.3. β-K₂SO₄, α-K₂SO₄, aphtitalite and glaserite-like compounds

Many $M^{II}M^{I}PO_4$ monophosphates were assigned to structure types of K_2SO_4 (the room-temperature form named arcanite or β-K_2SO_4 and the high-temperature form or α-K_2SO_4), $KNaSO_4$ (aphtitalite), and $K_3Na(SO_4)_2$ (glaserite), but in many cases simply on the basis of their unit-cell dimensions. Thus, many confusions can occur because the unit-cell dimensions of the high-temperature form of K_2SO_4 are very similar to those of glaserite and aphtitalite and that, in addition, for these basic structure types, the situation is far from being clear since the studies are not accurate yet. We must mention that for $KNaSO_4$ and $K_3Na(SO_4)_2$ we use the terminology employed by Wyckoff [98] for in the Dana's System of Mineralogy [99] the name aphtitalite is attributed to $K_3Na(SO_4)_2$.

As early as 1942, Bredig [100] discussed the polymorphism of $CaKPO_4$, $CaNaPO_4$, $SrNaPO_4$, and $BaNaPO_4$. He also investigated the possibility of stabilizing some of their forms by various adjuvants. Since this investigation, many data were published. With the exception of some convincing studies showing clearly that the two accurately investigated forms of $BaNaPO_4$ [101–102] are of glaserite type, that $BaKPO_4$ [103–104], $CaNaPO_4$ [105] and $BaRbPO_4$ [70] are arcanite-like salts etc..., much work remains to be done in order to verify numerous assumptions based mostly on the values of the unit-cell dimensions. We report in Tables 5.1.8 and 5.1.9 the main crystallographic data for the $M^{II}M^{I}PO_4$ monophosphates, belonging to the room-temperature form of K_2SO_4 (β-K_2SO_4), the high-temperature form of K_2SO_4 (α-K_2SO_4), $KNaSO_4$ (aphtitalite) and K_3NaSO_4 (glaserite) respectively. We shall describe below these different structure types in an attempt to compare them.

Table 5.1.8

$M^{II}M^{I}PO_4$ monophosphates reported as belonging to the room-temperature form of K_2SO_4. The common space group is *Pnma* and $Z = 4$.

Formula		a	b	c (Å)	Ref.
CaNaPO$_4$	(a)	6.83	5.215	9.32	[100]
SrNaPO$_4$		7.33	5.40	9.57	[100]
BaNaPO$_4$		7.11	5.59	8.82	[106]
SrKPO$_4$		7.22	5.51	9.76	[107]
BaKPO$_4$		7.709(4)	5.663(4)	9.972(5)	[103–104]
BaRbPO$_4$		7.812(2)	5.740(1)	10.056(2)	[70]

a: also given with $a = 20.397(10)$, $b = 5.412(4)$, $c = 9.161(5)$ Å, $Pn2_1a$ [105]

Table 5.1.9

$M^{II}M^{I}PO_4$ monophosphates which belong or possibly belong to α-K_2SO_4 ($P6_3/mmc$), $KNaSO_4$ ($P3m1$) or K_3NaSO_4 ($P\bar{3}m1$). In all cases, $Z = 2$.

Formula		a	c (Å)	Ref.
CaNaPO$_4$	(1173 K)	5.36	7.33	[100, 106–107]
SrNaPO$_4$	(1373 K)	5.45	7.67	[100, 106–107]
BaNaPO$_4$	(a)	5.622(5)	7.259(1)	[101]
CaKPO$_4$	(1273 K)	5.50	7.57	[100, 106–107]
SrKPO$_4$	(1373 K)	5.65	7.78	[106–107]
BaKPO$_4$	(1273 K)	5.80	8.15	[100, 106–107]

a: proved to be of the glaserite-type

b: also given with $a = 9.743(3)$, $b = 5.622(1)$, $c = 7.260(1)$ Å, $\beta = 90.10(3)°$, *C2/m* [102]

The temperatures of the measurements are given in parentheses.

The room-temperature form of K_2SO_4 (arcanite) – The room-temperature form of K_2SO_4 is othorhombic [108], *Pnma*, with $Z = 4$ and the following unit-cell dimensions:

$$a = 7.476(3), \quad b = 5.763(2), \quad c = 10.071(4) \text{ Å}$$

This pseudo-hexagonal arrangement is depicted in projection along the **a** direction in Figure 5.1.25. It can be described as being constructed by two types of rows both parallel to the **a** direction. The rows of the first kind, built by alternating SO_4 groups and K atoms, are arranged to create large hexagonal channels in which are inserted the rows of the second kind made by zigzag chains of potassium atoms. In this arrangement, the two crystallographically independent potassium atoms, the sulphur atom, and two oxygen atoms are located in the mirror planes in $y = 1/4$ and $3/4$.

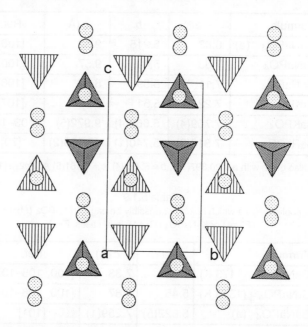

Figure 5.1.25. Projection, along the **a** axis, of the structure of β-K_2SO_4. The SO_4 groups are shown in polyhedral representation and dotted circles represent the potassium atoms.

The $K_3Na(SO_4)_2$ (glaserite), $KNa(SO_4)_2$ (aphtitalite), and β-K_2SO_4 structures – – Glaserite is trigonal [109], $P\bar{3}m1$, with $Z = 2$ and the following unit-cell dimensions:

$$a = 5.680(1), \quad c = 7.309(3) \text{ Å}$$

To describe the glaserite structure, one can use almost the same words as those used to describe arcanite. As shown by Figure 5.1.26, this arrangement is

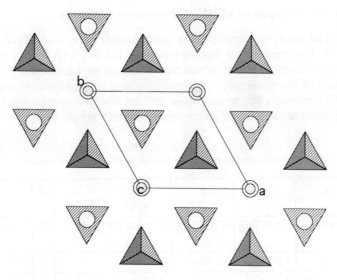

Figure 5.1.26. Projection, along the **c** direction, of the atomic arrangement of $K_3Na(SO_4)_2$. The potassium atoms are represented by the larger open circles, the sodium atoms by the smaller dotted circles. The SO_4 tetrahedra are shown by polyhedral representation.

built by the same kind of rows, but this time parallel to the ternary axis and whose building components are slightly different. Rows of the first type made by alternating potassium atoms and SO_4 groups are here disposed in a strict hexagonal way around the ternary axis. The second type of row is a linear array of alternating Na and K atoms. As in arcanite, these latter rows are inserted in the hexagonal channels created by the stacking of the first ones.

– Aphtitalite is trigonal [109], *P*3*m*1, with $Z = 2$ and the following unit-cell:

$$a = 5.067(1), \quad c = 7.177(1) \text{ Å}$$

The only difference between this arrangement and the precedent one is the replacement of one potassium atom sited on the internal threefold axis by a sodium atom.

– α-K_2SO_4 [110] is hexagonal, *P*6$_3$/*mmc*, with $Z = 2$ and unit-cell dimensions:

$$a = 5.947(2), \quad c = 8.375(3) \text{ Å } (1073 \text{ K})$$

are very similar to those of glaserite and aphtitalite. If the disorder of the SO_4 tetrahedra is not taken into account, this compound has the same type of structure that the two latter ones, the sodium atoms are being replaced by potassium atoms.

5.1.4.4. Other types of $M^{II}M^IPO_4$ monophosphates

As in all family of compounds, a good number of representatives cannot be classified. The crystal data that we gather in Table 5.1.10 correspond to some $M^{II}M^IPO_4$ monophosphates which cannot clearly be assigned or even compared to one of the previously described structure types. However, among them, $HgNaPO_4$ presents strong analogies with an olivine arrangement. Most of the investigations reported in this table are recent and accurate.

Table 5.1.10
Crystallographic data for various $M^{II}M^IPO_4$ monophosphates.

Formula	a α	b β	c (Å) γ (°)	S. G.	Z	Ref.
α-ZnLiPO$_4$	17.250(6)	9.767(3)1 10.9(3)	17.106(6)	Cc	32	[111–112]
PbLiPO$_4$	7.945(6)	18.46(2)	4.928(4)	$Pna2_1$	8	[70, 113–115]
α-CuNaPO$_4$	9.708(5)	4.805(2)	7.166(4)	$P2_12_12_1$	4	[116]
β-CuNaPO$_4$	6.944(1)	8.851(1) 97.25(1)	5.014(1)	$P2_1/n$	4	[117]
CdNaPO$_4$	5.226(3)	11.66(1)	6.549(3)	$Pnma$	4	[95]
HgNaPO$_4$	5.883(1)	9.401(3)	6.448(1)	$Cmcm$	4	[118]
CoAgPO$_4$	9.516(2) 102.2(1)	5.574(1) 106.2(1)	6.572(2) 80.1(1)	$P\bar{1}$	4	[119]
β-CuAgPO$_4$	7.500(1)	15.751(2)	5.702(1)	$Pbca$	8	[120]
CuKPO$_4$	8.278(1)	9.720(1) 92.13(1)	4.942(1)	$P2_1$	4	[121]
CuKPO$_4$	17.94(2)	6.742(6)	6.795(6)	$Pbca$	8	[122]
MgKPO$_4$	8.582(1)	5.080(1) 91.68(1)	19.020(2)	$P2_1/c$	8	[65]
NiKPO$_4$	8.64	9.28	4.93	$Pna2_1$	4	[123]
BeNH$_4$PO$_4$	12.80	12.80	9.65	$P\bar{4}21c$	16	[124]

5.1.5. Trivalent-Cation Monophosphates

The chemistry of trivalent-cation monophosphates is very rich. For instance, let us examine the presently known aluminum phosphates. Eleven natural compounds and nine synthetic ones are well characterized in the $Al_2O_3–P_2O_5–$

H_2O system and their list given in Table 5.1.11, but established some months ago, is probably not exhaustive today.

Table 5.1.11
Aluminum monophosphates.

Natural compound	Mineral name	Synthetic compound
$AlPO_4$	Berlinite	$Al(H_2PO_4)_3$
$AlPO_4.2H_2O$ (I)	Variscite	$[Al_3(H_2PO_4)_6(HPO_4)_2](H_3O).4H_2O$
$AlPO_4.2H_2O$ (II)	Metavariscite	$Al(H_2PO_4)(HPO_4).H_2O$
$Al_2(OH)_3(PO_4)$	Augelite	$Al_2(HPO_4)_3.7/2H_2O$
$Al_2(OH)_3(PO_4).H_2O$	Senegalite	$Al_2(HPO_4)_3.4H_2O$
$Al_2(OH)_3(PO_4).19/4H_2O$	Bolivarite	$Al(H_2PO_4)(HPO_4).5/2H_2O$
$Al_3(OH)_3(PO_4)_2.5H_2O$	Wavellite	$Al_2(H_{2-x}PO_4)_3(H_{3x}PO_4).6H_2O$
$Al_3(OH)_6(PO_4).6H_2O$	Evansite	$Al_2(HPO_4)_3.13/2H_2O$
$Al_3(OH)_3(PO_4)_2.9H_2O$	Kingite	$Al_2(HPO_4)_3.8H_2O$
$Al_4(OH)_3(PO_4)_3$	Trolleite	
$Al_4(OH)_3(PO_4)_3.11H_2O$	Vashegyite	

The interest in this system for various areas as mineralogy, chemical industry, and medicine, can probably explain the numerous investigations made in this field and consequently the abundance of derivatives. Moreover one can note that in most of cases the corresponding isotypic iron salts exist as well as a continuous series of solid solutions between the two salts. Among the eleven natural compounds, only three of them are, in fact, monophosphates, berlinite, variscite, and metavariscite, the other eight are hydroxide-monophosphates and should strictly speaking be classified in the category of addition compounds.

Nevertheless, the family of trivalent-cation monophosphates is mainly dominated by the $M^{III}PO_4$ compounds that we have selected to illustrate in this

Table 5.1.12
Melting temperature of some $M^{III}PO_4$ monophosphates.

Formula	mp (K)	Formula	mp (K)	Formula	mp (K)
$AlPO_4$	1873	$GaPO_4$	1943	$FePO_4$	1503
$GdPO_4$	2223	$LaPO_4$	2323	$CePO_4$	2318
YPO_4	2253	$NdPO_4$	2373	$SmPO_4$	2313

section. A restricted selection of their hydrates will be also examined.

These compounds are very refractory and evidently not water soluble. We report in Table 5.1.12 the melting temperatures of some of them. For lanthanide monophosphates, these temperatures are relatively inaccurate because of the experimental difficulties of measurement in this range of temperatures.

These monophosphates can be classified in three categories according to the nature of the oxygen coordination of their associated cation.

– The $M^{III}PO_4$ compounds, (observed for M^{III} = B, Al, Ga, Fe), in which the trivalent cation has tetrahedral oxygen coordination, adopt atomic arrangements closely related to the various forms of silica.

– The $M^{III}PO_4$ phosphates (M^{III} = Cr, Ti, In, V, Rh), in which the trivalent cation has octahedral oxygen coordination and generally have atomic arrangements related to the $MgSO_4$ structure type.

– When M^{III} = Sc, Y, Ln, Bi, the corresponding $M^{III}PO_4$ compounds adopt structures similar to the zircon or monazite forms of the $M^{IV}SiO_4$ silicates.

We will examine successively these three classes of derivatives and some hydrates, and illustrate this review by some structural descriptions.

5.1.5.1. $M^{III}PO_4$ compounds with M^{III} = B, Al, Ga, Fe

As early as 1934, Schulze [125] performed the crystal structure of BPO_4 and recognized it as a cristobalite-like compound. Later, Mackenzie *et al.* [126] and Dachille and Glasser [127] identified for this salt as a quartz-like form prepared under high pressure. Since the work of Schulze, many other derivatives were characterized. Most of them adopt atomic arrangements very similar to those observed for the most known forms of silica as quartz, tridymite, and cristobalite, but, to date, no $M^{III}PO_4$ monophosphate was observed as an analogue of some less classical forms of silica, like coesite, keatite, or stishovite.

The way from SiO_2 to $M^{III}PO_4$ is illustrated by the following schemes:

$$2Si^{IV} \longrightarrow P^V + M^{III} \quad [1]$$

$$2SiO_2 \longrightarrow M^{III}PO_4 \quad [2]$$

As silica, most of these monophosphates are polymorphic. Among them,

Table 5.1.13
Comparison of the transition temperatures (K) in $AlPO_4$ and SiO_2.

Transformation	$AlPO_4$	SiO_2
low <——> high-quartz	853	856
low <——> high-tridymite	363	436
low <——> high-cristobalite	493	541

the most investigated is *AlPO₄*. For this compound, isoelectronic of silica, all the classical forms of silica were observed: two quartz-like, two cristobalite-like and two tridymite-like. Moreover, as can be seen in Table 5.1.13, the transition temperatures observed for AlPO₄ are close to those of silica. For some of these temperatures, the values reported in literature are ranging in a wide interval.

Nevertheless, from a structural point of view, things are not quite established. If we can compare, in Table 5.1.14, the unit-cell dimensions of the high and low form of quartz with some of the quartz-like $M^{III}PO_4$ monophosphates and make, in Table 5.1.15, the same comparison for some cristobalite-like com-

Table 5.1.14
Unit-cell dimensions for the two forms of quartz and of the quartz-like $M^{III}PO_4$ monophosphates. Space group is $P3_121$ for low-quartz and $P6_422$ for high-quartz derivatives.

Formula	low-quartz form	Ref.	high-quartz form	Ref.
SiO₂	$a = 4.913$ (Å) $c = 5.405$	[128]	$a = 5.02$ (Å) $c = 5.48$	[128]
BPO₄	$a = 4.470(5)$ $c = 9.93(1)$	[127]		
AlPO₄	$a = 4.983$ $c = 10.93$	[129]	$a = 5.029(5)$ $c = 11.05(2)$	[130]
GaPO₄	$a = 4.899(1)$ $c = 11.034(2)$	[130– 131]	$a = 4.92(1)$ $c = 11.00(2)$ (a)	[132]
FePO₄	$a = 5.051(3)$ $c = 11.27(2)$	[130]	$a = 5.170(5)$ $c = 11.37(3)$	[130]
MnPO₄			$a = 4.94$ $c = 10.96$ (a)	[132]

a: from a recent investigation [130], the existence of the high-quartz form of GaPO₄ and MnPO₄ reported in [132] is questionable.

pounds, one must confess that owing to a lack of data a similar comparison is almost impossible for the tridymite-like derivatives.

Table 5.1.14 shows clearly the relationships between the unit-cells dimensions of the quartz forms and those of the *quartz-like* $M^{III}PO_4$ monophosphates:

$$a = a_q, \quad c = 2c_q$$

In this case, the doubling of the c dimension shows the existence of the 1/1 order between the PO₄ and $M^{III}O_4$ tetrahedra in the atomic arrangements, suggested by the substitution schemes [1] and [2]. The structure determinations confirm this assumption. In these silica-like compounds, each PO₄ shares its four corners with the four adjacent $M^{III}O_4$ tetrahedra and vice versa.

Table 5.1.15
Unit-cell dimensions of the two forms of cristobalite and of some cristobalite-like $M^{III}PO_4$ monophosphates.

Formula	low-cristobalite form	Ref.	high-cristobalite form	Ref.
SiO_2	$a = 4.97$ $c = 6.94$ $P4_12_12$	[133]	$a = 7.131$ $Fd3m$	[133]
BPO_4	$a = 4.46(3)$ $c = 6.65(1)$ $I\bar{4}$	[125]		
$AlPO_4$	$a = 7.09(1)$ $b = 7.09(1)$ $c = 6.99(1)$ $C222_1$	[133]	$a = 7.21(1)$ $F\bar{4}3m$	[130]
$GaPO_4$	$a = 6.963(4)$ $b = 6.963(4)$ $c = 6.88(1)$ $C222_1$	[130, 133]	$a = 7.148(4)$ $F\bar{4}3m$	[130]
$MnPO_4$	$a = 4.97$ $c = 6.97$ Tetra.	[132]		

$GaPO_4$ is also given with a = 5.06, c = 7.16 Å, $I\bar{4}$, [133].

The comparison of the unit-cell relationships between the two forms of cristobalite and the *cristobalite-like* $M^{III}PO_4$ monophosphates (Table 5.1.15) is more difficult. In the case of the high-temperature form, the analogues crystallize with very similar cubic unit-cell dimensions and the order M^{III}/P induces a change of space group. For the low-temperature analogues, the same phenomenon occurs, but in some reported examples, the observed unit cells are orthorhombic pseudotetragonal. In this case, the relationships between low cristobalite and the $M^{III}PO_4$ analogues are given by:

$$a = a_c\sqrt{2}, \quad b = a_c\sqrt{2}, \quad c = c_c$$

If the existence of *tridymite-like* $M^{III}PO_4$ monophosphates seems well established [130] from various measurements, crystallographic data are rare and only based on powder-diffraction experiments. One can, nevertheless, report a recent investigation by Debnath and Chauduri [134] who measured the following orthorhombic cell for a tridymite-like form of $AlPO_4$:

$$a = 9.64, \quad b = 8.66, \quad c = 18.28 \text{ Å}$$

These values are in agreement with those observed in a natural two-layer polytype sample of low-tridymite [128].

The $GaPO_4$ atomic arrangement – Owing to its interesting piezoelectric properties, the trigonal low-quartz form of $GaPO_4$ is presently the object of many investigations. Thus, we have selected it to illustrate this family of $M^{III}PO_4$ monophosphates. For that we use a recent investigation performed by Goiffon *et al.* [131]. As shown by Figure 5.1.27, planes containing the gallium and planes containing the phosphorus atoms, separated by a distance of $c/6$, alternate perpendicularly to the **c** direction. Each GaO_4 tetrahedron shares its four corners with the four adjacent PO_4 tetrahedra and reversely. This planar order between the PO_4 groups and the GaO_4 groups explain the doubling of the c value. Inside the GaO_4 tetrahedron, the Ga–O distances are 1.810 ($\times 2$) and 1.821 ($\times 2$) Å and in the PO_4 tetrahedron, the P–O distances are 1.523 ($\times 2$) and 1.527 ($\times 2$) Å.

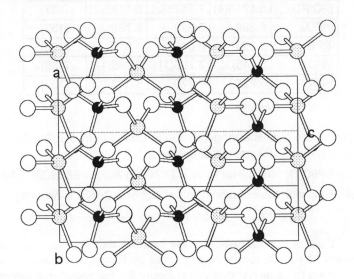

Figure 5.1.27. Orthogonal projection on the (b, c) plane of the atomic arrangement of $GaPO_4$. Black circles represent the phosphorus atoms and dotted circles the gallium atoms.

Other applications – BPO_4 is commonly used for the preparation of mono- or diphosphates by a metathesis reaction with oxides or fluorides. This type of reaction mainly employed to convert the refractory oxides present in nuclear wastes into phosphates will be examined in the section devoted to chemical preparations

5.1.5.2. $M^{III}PO_4$ compounds with M^{III} = Cr, Ti, Fe, In, V, Rh, Sb, Tl

These compounds can be roughly classified into two categories commonly denominated by: α- and β-$CrPO_4$.

There are only two representatives of the α-$CrPO_4$ structure type, α-$CrPO_4$

itself and RhPO$_4$. Their unit cells are orthorhombic, *Imma*, with Z = 12:

a = 10.403, b = 12.898, c = 6.299 Å   for α-CrPO$_4$ [135]

a = 10.397(1), b = 13.112(1), c = 6.3929(5) Å   for RhPO$_4$ [136]

Four M^{III}PO$_4$ phosphates are isotypic with β-CrPO$_4$. Their crystallographic data are gathered in Table 5.1.16.

Table 5.1.16
Unit-cell dimensions of the β-CrPO$_4$-like monophosphates. The space group is *Cmcm* and Z = 4.

Formula	a	b	c (Å)	Ref.
β-CrPO$_4$	5.1710(4)	7.7573(2)	6.1183(2)	[137]
InPO$_4$	5.320	7.993	6.785	[133]
TlPO$_4$	5.395	8.010	7.071	[133]
VPO$_4$	5.245(1)	7.795(1)	6.285(1)	[138]
FePO$_4$ (a)	5.227	7.770	6.322	[139]

a: high-pressure form

TlPO$_4$ has an atomic arrangement very similar to that of β-CrPO$_4$, but its unit cell is monoclinic, P2$_1$/m, with Z = 8, and:

a = 4.760(1), b = 6.349(1), c = 17.760(2) Å, β = 97.39(5)° [140]

The relationship with the orthorhombic unit cell of β-CrPO$_4$ is given by:

$$a_m = b_0/2 - a_0/2, \quad b_m = c_0, \quad c_m = 3a_0 + b_0$$

The α-CrPO$_4$ atomic arrangement – We examine this structure type using the arrangement of RhPO$_4$ recently determined by Rittner and Glaum [136]. The projection along the **c** direction, given in Figure 5.1.28, shows that this structure can be described as built by RhO$_6$–PO$_4$–RhO$_6$–PO$_4$ rows located in planes x = 0 and 1/2. These rows made by corner-sharing polyhedra are not linked together and connect along the **a** direction complex layers centered by planes x = 1/4 and 3/4. One of these layers is represented in projection along the **a** direction in Figure 5.1.29. This drawing shows that such a layer is built by pairs of edge-sharing RhO$_6$ octahedra interconnected by PO$_4$ tetrahedra sharing either edges or corners with the Rh$_2$O$_{10}$ clusters. Within the two crystallographically independent RhO$_6$ octahedra, the Rh–O distances range between 2.002 and 2.044 Å.

The β-CrPO$_4$ atomic arrangement – This structure type was determined and discussed by Attfield *et al.* [137]. The atomic arrangement of β-CrPO$_4$ that we

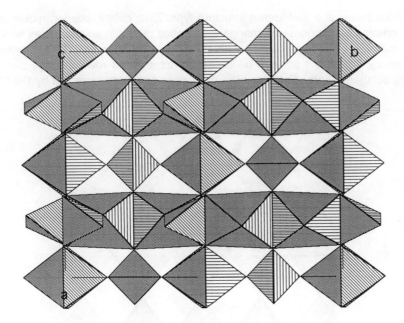

Figure 5.1.28. Projection, along the **c** direction, of the atomic arrangement in RhPO$_4$. Both RhO$_6$ octahedra and PO$_4$ tetrahedra are shown in polyhedral representation.

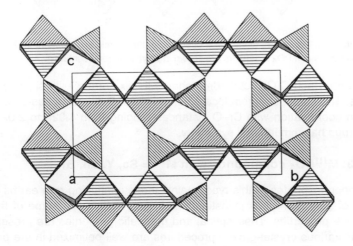

Figure 5.1.29. Projection, along the **a** direction, of the arrangement of an isolated layer located around the plane $x = 1/4$ in the atomic arrangement of RhPO$_4$.

describe below is a well-known structure type. Zinc, cobalt, copper, nickel, and zinc chromates, magnesium, manganese, cobalt, and iron sulphates, as well as chromium vanadate crystallize with the same arrangement. This structure, shown by Figure 5.1.30, can be described as isolated rows of edge-sharing CrO_6 octahedra spreading along the **c** direction and interconnected by the PO_4

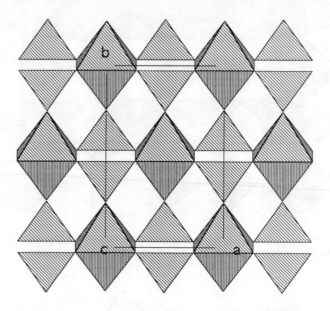

Figure 5.1.30. Projection, along the **c** direction, of the atomic arrangement in β-$CrPO_4$. Both CrO_6 octahedra and PO_4 tetrahedra are shown in polyhedral representation.

tetrahedra. Each CrO_6 octahedron shares its six corners with six adjacent PO_4 groups. The chromium atom located on a center of inversion has a very regular oxygen coordination with Cr–O distances ranging from 1.925 to 2.021 Å. The PO_4 group has mm internal symmetry.

5.1.5.3. $M^{III}PO_4$ compounds with M^{III} = Sc, Y, Ln, Bi

Monophosphates of this type, mainly those containing rare earths were the object of very numerous investigations in connection with some of their luminescence properties. Their optical and magnetic properties are presently well-known. Various crystal-growth procedures are well optimized in the production of large single crystals today.

They belong to two structure types:

– for M^{III} = Sc, Y, Tb, Dy, Ho, Er, Tm, Yb, Lu, the corresponding phosphates are tetragonal, isotypic with *zircon* ($ZrSiO_4$),
– for M^{III} = La, Ce, Pr, Nd, Pm, Sm, Eu, Gd, Tb, Dy, they are monoclinic, isotypic with *monazite*, (La, Ce, ...)PO_4.

In Table 5.1.17 are reported the main crystallographic data of the zircon-like derivatives and in Table 5.1.18 those of the monazite-like compounds. From these tables, one can observe that two of these derivatives, $DyPO_4$ and $TbPO_4$, are dimorphous.

Table 5.1.17
Unit-cell dimensions of the zircon-like $M^{III}PO_4$ compounds. Space group is $I4_1/amd$ and $Z = 4$.

Formula	a	c (Å)	Ref.
$ScPO_4$	6.574(1)	5.791(1)	[141]
YPO_4	6.882(1)	6.018(1)	[141]
$TbPO_4$	6.941	6.070	[142]
$DyPO_4$	6.917	6.053	[142]
$HoPO_4$	6.901	6.031	[142]
$ErPO_4$	6.863(3)	6.007(3)	[143–144]
$TmPO_4$	6.848	5.994	[142, 144]
$YbPO_4$	6.824	5.980	[142, 144]
$LuPO_4$	6.792(2)	5.954(2)	[141]

We illustrate this class of compounds with the description of these two structure types.

The zircon atomic arrangement – The structure of zircon, $ZrSiO_4$, was first solved, as early as 1926, by Vegard [145–146]. The structures of isotypic silicates, phosphates, arsenates or vanadates have also been reported.

For the description below, we have used the results obtained by Milligan *et al.* [141] when they investigated YPO_4. This arrangement is a stacking of PO_4 tetrahedra and YO_8 dodecahedra, sharing both edges and corners. All the Y and P atoms are located in planes y = 1/4 and 3/4. In Figure 5.1.31, we present a projection made approximately along the [120] direction of the layer centered by the plane y = 1/4. In this layer, each YO_8 polyhedron is surrounded by four PO_4 tetrahedra and shares with them two edges and two corners. The cohesion along the **a** direction is performed by corner sharing and along **c** by edge sharing. The YO_8 polyhedron shares its last two oxygen atoms with two PO_4 groups of the two adjacent layers. Figure 5.1.32 represents the surrounding of

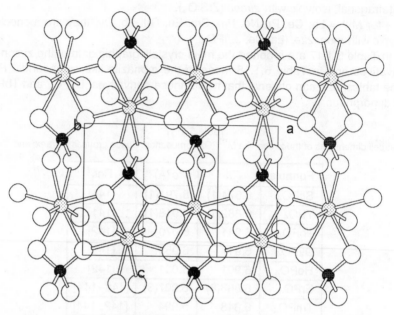

Figure 5.1.31. Projection, approximately along the [120] direction, of a layer of PO_4 tetrahedra and YO_8 dodecahedra in YPO_4.

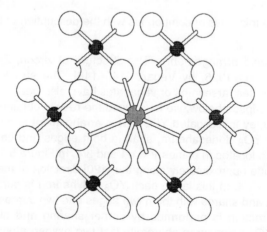

Figure 5.1.32. Projection, along the [1$\bar{1}$0] direction, of an yttrium atom and of its six PO_4 neighbors in YPO_4.

Table 5.1.18
Unit-cell dimensions of the monazite-like $M^{III}PO_4$ compounds. Space group is $P2_1/n$ and $Z = 4$.

Formula	a	b	c (Å)	β (°)	Ref.
$LaPO_4$	6.85	7.07	6.47	103.63	[147]
$CePO_4$	6.78	7.01	6.45	103.83	[147–148]
$PrPO_4$	6.75	6.98	6.43	103.85	[147]
$NdPO_4$	6.72	6.96	6.40	103.61	[147]
$PmPO_4$	6.72	6.89	6.36	104.28	[149]
$SmPO_4$	6.67	6.88	6.36	103.78	[147]
$EuPO_4$	6.65	6.84	6.33	103.83	[147]
$GdPO_4$	6.64	6.83	6.32	103.90	[147]
$TbPO_4$	6.60	6.82	6.31	103.93	[147]
$DyPO_4$	6.58	6.78	6.30	103.93	[147]
$BiPO_4$	6.74	6.92	6.46	103.30	[150]

an yttrium atom. The Y–O distances inside the YO_8 polyhedron are 2.300 (x 4) and 2.373 (x 4) Å.

The monazite structure – The monazite structure is a simple monoclinic distortion of that of zircon. In spite of that, it is difficult to describe its atomic arrangement as clearly as that of zircon. The decrease of symmetry induces an irregular ninefold coordination around the metal atom. For instance, in $CePO_4$ [148], the Ce–O distances vary from 2.445 to 2.779 Å, with an average of 2.550 Å and each cerium atom has seven PO_4 neighbors instead of six in the zircon structure.

$BiPO_4$ has a particular behavior. It is dimorphous with a room-temperature form of the monazite type [150] and a high-temperature form [151] which is a slight distortion of the first one. The unit-cell dimensions of the high-temperature form are:

$$a = 4.871(3), \quad b = 7.073(3), \quad c = 4.709(3) \text{ Å}, \quad \beta = 96.24(8)°$$

The space group is $P2_1/m$ and $Z = 2$. The relation with the unit cell of the monazite form is:

$$a = a_m\sqrt{2}/2, \quad b = b_m, \quad c = c_m\sqrt{2}/2$$

Hydrated LnPO₄ compounds – Mooney [152] was the first to describe a series of $LnPO_4$ hemihydrates and to perform the crystal structure determination of the hexagonal cerium salt, $CePO_4 \cdot 1/2H_2O$. Later, various investigators characterized several other isotypic monophosphates. Their unit cells are given in Table 5.1.19.

This type of structure, determined by Mooney [152] with $CePO_4.1/2H_2O$ and represented in projection along the **c** direction in Figure 5.1.33, can be descri-bed as built by arrays of alternating PO_4 groups and Ce atoms extending along

Table 5.1.19
Unit-cell dimensions for the $M^{III}PO_4.1/2H_2O$ monophosphates. The common space group is $P6_222$ and $Z = 3$.

Formula	a	c (Å)	Ref.
$LaPO_4.1/2H_2O$	7.081(5)	6.468(8)	[153]
$CePO_4.1/2H_2O$	7.055(3)	6.439(5)	[153]
$PrPO_4.1/2H_2O$	7.00	6.43	[154]
$NdPO_4.1/2H_2O$	7.001(1)	6.39(3)	[153]
$EuPO_4.1/2H_2O$	6.91	6.34	[154]
$GdPO_4.1/2H_2O$	6.89(1)	6.32(2)	[153–154]
$TbPO_4.1/2H_2O$	6.87	6.33	[154]
$DyPO_4.1/2H_2O$	6.80	6.29	[154]

the **c** axis and disposed around the 6_2 helical axis to create large hexagonal channels. The location of the water molecules was suggested by this author from geometrical considerations. They are probably located on the 6_2 axes, in the large channels created by the stacking of the $-PO_4-Ce-PO_4-Ce-$ rows. Figure 5.1.34 represents one of these rows in projection along the **b** direction. As shown by this figure, each cerium atom is surrounded by six PO_4 tetrahedra and has eightfold oxygen coordination.

5.1.6. Tetravalent-Cation Monophosphates

Among many investigations of various $MO_2-P_2O_5-H_2O$ systems, those dealing with zirconium salts are the most numerous. Since 1964, the interesting ion-exchange and intercalation properties of $Zr(HPO_4)_2$ and of some of its hy-drates were the subject of numerous studies. There is some confusion in the chemical literature dealing with these compounds, often abbreviated as α-ZrP, β-ZrP, and γ-ZrP, giving the impression that this compound is polymorphic. In fact, from the results of various investigations, it appears that α-ZrP corresponds to the monohydrate, $Zr(HPO_4)_2.H_2O$, whereas γ-ZrP is the dihydrate, and β-ZrP the anhydrous salt [155].

These compounds are difficult to crystallize. Under normal conditions, the

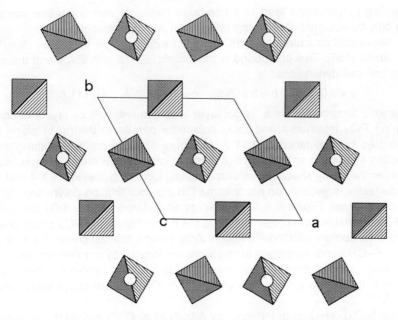

Figure 5.1.33. Projection, along the **c** direction, of the atomic arrangement of CePO$_4$.1/2H$_2$O. PO$_4$ groups are shown in polyhedral representation and cerium atoms represented by open circles. The suggested positions of the water molecules are not represented.

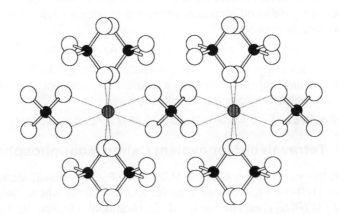

Figure 5.1.34. Projection, along the **b** direction, of an array of PO$_4$ and CeO$_8$ polyhedra spreading along the **c** direction. Black circles represent the phosphorus atoms and grey circles cerium atoms.

chemical preparations lead to amorphous materials and crystalline samples can only be obtained after a long refluxing in phosphoric acid.

The crystal structure of $Zr(HPO_4)_2.H_2O$ (α-ZrP) was performed by Clearfield and Smith [156]. This compound is monoclinic, $P2_1/c$, with $Z = 4$ and the following unit-cell dimensions:

$$a = 9.076(3), \quad b = 5.298(6), \quad c = 16.22(2) \text{ Å}, \quad \beta = 111.5(1)°$$

Its atomic arrangement is a typical layer arrangement, built by layers of corner sharing PO_4 tetrahedra and ZrO_6 octahedra parallel to the (a, b) plane and separated by a distance of 8.11 Å. In Figure 5.1.35 a projection along the **b** direction shows the arrangement of these layers and how they are intercalated, between them, the sheets of water molecules, located in planes $z = 1/4$ and $3/4$. These water molecules do not take part in the coordination polyhedron of the zirconium atom. Figure 5.1.36 represents an isolated layer of PO_4 tetrahedra and ZrO_6 octahedra in projection along the **c** direction. Each PO_4 group shares three of its corners with three adjacent ZrO_6 groups, the last corner corresponds to the P–OH bond. Hydrogen atoms were not located by the authors, but, from the examination of the O–O distances, they suggest a possible hydrogen-bond scheme. Inside the ZrO_6 octahedron, the Zr–O distances range from 2.04 to 2.11 Å.

$Ti(HPO_4)_2.H_2O$, characterized by Alberti *et al.* [157] and later investigated by Christensen *et al.* [158] is isotypic. It also has exchange and intercalation properties.

There is little structural information concerning β-ZrP and γ-ZrP. Nevertheless, one can report that according to Yamanaka and Tanaka [159], who prepared the single crystals and proposed a possible structural model, the dihydrate (γ-ZrP), is monoclinic with:

$$a = 5.376, \quad b = 6.636, \quad c = 24.56 \text{ Å}, \quad \beta = 93.94°$$

The dihydrate of the titanium salt, formulated as $Ti(H_2PO_4)(PO_4).2H_2O$ by Christensen *et al.* [158] seems to have a different arrangement as described by these authors with the following monoclinic unit cell:

$$a = 5.181(1), \quad b = 6.347(1), \quad c = 11.881(1) \text{ Å}, \quad \beta = 102.59(1)°$$

5.1.7. Tetravalent-Monovalent Cation Monophosphates

During the investigation of some MO_2–M_2O–P_2O_5 systems, Matkovic and co-workers [160–166] discovered a series of compounds whose general formula is $M_2^{IV}M^I(PO_4)_3$, with M^{IV} = Th, U, Zr, Hf and M^I = Li, Na, Ag, K, Rb, Cs. They mentioned that some of these compounds were characterized as early as 1883 by Wallroth and by Colani in 1907. Later Hagman and Kierkgaard [167] identified similar compounds for M^{IV} = Ge and Ti.

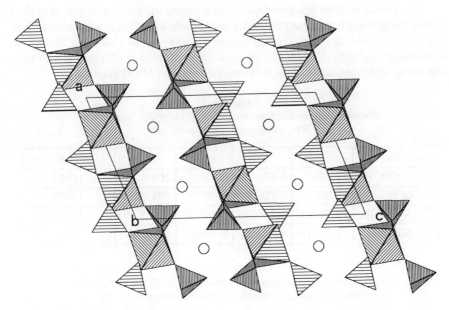

Figure 5.1.35. Projection, along the **b** direction, of the atomic arrangement of Zr(HPO$_4$)$_2$.H$_2$O. The PO$_4$ tetrahedra and the ZrO$_6$ octahedra are shown in polyhedral representation. The open circles represent the water molecules.

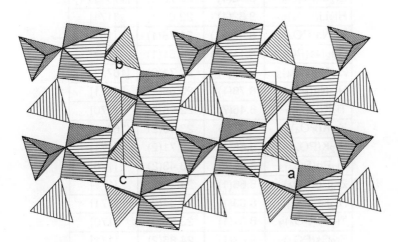

Figure 5.1.36. Projection, along the **c** direction, of an isolated layer of PO$_4$ tetrahedra and ZrO$_6$ octahedra in Zr(HPO$_4$)$_2$.H$_2$O.

The earlier studies showed that these compounds belong to two structural types, a trigonal (rhombohedral) and a monoclinic one. The crystal structure of the rhombohedral $Zr_2K(PO_4)_3$ was solved by Sljukic *et al.* [163] and that of the monoclinic $Th_2K(PO_4)_3$ by Matkovic *et al.* [162]. Tables 5.1.20 and 5.1.21 contain some crystallographic data for these two classes of compounds.

Table 5.1.20
Unit-cell dimensions of some monoclinic $M_2^{IV}M^I(PO_4)_3$ compounds.
The space group is *C2/c* and $Z = 4$.

Formula	a	b	c (Å)	β (°)	Ref.
$Th_2Na(PO_4)_3$	17.39	6.80	8.13	101.13	[160]
$U_2Na(PO_4)_3$	17.23	6.70	8.03	101.36	[160]
$Th_2Ag(PO_4)_3$	17.385(5)	6.815(3)	8.148(5)	101.10(5)	[166]
$Th_2K(PO_4)_3$	17.57(1)	6.863(3)	8.138(4)	101.57(1)	[161–162, 165]

Table 5.1.21
Unit-cell dimensions of some rhombohedral $M_2^{IV}M^I(PO_4)_3$ compounds.
The space group is $R\bar{3}c$ and $Z = 6$ in the hexagonal setting.

Formula	a	c (Å)	Ref.
$Ti_2Li(PO_4)_3$	8.53(5)	20.88(3)	[170]
$Zr_2Li(PO_4)_3$	8.72(1)	22.72(2)	[170–171]
$Hf_2Li(PO_4)_3$	8.82(1)	22.03(1)	[171]
$Ti_2Na(PO_4)_3$	8.492(1)	21.78(1)	[167]
$Ge_2Na(PO_4)_3$	8.112(1)	21.51(1)	[167]
$Zr_2Na(PO_4)_3$	8.78(1)	22.71(2)	[167, 171]
$Hf_2Na(PO_4)_3$	8.76(1)	22.68(2)	[171]
$Ti_2K(PO_4)_3$	8.46(7)	22.96(8)	[170]
$Zr_2K(PO_4)_3$	8.71(1)	23.89(2)	[163, 171]
$Hf_2K(PO_4)_3$	8.68(1)	23.71(2)	[171]
$Ti_2Rb(PO_4)_3$	8.35(5)	23.43(2)	[170]
$Zr_2Rb(PO_4)_3$	8.66(1)	24.38(2)	[171]
$Hf_2Rb(PO_4)_3$	8.63(1)	24.34(2)	[171]
$Ti_2Cs(PO_4)_3$	8.31(1)	23.86(2)	[170]
$Zr_2Cs(PO_4)_3$	8.62(1)	24.86(2)	[171]
$Hf_2Cs(PO_4)_3$	8.54(1)	24.81(2)	[171]

Ammonium derivatives are rare in this class of compounds. $Zr_2NH_4(PO_4)_3$ was prepared under hydrothermal conditions by Clearfield *et al.* [168]. The same authors have been able to prepare $Zr_2H(PO_4)_3$ by heating the ammonium derivative between 723 and 923 K:

$$Zr_2NH_4(PO_4)_3 \longrightarrow Zr_2H(PO_4)_3 + NH_3$$

When heated at temperatures higher than 923 K, $Zr_2H(PO_4)_3$ decomposes according to the following reaction:

$$2Zr_2H(PO_4)_3 \longrightarrow 3ZrP_2O_7 + ZrO_2 + H_2O$$

Both $Zr_2NH_4(PO_4)_3$ and $Zr_2H(PO_4)_3$ belong to the rhombohedral form.

Owing to their ionic conduction properties, the $M_2^{IV}M^I(PO_4)_3$ monophosphates provoked considerable interest in the last twenty years. A good number of cation substitutions performed in the basic framework were reported in hundreds of publications. We report the following example in order to illustrate the richness of this class of compounds. In one recent publication on the $M^{IV}M^{III}M^{II}(PO_4)_3$ compounds by Sugantha *et al.* [169], twenty-two new derivatives were described.

The $M_2^{IV}M^I(PO_4)_3$ monophosphates and the numerous compounds obtained through various cation substitutions are popularly known today as NZP phases. Moreover, one must note that the atomic arrangement in this family of monophosphates is very similar to that found in the Nasicon ($Zr_2Na_3Si_2PO_{12}$).

The structure of $Zr_2K(PO_4)_3$ using the data reported by Sljukic *et al.* [163] is illustrated in this section. Unit-cell dimensions and space group are given in Table 5.1.21.

This atomic arrangement can be described by examining the situation of its

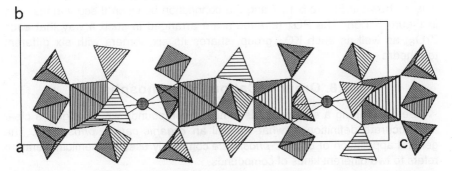

Figure 5.1.37. Arrangement, around a threefold axis, of the various components of the atomic arrangement in $Zr_2K(PO_4)_3$, viewed along the **a** direction. Both ZrO_6 and PO_4 polyhedra are shown in polyhedral representation, whereas the grey circles represent the potassium atoms.

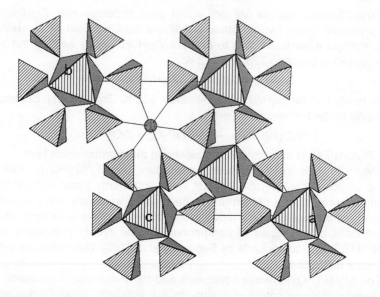

Figure 5.1.38. Projection, along the **c** axis, of a portion of the atomic arrangement in $Zr_2K(PO_4)_3$. ZrO_6 and PO_4 groups are shown by polyhedral representation, whereas the grey circles are potassium atoms.

various components along a ternary axis. The zirconium and potassium atoms in octahedral oxygen coordination are located on this axis and along it, each KO_6 polyhedron shares two opposite faces with its two adjacent ZrO_6 octahedra to form a Zr_2KO_{12} unit. These units are connected through the PO_4 tetrahedra disposed around the ternary axis. This alignment along a ternary axis is shown in Figure 5.1.37 and the connection between these various rows in Figure 5.1.38. The PO_4 tetrahedra are arranged in such a way that each ZrO_6, as well as each KO_6 group, shares its six corners with six different tetrahedra.

5.1.8. Organic-Cation Monophosphates

Before starting a survey of this family of compounds we must give a clear and accurate definition of what we call an *organic-cation phosphate*. The general appellation of *organic phosphate* commonly used in chemical literature refers to two different kinds of compounds.

– The first category corresponds to a wide family of compounds including P–O–C bonds in their atomic arrangements. These compounds are in fact *phosphoric esters* and must simply be called organic phosphates

– The second category corresponds to compounds which include in their atomic arrangements a phosphoric anion whether condensed or not, and a protonated organic entity. In our opinion, the term *organic-cation phosphate* must apply to the members of this second category.

Two examples are:

– Ethylenediammonium bis(dihydrogenmonophosphate), recently described by Kamoun *et al.* [172] and formulated as:

$$[NH_3-(CH_2)_2-NH_3]^{2+}.2[H_2PO_4]^-$$

is an *organic-cation monophosphate*.

– Dimethyl ammonium phosphate, studied by Giarda *et al.* [173] and formulated as:

$$[NH_4]^+[(CH_3)_2O_2PO_2]^-$$

is an *organic phosphate* for it includes two P–O–C bonds.

Table 5.1.22
Main crystal data for a selection of organic-cation monophosphates.

Formula	a α	b β	c (Å) $\gamma(°)$	S. G.	Z	Ref.
(eda)Na$_2$(HPO$_4$)$_2$.6H$_2$O	11.699(9)	10.164(9) 105.00(5)	6.835(4)	$P2_1/c$	2	[174]
(eda)HPO$_4$	8.059(3)	11.819(5) 110.12(5)	7.513(3)	$P2_1/a$	4	[175]
(sar)H$_2$PO$_4$	13.180(8)	9.275(3)	6.245(2)	$P2_12_12_1$	4	[176]
(L-his)H$_2$PO$_4$.H$_2$O	8.923(5)	14.383(9)	8.362(5)	$P2_12_12_1$	4	[177]
(gly)H$_2$PO$_4$	9.580(2)	7.840(2) 114.7(2)	9.249(3)	$P2_1/c$	4	[178]
(eda)H$_2$PO$_4$.H$_5$P$_2$O$_8$	13.03(1)	7.923(7) 107.43(5)	13.22(1)	$C2/c$	4	[179]
(amp)H$_2$PO$_4$.H$_2$O	8.951(9) 113.47(5)	7.982(8) 108.75(5)	7.398(8) 77.18(5)	$P\bar{1}$	2	[180]
(iso)H$_2$PO$_4$	5.769(2)	15.39(1) 109.43(3)	8.504(3)	$P2_1/c$	4	[181]

eda	= ethylenediammonium or	$[NH_3-(CH_2)_2-NH_3]^{2+}$
sar	= sarcosinium or	$[CH_3-NH_2-CH_2-COOH]^+$
L-his	= L-histidinium or	$[(C_3H_4N_2)-CH_2-CH-(NH_3)(CO_2)]^+$
gly	= glycinium or	$[NH_3-CH_2-COOH]^+$
amp	= 4-aminopyridinium or	$[C_5H_7N_2]^+$
iso	= isopropylammonium or	$[(CH_3)_3-NH]^+$

From the definition given above, it is clear that in the organic-cation phosphates, the two constituting entities can only be connected by H–bonds.

During the last twenty years, many organic-cation monophosphates were investigated. We will describe in the section "Chemical Preparation of Mono-phosphates", the general procedures used for their syntheses, which are not fundamentally different from those used for the preparation of the inorganic derivatives.

In Table 5.1.22 we give the main crystallographic features for some orga-nic-cation monophosphates and illustrate this section with the descriptions of the atomic arrangements of two of them that we consider as well representative.

The ethylenediammonium monohydrogenmonophosphate structure – The main crystallographic data for [NH$_3$–(CH$_2$)$_2$–NH$_3$]HPO$_4$ were given in Table 5.1.22. The atomic arrangement of this salt [175] can be described as built by thick layers of ethylenediammonium groups alternating with layers of phosphoric anions, both of which are perpendicular to the **c** direction, in $z = 0$ and 1/2 respectively. Figure 5.1.39 represents this arrangement in projection along the

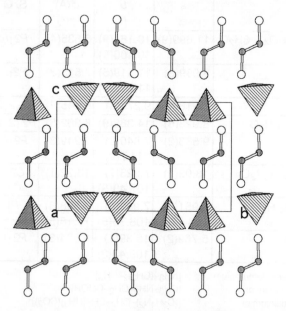

Figure 5.1.39. Projection of the structure of [NH$_3$–(CH$_2$)$_2$–NH$_3$]HPO$_4$, along the **a** axis, showing its layer arrangement. A polyhedral representation is used for the phosphoric groups. The grey circles are the N atoms and the open ones the C atoms. The H atoms are omitted.

a direction. Inside a layer of phosphoric anions, the [HPO4]2- entities are assembled by strong hydrogen bonds (O–O = 2.578 Å) to form corrugated infinite chains extending along the **a** direction. In Figure 5.1.40 a projection along the **c** direction illustrates the main features of these chains. The three-dimensional cohesion of the arrangement is performed by a network of strong hydrogen bonds established between the hydrogen atoms of the NH$_3$ radicals and the oxygen atoms of the HPO$_4$ groups.

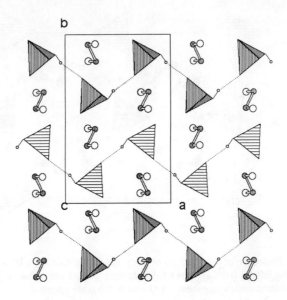

Figure 5.1.40. Projection of the structure of [NH$_3$–(CH$_2$)$_2$–NH$_3$]HPO$_4$, along the **c** axis, showing the infinite (HPO$_4$)$_n$ chains. The drawing conventions are similar to those used in Figure 5.1.39. The H atoms belonging to the organic entities are omitted. The H–bonds are represented by solid and dotted lines.

The sarcosinium dihydrogenmonophosphate atomic arrangement – The main crystallographic data for [CH$_3$–NH$_2$–CH$_2$–COOH]H$_2$PO$_4$ were reported in Table 5.1.22. As in the latter example, this structure [176] is a layer arrangement between slabs of phosphoric anions centered around the planes $x = 0$ and 1/2 and thick layers of organic entities. Figure 5.1.41 represents the projection of this atomic arrangement along the **c** axis. Inside a phosphoric layer, the H$_2$PO$_4$ groups connected by strong H–bonds (O–O = 2.566 Å) are assembled to form infinite chains extending along the **b** direction. Only one of the H atoms of the phosphoric anion is involved in the chain formation. The second H atom of this group is connected by H–bond to one of the O atoms of the carboxylic radical of the sarcosinium. The three-dimensional cohesion of the arrangement is completed by a network of strong H–bonds established between the H atoms of the NH_2 radicals, the OH radicals of the sarcosinium group and the O atoms of the H_2PO_4 groups. The main geometric features of the various H–bonds in the chain are reported in Chapter 6.

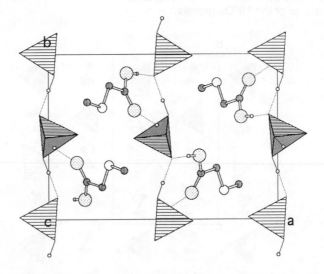

Figure 5.1.41. Projection of the structure of $[CH_3–NH_2–CH_2–COOH]H_2PO_4$ along the **c** direction, showing the infinite $(HPO_4)_n$ chains. Phosphoric groups are shown in polyhedral representation, the spotted circles represent the O atoms, the open circles the N atoms and the grey ones the C atoms. The H atoms belonging to the organic entities are omitted. The H–bonds are represented by solid and dotted lines.

General structural considerations – All the organic-cation monophosphates exhibit features similar to those that we could observe in the two preceding examples.

– Except for a very restricted number of examples, the phosphoric anions $[HPO_4]^{2-}$ or $[H_2PO_4]^-$ are acidic. Generally, these anions are assembled by strong hydrogen bonds to build infinite networks. The various geometries of the networks generated by this process will be examined in detail in Chapter 6. In this same chapter will be reported the main numerical data for the hydrogen bonds corresponding to the two arrangements described above.

– A strong set of hydrogen bonds interconnects the various components of the organic cations (NH_2, NH_3 or OH radicals) to the oxygen atoms of the phosphoric entities and reversely.

These compounds are intensively investigated today in order to introduce chiral organic groups in atomic arrangements for the production of non-linear optics materials.

5.1.9. Substituted Monophosphates

5.1.9.1. Phosphites

Phosphites constitute a well-known class of compounds including substituted tetrahedra, PO_3H, in their atomic arrangements. The confusing formula, $P(OH)_3$, commonly used for phosphorus acid, should actually be written as $H_2(PO_3H)$. The preparation of this acid and its main properties were described in the Chapter 3. The corresponding salts, known since the last century, are commonly denominated phosphites.

Their preparation is generally based on the direct action of the acid on the required carbonate or metal. In some cases, hydrothermal processes were also used. For example, Attfield *et al.* [182] prepared two hydroxyphosphites, $Fe(HPO_3)_8(OH)_6$ and $Mn(HPO_3)_8(OH)_6$, by heating a mixture of $FeCl_3$, NaOH and H_2PO_3H in the case of the iron salt at 485 K for 100 hours in autoclaves.

A recent survey of the crystal chemistry of inorganic phosphites was published by Loub [183]. The main geometrical features of the PO_3H tetrahedron are carefully analyzed by Loub through a large number of accurate atomic arrangements.

In order to illustrate with a numerical example the geometry of a PO_3H group, we give, in Table 5.1.23, the geometrical features of this anion as observed after the accurate structural investigation of tris(hydroxymethylaminomethane) monohydrogenphosphite, $[CH_3N(CH_2OH)_3]HPO_3H$, recently performed by Averbuch-Pouchot [184].

In the above example, the largest P–O distance found for P–O(3) corresponds to the P–OH bond. The P–H distance observed here (1.45 Å) is significantly longer that the average value 1.30 Å given by Loub [183], confirming his observation on the wide range of values observed for this distance.

Table 5.1.23

Geometrical features in the PO_3H tetrahedron observed in $CH_3N(CH_2OH)_3HPO_3H$. The distances are in Å and bond angles in °.

P	O(1)	O(2)	O(3)	H
O(1)	1.503(1)	2.556(2)	2.525(2)	2.41(3)
O(2)	117.19(9)	1.492(1)	2.466(3)	2.39(3)
O(3)	110.65(9)	107.37(9)	1.568(2)	2.36(3)
H	109.5(9)	108(1)	103(1)	1.45(3)

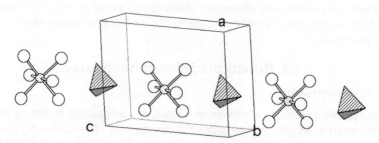

Figure 5.1.42. Perspective view of an isolated $-HPO_3-Mg(H_2O)_6-HPO_3-$ row in $MgPO_3H.6H_2O$. The HPO_3 groups are shown in polyhedral representation. Dotted circles represent the magnesium atoms and the open ones the water molecules of the $Mg(H_2O)_6$ octahedra.

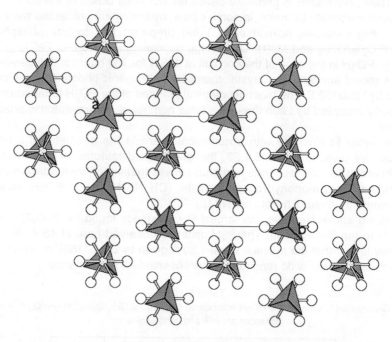

Figure 5.1.43. Projection along the [001] direction of the atomic arrangement in $MgPO_3H.6H_2O$. PO_3H tetrahedra are shown in polyhedral representation. Open circles represent the water molecules and dotted ones the magnesium atoms.

The $MgPO_3H.6H_2O$ atomic arrangement – This compound is trigonal (rhombohedral), with the space group $R3$ and $Z = 3$ in the hexagonal setting. Its unit-cell dimensions are:

$$a = 8.828, \quad c = 9.104 \text{ Å}$$

The atomic arrangement was first determined by Corbridge [185] and later refined by Powell *et al.* [186]. The crystal structure is a stacking of PO$_3$H groups and Mg(H$_2$O)$_6$ octahedra. These two entities located on the ternary axes, alternate along these axes. Figure 5.1.42 is a perspective view of an isolated –HPO$_3$–Mg(H$_2$O)$_6$–HPO$_3$– row. In Figure 5.1.43 a projection along the **c** direction shows the stacking of these rows and in Figure 5.1.44 a projection along the [110] direction illustrates another aspect of the same stacking. The three-dimensional cohesion is performed through hydrogen bonds interconnecting the water molecules to the oxygen atoms of the PO$_3$H groups. The Mg(H$_2$O)$_6$ octahedron is rather regular with Mg–O distances of 2.056 and 2.102 Å and O–Mg–O angles close to the ideal values.

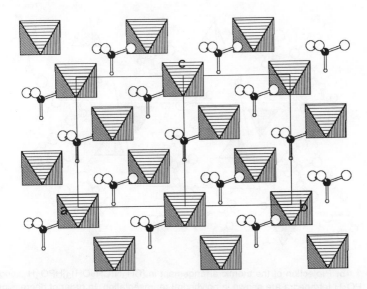

Figure 5.1.44. Projection, along the [110] direction, of the structure of in MgPO$_3$H.6H$_2$O. Mg(H$_2$O)$_6$ octahedra are shown in polyhedral representation. In order of decreasing size, the circles represent the oxygen, the phosphorus and the hydrogen atoms of the PO$_3$H tetrahedra.

The tris(hydroxymethylaminomethane) monohydrogenphosphite structure – In some rare cases, a strong structural analogy exists between a given phosphite and the corresponding phosphate. Recently Averbuch-Pouchot [184] determined the crystal structure of [CH$_3$N(CH$_2$OH)$_3$]HPO$_3$H and observed that for this salt the atomic arrangement is similar to that found for [CH$_3$N(CH$_2$OH)$_3$]H$_2$PO$_4$ previously determined by Jingxian *et al.* [187]. This analogy could be guessed by comparing the two sets of unit-cell dimensions that we report below.

a = 9.576, b = 6.170, c = 8.179 Å, β = 106.32°, for the phosphate and

a = 9.594, b = 6.181, c = 8.191 Å, β = 106.26°, for the phosphite

For the two compounds, the space group is $P2_1$ and Z = 2.

The atomic arrangement in [CH$_3$N(CH$_2$OH)$_3$]HPO$_3$H is a typical layer packing (Figure 5.1.45). Parallel to the (b, c) plane are alternate thick layers of the organic cations and layers containing the phosphoric groups. Inside a phosphoric layer, the [HPO$_3$H]$^-$ entities connected by strong hydrogen bonds are assembled to build [HPO$_3$H]$_n$ infinite chains parallel to the **b** direction (Figure 5.1.46). Each phosphoric group is connected to its adjacent neighbours by a strong H–bond (O–O = 2.580 Å). The details of the geometry of the PO$_3$H

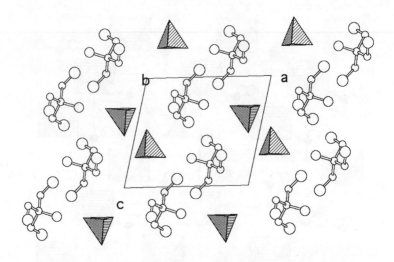

Figure 5.1.45. Projection of the atomic arrangement in [CH$_3$N(CH$_2$OH)$_3$]HPO$_3$H along the **b** direction. PO$_3$H tetrahedra are shown in polyhedral representation. In order of decreasing size, the circles denote oxygen, nitrogen, carbon atoms. The hydrogen atoms are omitted.

group were given in Table 5.1.23. Within the organic layer, each entity is connected to its three neighbours by hydrogen bonds, all of them establishing connections along the **b** direction. Each organic group is moreover bonded to three different [HPO$_3$H]$^-$ groups belonging to two different phosphoric chains.

The only difference between the present atomic arrangement and the structure of the corresponding monophosphate concerns the hydrogen-bond net-

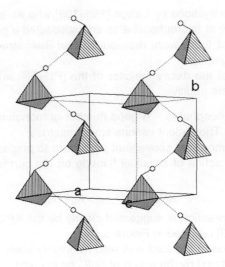

Figure 5.1.46. Perspective view of the two phosphoric $[HPO_3H]_n$ chains crossing the unit cell of $[CH_3N(CH_2OH)_3]HPO_3H$.

work connecting the acidic anions. In both cases, they are assembled to build chains, but in the phosphate, each anionic entity is connected to the adjacent one by two H–bonds, while in the phosphite only one performs this linkage.

Condensation of phosphites – As in phosphoric acid, phosphorous acid can also be condensed. The simplest condensation scheme:

$$2H_2PO_3H \longrightarrow H_2(O_2H)–P–O–P–(O_2H) + H_2O$$

leads to an acid with similarities to $H_4P_2O_7$ and whose corresponding salts are known as diphosphites or pyrophosphites. They must be classified as substituted diphosphates.

5.1.9.2. Fluoromonophosphates

One or two oxygen atoms of each PO_4 tetrahedron can be substituted by fluorine atoms, leading to two categories of substituted monophosphates: the mono- and difluoromonophosphates, corresponding respectively to the $[PO_3F]^{2-}$ and $[PO_2F_2]^-$ anions. Both the $[PO_3F]^{2-}$ and $PO_2F_2]^-$ anions are stable in aqueous solutions.

Fluoromonophosphoric acids can be prepared by the action of HF on monophosphoric acid according to the following schemes:

$$H_3PO_4 + HF \longrightarrow H_2PO_3F + H_2O$$

$$H_3PO_4 + 2HF \longrightarrow HPO_2F_2 + 2H_2O$$

Since the investigations by Lange [188–189] who as early as 1929 prepared a good number of fluorophosphates and suspected a possible analogy with sulphates, many of them were reproduced and their structural investigations confirmed his assumption.

The property of the decay inhibitor of the $[PO_3F]^{2-}$ anion is commercially used as a toothpaste additive.

Monofluoromonophosphates – A good number of monofluoromonophosphates are known to date. They adopt various stoichiometries.

The most common and convenient procedure to prepare monofluoromonophosphates is the action of an alkali fluoride on the corresponding long-chain polyphosphate:

$$KPO_3 + KF \longrightarrow K_2PO_3F$$

This possibility of reaction is suggested clearly by the KPO_3–KF phase-equilibrium diagram [190] reported in Figure 5.1.47.

The direct action of the acid was also frequently used. Thus, $SnPO_3F$ was prepared by Bernt [191] by the action of SnF_2 on the acid:

$$SnF_2 + H_2PO_3F \longrightarrow SnPO_3F + 2HF$$

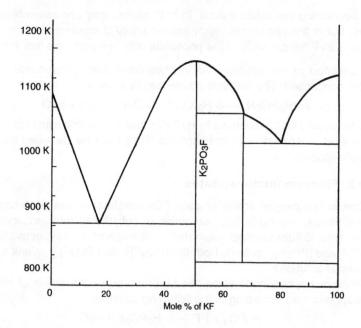

Figure 5.1.47. The KPO_3–KF phase-equilibrium diagram.

The silver salt, Ag_2PO_3F, was also used as a starting material in some syntheses described by the scheme:

$$Ag_2PO_3F + 2MCl \longrightarrow M_2PO_3F + 2AgCl$$

Among the salts whose main crystallographic features are gathered in Table 5.1.24, all the mixed-cation compounds were prepared by the crystallization of aqueous solutions containing the stoichiometric amounts of the two fluorophosphates. Some of them belong to well-known structures types. Thus, K_2PO_3F is isotypic with α-K_2SO_4, $NaK_3(PO_3F)_2$ and $NaRb_3(PO_3F)_2$ with glaserite, while $CaPO_3F.2H_2O$ has close analogies with gypsum.

Table 5.1.24

Main crystal data for some monofluoromonophosphates.

Formula	a α	b β	c (Å) γ (°)	S. G.	Z	Ref.
α-Na_2PO_3F	5.750	9.280	6.854	$Cmcm$	4	[192]
β-Na_2PO_3F	5.416(3)	6.929(4)	19.08(1)	$P2_12_12_1$	8	[193]
K_2PO_3F	7.543(5)	10.16(1)	5.953(5)	$Pnma$	4	[194]
$(NH_4)_2PO_3F.H_2O$	6.3042(5)	8.294(1) 98.42(1)	12.76(1)	$P2_1/c$	4	[195–196]
$LiNaPO_3F.H_2O$	7.428 119.16	9.323 117.16	6.934 98.15	$P\bar{1}$	2	[197]
$LiKPO_3F$	5.438	4.972 91.24	13.869	$P2/m$	4	[197]
$LiRbPO_3F$	5.484	5.060 90.77	14.360	$P2/m$	4	[197]
$LiCsPO_3F$	5.591	5.169	15.127	ortho.	4	[197]
$LiKPO_3F.H_2O$	5.426(5)	7.474(7) 109.55(5)	12.54(1)	$P2_1/c$	4	[198]
$LiNH_4PO_3F$	5.463(3)	5.049(3) 90.82(3)	14.362(9)	$P2_1/c$	4	[197, 199]
$NaNH_4PO_3F.H_2O$	4.946	6.051	9.010	$Pmmn$	2	[192]
$NaK_3(PO_3F)_2$	5.761(6)	5.761(6)	7.374(7)	$P\bar{3}$	1	[200]
$NaRb_3(PO_3F)_2$	5.909	5.909	7.661	$P\bar{3}$	1	[192]
$CaPO_3F.2H_2O$	8.650(1) 119.00(1)	6.461(1) 110.85(1)	5.7353(4) 94.15(1)	$P\bar{1}$	2	[195]

We illustrate this class of compounds with some examples, but before describing their structures, it is useful to show a numerical example of the kind of distortion introduced within a phosphoric anion by the substitution of one oxygen atom with a fluorine atom. Table 5.1.25 provides the main geometrical features within the PO_3F group of $CaPO_3F.2H_2O$ observed by Perloff [195]. The P–F distance (1.583 Å) is significantly longer than the three P–O distances ranging from 1.503 to 1.515 Å, but the various O–P–F angles are quite similar to those commonly found for O–P–O angles in a non substituted tetrahedron.

Table 5.1.25

An example of PO_3F tetrahedron geometry as observed by Perloff [195] in $CaPO_3F.2H_2O$.

P	O(1)	O(2)	O(3)	F
O(1)	1.506(1)	115.68(6)	116.48(7)	102.68(6)
O(2)	2.557(2)	1.515(1)	109.52(8)	105.03(6)
O(3)	2.558(2)	2.465(2)	1.503(2)	106.13(7)
F	2.413(1)	2.459(1)	2.468(2)	1.583(1)

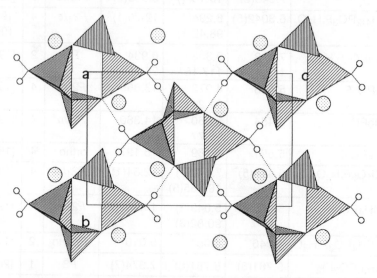

Figure 5.1.48. Projection of the $LiKPO_3F.H_2O$ atomic arrangement along the **a** direction. The larger tetrahedra represent the $LiO_3(H_2O)$ groups, the smaller ones the PO_3F groups. The large dotted circles represent the potassium atoms, the smaller open ones the hydrogen atoms. Hydrogen bonds are represented by dotted and solid lines.

– The LiKPO$_3$F.H$_2$O atomic arrangement. This atomic arrangement [198] can be described as a packing of infinite chains built by corner sharing LiO$_3$(H$_2$O) and PO$_3$F tetrahedra. These chains, spreading along the **a** axis, are interconnected by a set of hydrogen and K–O bonds. Figure 5.1.48 represents the atomic arrangement in projection along the **a** axis, the direction of the chain axis. Within a chain, as represented by Figure 5.1.49, each PO$_3$F shares its three oxygen atoms with the three adjacent LiO$_3$(H$_2$O) and each LiO$_3$(H$_2$O) tetrahedron its three oxygen atoms with three adjacent phosphoric groups. In the phosphoric group, the three P–O distances range from 1.506 to 1.511 Å, while the P–F distance is significantly longer (1.589 Å). Li–O or Li–H$_2$O distances are within a range of 1.870–1.975 Å.

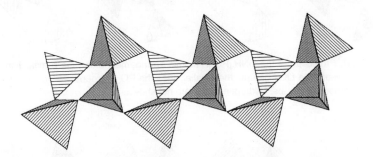

Figure 5.1.49. Projection of an isolated chain of LiO$_3$(H$_2$O) and PO$_3$F tetrahedra in LiKPO$_3$F.H$_2$O. The larger tetrahedra represent the LiO$_3$(H$_2$O) groups, the smaller ones the PO$_3$F groups. The unshared corners of the tetrahedra are the fluorine atoms in the case of PO$_3$F and the water molecules in LiO$_3$(H$_2$O).

– The LiNH$_4$PO$_3$F atomic arrangement. For this compound, one could expect an arrangement similar to that found in LiNH$_4$SO$_4$, in which a tridymite-like LiO$_4$–SO$_4$ framework exists. But, the impossibility of the existence of Li–F–P bonds leads to a rather different stacking.

This structure [199], as shown by Figure 5.1.50, is a layered arrangement. Thick layers, built by LiO$_4$ and PO$_3$F tetrahedra, alternate perpendicular to the **c** direction with slabs of ammonium groups. In order to show the arrangement within these layers, we report in Figure 5.1.51, the projection of one of them along the **c** direction. One can describe such a slab as composed of infinite chains of corner-sharing LiO$_4$ tetrahedra, zigzagging parallel to the **b** direction, and interconnected by PO$_3$F groups. This layer is built in such a way that, among the four oxygen atoms of a LiO$_4$ tetrahedron, two of them are each shared with another LiO$_4$ tetrahedron and a PO$_3$F group, the last two are common with only one PO$_3$F group. Reversely, one can say also that a PO$_3$F group

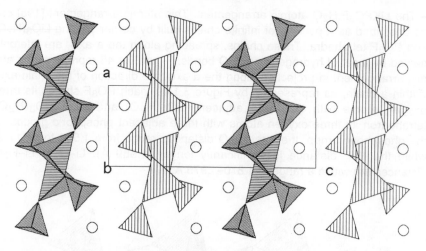

Figure 5.1.50. Projection of the LiNH$_4$PO$_3$F atomic arrangement along the **b** axis. The larger tetrahedra represent the LiO$_4$ groups, the smaller ones the PO$_3$F groups. The open circles represent the ammonium groups. Hydrogen atoms are not shown.

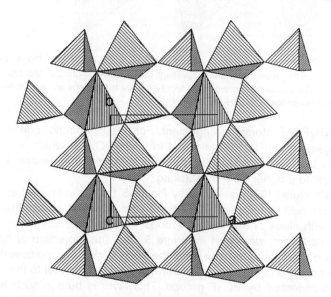

Figure 5.1.51. Projection, along the **c** axis, of an isolated layer of LiO$_4$ and PO$_3$F tetrahedra as observed in LiNH$_4$PO$_3$F. The larger tetrahedra represent the LiO$_4$ groups, the smaller ones the PO$_3$F groups. The unshared corners of the PO$_3$F tetrahedra are the fluorine atoms.

shares one of its oxygen atoms with two adjacent LiO_4 tetrahedra, while the last two are each shared with only one LiO_4. The fluorine atom remains unshared. The PO_3F tetrahedron, as commonly observed in monofluorophosphate groups, has a P–F distance (1.592 Å) significantly longer than the three P–O distances ranging from 1.506 to 1.512 Å. The LiO_4 tetrahedron is almost regular, with Li–O distances ranging from 1.902 to 2.032 Å and O–Li–O angles from 103.9 to 115.2°.

Difluoromonophosphates – Probably because their chemical preparation is more difficult, difluoromonophosphates are less numerous than monofluoromonophosphates.

The first reported preparation of a difluoromonophosphate was performed using NH_4F and P_4O_{10}. Schematically the reaction is:

$$4NH_4F + P_2O_5 \longrightarrow 2(NH_4)PO_2F_2 + 2NH_3 + H_2O$$

This reaction produces both monofluoro- and difluoromonophosphates which must be separated by fractional crystallization.

Other ways of preparation were reported. For example:

$$NaPF_6 + 2NaPO_3 \longrightarrow 3NaPO_2F_2$$

$$3KPF_6 + 2B_2O_3 \longrightarrow 4BF_3 + 3KPO_2F_2$$

In fact, most of the difluoromonophosphates were obtained by the action of the acid, HPO_2F_2, on various salts. This acid can be prepared by the action of gaseous POF_3 on the monofluoromonophosphoric acid:

$$H_2PO_3F + POF_3 \longrightarrow 2HPO_2F_2$$

or by the action of HF on P_4O_{10}:

$$P_4O_{10} + 8HF \longrightarrow 4HPO_2F_2 + 2H_2O$$

By the action of the difluoromonophosphoric acid on the corresponding chlorides, Shihada *et al.* [201] prepared $Al(PO_2F_2)_3$ and $Ga(PO_2F_2)_3$:

$$MCl_3 + 3HPO_2F_2 \longrightarrow M(PO_2F_2)_3 + 3HCl$$

The three isotypic compounds, KPO_2F_2, $RbPO_2F_2$, and $CsPO_2F_2$, whose the structural type is described below, were also prepared by this procedure. In the case of the potassium salt, Tul'chinski *et al.* [202] also reported an optimized procedure using the reaction of KPF_6 on KPO_3:

$$KPF_6 + 2KPO_3 \xrightarrow{673\,K} 3KPO_2F_2$$

Fluorination of the corresponding dichloromonophosphates was also used by Weidlein [203] to produce $Fe(PO_2F_2)_3$ and $In(PO_2F_2)_3$:

$$M(PO_2Cl_2)_3 + 3F_2 \longrightarrow M(PO_2F_2)_3 + 3Cl_2$$

The existence of an adduct between the zinc derivative and the acid,

$Zn(PO_2F_2)_2 \cdot 2HPO_2F_2$, is reported by Shihada *et al.* [201].

We have selected a series of three compounds, KPO_2F_2, $RbPO_2F_2$, and KPO_2F_2, all isotypic with $BaSO_4$ to illustrate this section. We report in Table 5.1.26, their unit-cell dimensions and that of $BaSO_4$.

Table 5.1.26
Unit-cell dimensions of the MPO_2F_2 (M = K, Rb, Cs) compounds isotypic with $BaSO_4$.
The space group is *Pnma* and $Z = 4$.

Formula	*a*	*b*	*c* (Å)	Ref.
$BaSO_4$	8.882	5.456	7.155	
KPO_2F_2	8.03	6.20	7.63	[204]
$RbPO_2F_2$	8.15	6.45	7.79	[205]
$CsPO_2F_2$	8.43	6.79	8.06	[205]

A structural investigation of KPO_2F_2 by Harrison *et al.* [204] confirmed its isotypy with $BaSO_4$ as was suggested by the analogy of their unit cells. In Figure 5.1.52 a projection along the **a** direction shows the atomic arrangement of KPO_2F_2 using the data published by these authors. Rows of PO_2F_2 tetrahedra, parallel to the **a** direction are arranged to create large diamond-like channels in which zigzag rows of potassium atoms are located.

Table 5.1.27 presents the geometrical features of the PO_2F_2 tetrahedron observed in this compound. This tetrahedron has mirror symmetry, with the phosphorus atom and the two crystallographically independent fluorine atoms in the mirror.

Table 5.1.27
Interatomic (Å) distances and bond angles (°) in the PO_2F_2 tetrahedron in KPO_2F_2. Some
numerical values are duplicated because the tetrahedron has mirror symmetry.

P	O(1)	O(1)	F(1)	F(2)
O(1)	1.457(5)	2.553	2.439	2.448
O(1)	122.4(4)	1.457(5)	2.439	2.448
F(1)	108.3(4)	108.3(4)	1.552(7)	2.326
F(2)	108.9(4)	108.9(4)	97.1(4)	1.551(7)

Within a range of 3.5 Å, the potassium atom sited in the mirror plane has tenfold coordination built by six oxygen and four fluorine atoms. This coordination is provided by its seven PO_2F_2 neighbors with which the KO_6F_4 polyhedron shares either edges or corners. Figure 5.1.53 shows a perspective view of

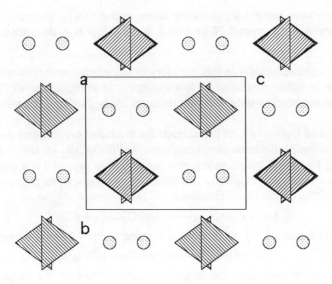

Figure 5.1.52. Projection of the atomic arrangement in KPO_2F_2 along the **a** axis. The PO_2F_2 tetrahedra are shown in polyhedral representation. The dotted circles represent the K atoms.

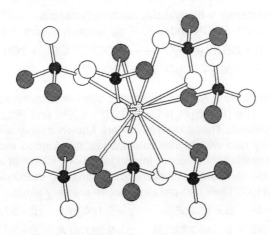

Figure 5.1.53. Perspective view of a part of the structure of KPO_2F_2 showing how a K atom is connected to seven PO_2F_2 groups within a range of 3.5 Å. The K atom is represented by a dotted circle, the P atoms by black, the oxygen atoms by white, and the F atoms by grey circles.

this arrangement around a potassium atom. In the KO_6F_4 polyhedron, the six K–O distances range from 2.76 to 3.08 Å and the four K–F distances from 2.83 to 3.34 Å.

Chloromonophosphates – In this category of substituted monophosphates, our knowledge is rather restricted. A few examples of compounds with $[PO_2Cl_2]^-$ anions are reported to date, but no structural study confirms the nature of the anions.

Müller and Dehnicke [206] described the chemical preparation and characterized a series of dichloromonophosphates, $M^{III}(PO_2Cl_2)_3$ for M^{III} = Al, Ga, In, Fe. During the same investigation, the authors also reported the existence of the beryllium salt, $Be(PO_2Cl_2)_2$. All these compounds were prepared by the action of $POCl_3$ on the metal chlorides:

$$MCl_3 + 3POCl_3 \longrightarrow M(PO_2Cl_2)_3 + 6Cl_2$$

When heated, these salts decompose according to the following scheme:

$$M(PO_2Cl_2)_3 \longrightarrow 2POCl_3 + MPO_4$$

According to infrared data, these authors suspected that the trivalent-cation derivatives could have polymeric structures.

Later, Weidlein [207] investigated the properties of these compounds and showed that the corresponding difluoromonophosphates can be obtained from the indium and iron derivatives by direct fluorination:

$$M(PO_2Cl_2)_3 + 3F_2 \longrightarrow M(PO_2F_2)_3 + 3Cl_2$$

Weidlein prepared a titanyldichloromonophosphate, $TiO(PO_2Cl_2)_2$, by the reaction of $TiCl_4$ with $POCl_3$ and Cl_2O. This reaction occurs in two steps:

$$TiCl_4 + 2POCl_3 + Cl_2O \longrightarrow TiOCl_2.2POCl_3 + 2Cl_2$$

$$TiOCl_2.2POCl_3 + 2Cl_2O \longrightarrow TiO(PO_2Cl_2)_2 + 4Cl_2$$

Thiomonophosphates – A good number of sulphur-substituted monophosphates are known. The $[PO_3S]^{3-}$, $[PO_2S_2]^{3-}$, $[POS_3]^{3-}$, and $[PS_4]^{3-}$ anions were clearly characterized. These derivatives are known mainly with alkali metals, but are relatively rare with higher valency cations. A good example of such a compound is provided by a recent investigation of Palkina *et al.* [208], dealing with two isotypic lanthanide dithiomonophosphates, $LaPO_2S_2.11/2H_2O$ and $NdPO_2S_2.11/2H_2O$. Their unit-cell dimensions are very similar:

$a = 7.028(4)$, $b = 15.208(5)$, $c = 9.164(6)$ Å, $\beta = 97.32(5)°$ (La)

$a = 7.157(4)$, $b = 15.283(13)$, $c = 9.281(6)$ Å, $\beta = 97.29(6)°$ (Nd)

The space group is $P2_1/n$ and $Z = 4$. We will not represent the atomic arrangement in this category of salts, but simply give some information on the deformations induced in the PO_2S_2 group by the oxygen-sulphur substitution. In Table 5.1.28 the main numerical values observed in this tetrahedron are gathered. The two P–O distances, 1.558 and 1.531 Å, are those normally found in any

Table 5.1.28
Main interatomic distances (Å) and bond angles (°) in the PO_2S_2 tetrahedron observed in $NdPO_2S_2.11/2H_2O$.

P	O(1)	O(2)	S(1)	S(2)
O(1)	1.558(9)	2.41(1)	2.924(8)	2.94(1)
O(2)	102.5(5)	1.531(9)	2.949(8)	2.944(9)
S(1)	109.8(4)	112.4(4)	2.002(4)	3.322(5)
S(2)	109.7(4)	110.9(4)	111.1(2)	2.026(5)

kind of phosphoric anion, but the two P–S distances are, as can be expected, considerably longer, 2.022 and 2.026 Å. In spite of this very significant distortion, an examination of the various O–P–O, S–P–S, and S–P–O angles shows

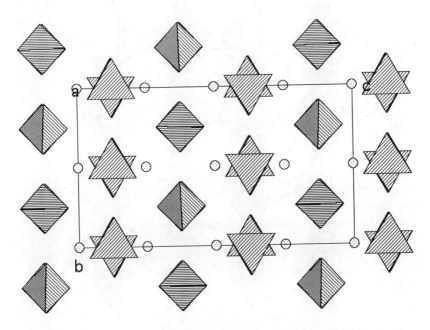

Figure 5.1.54. The atomic arrangement in $SmPS_4$ in projection along the **a** direction. The PS_4 groups are shown in polyhedral representation and the grey circles represent the Sm atoms.

that they do not depart from the values ordinarily measured in a PO_4 group. In the present example, they are within a range of 102.5 to 112.4°, with an average value of 109.4°.

Totally substituted thiomonophosphates of trivalent cations, as BPS_4, $AlPS_4$, $GaPS_4$, $InPS_4$, were extensively studied. Recently, $LnPS_4$, (Ln = La, Pr, Nd, Sm, Gd, Tb, Dy, Ho, Er, Yb), were characterized by Palkina *et al.* [209]. We report below the main crystallographic features and the atomic arrangement of *SmPS₄* as described by these authors. $SmPS_4$ is tetragonal, $I4_1/acd$, with a large unit cell:

$$a = 10.790(2), \quad c = 19.181 \text{ Å}$$

containing sixteen formula units. The atomic arrangement, represented in Figure 5.1.54 can be described as built by planes of samarium atoms alternating with layers of PS_4 tetrahedra, both perpendicular to the **c** axis. Another interesting aspect of this structure is given in Figure 5.1.55. This drawing shows how the PS_4 groups arrange themselves around the 4_1 axis to generate chan-

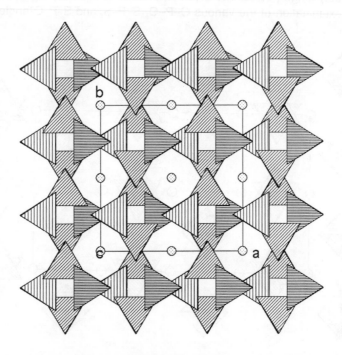

Figure 5.1.55. The atomic arrangement in SmPS₄ in projection along the **c** axis. The PS₄ groups are shown in polyhedral representation and the grey circles represent the samarium atoms.

nels containing rows of samarium atoms. In the PS_4 tetrahedron which has binary symmetry, the two P–S distances are 2.038 and 2.044 Å. The two crystallographically independent samarium atoms have eightfold coordination, with Sm–S distances ranging from 2.851 to 3.046 Å.

Nitromonophosphates – The chemistry of phosphoric anions that are partly or totally substituted by nitrogen is presently under development but are not clear enough to deserve a survey. We examine one example of such a substitution in the section devoted to heteropolyphosphates.

5.1.10. Chemical Preparations of Monophosphates

Many processes were used for the synthesis of monophosphates and it seems hopeless to report an exhaustive examination of these methods. Nevertheless, we attempt a survey of the most commonly used. In this survey, it will soon appear that some processes, like flux methods or hydrothermal syntheses, are specific to single-crystal production, whereas other more classical ones are used for the production of polycrystalline specimens.

The synthesis of monophosphates is greatly facilitated by the existence of a good number of commercially available starting materials. Phosphoric acid, sodium, potassium, and ammonium monophosphates as well as many other derivatives are common compounds.

The chemical preparations of substituted monophosphates are most of time very specific and were described in the corresponding sections.

5.1.10.1. Aqueous chemistry processes

The great majority of the monophosphates produced on a large scale for industrial purposes as sodium, potassium, and calcium derivatives, are prepared simply by the action of phosphoric acid on the corresponding hydroxides or carbonates.

Classical methods of aqueous chemistry are extensively used for the preparation of monophosphates. In spite of an apparent simplicity, these reactions are dependent on a good number of parameters: pH, relative ratio of the components, nature of some components, concentration, duration, and temperature, often difficult to optimize properly. For instance, the procedure described by Mathew and Schroeder [210] for the synthesis of $MgKPO_4.6H_2O$ and we report below, may appear as very logical but is, in fact, the fruit of a previous careful investigation of the $MgO–K_2O–P_2O_5–H_2O$ system by Lehr *et al.* [211]. To prepare $MgKPO_4.6H_2O$, a diluted KOH aqueous solution is added dropwise to a solution of 5 g of $MgCl_2$ and 25 g KH_2PO_4 in 100 cm^3 of water till the pH is 7.5. After a few days, crystals of $MgKPO_4.6H_2O$ will appear in the solution.

The same words can be used to comment on the preparation of the manganese analogue of newberite, *MnHPO$_4$.3H$_2$O*, reported by Durif [212]. In an aqueous solution containing 16 g of Na$_2$HPO$_4$.12H$_2$O in 300 cm^3 of water, one adds 50 cm^3 of a solution containing 20 g of MnCl$_2$.4H$_2$O. The voluminous gelatinous precipitate which forms immediately is kept under strong mechanical stirring and added with about 300 cm^3 of water. This mechanical stirring is maintained for at least two hours. After two days of standing at room temperature, the precipitate is totally crystallized and crystals of MnHPO$_4$.3H$_2$O, up to 1/2 mm can be separated by filtration.

In the same manner, the preparation of pure *LiH$_2$PO$_4$*, an important starting material for the synthesis of cyclohexaphosphates, cannot be achieved if the initial mixture of Li$_2$CO$_3$ and H$_3$PO$_4$ is stoichiometric. A 10% excess of the acid is necessary.

Pure *Cd(H$_2$PO$_4$)$_2$.2H$_2$O*, in a good crystalline state, can only be obtained conveniently if solid cadmium carbonate is slowly added to warm monophosphoric acid till is formed a solid phase made of well crystallized pure Cd(H$_2$PO$_4$)$_2$.2H$_2$O.

Most of the apparently simplest preparations of monophosphates are tributary of procedures which were, in many cases, difficult to optimize.

5.1.10.2. Thermal processes

Many anhydrous monophosphates can be prepared as polycrystalline specimens by heating various stoichiometric mixtures at relatively moderate temperatures. For instance, most of the M^{I}M^{II}PO$_4$ monophosphates are easily synthesized by using the following types of reactions:

$$CaCO_3 + NaH_2PO_4 \longrightarrow CaNaPO_4 + CO_2 + H_2O$$

$$MgO + NaPO_3 \longrightarrow MgNaPO_4$$

Through a very similar scheme, divalent-cation monophosphates are also obtained:

$$2(NH_4)_2HPO_4 + 3BaCO_3 \longrightarrow Ba_3(PO_4)_2 + 3CO_2 + 4NH_3 + 3H_2O$$

The temperature of the reaction and its duration must be optimized in function of the nature of the starting materials. Frequent homogeneization grinds shorten significantly the time required for heating.

5.1.10.3. Gel diffusion technique

This technique widely used in certain domains of chemistry is seldom used for the production of phosphates. One can, nevertheless, report the synthesis of *MgKPO$_4$.6H$_2$O* by Banks *et al.* [213]. Some attempts to prepare cobalt and nickel monophosphates are reported by Henisch [214].

5.1.10.4. Hydrothermal synthesis

A good number of new monophosphates were characterized during hydro-thermal experiments. These high-pressure processes are mainly used by crys-tallographers for the production of single crystals with a suitable size and also for the systematic investigation of a $MO-P_2O_5-H_2O$ or other more complex systems. We report two examples of such studies carried out in the $CuO-Na_2O-P_2O_5-H_2O$ system.

Single crystals of a new form of $CuNaPO_4$ were obtained by heating a stoichiometric mixture of $CuSO_4.5H_2O$ and $Na_3PO_4.12H_2O$ at 773 K for three days, under a pressure of 2000 Kg.cm^{-2} by Kawahara *et al.* [215]. At the end of the experiment, the vessel was slowly cooled at the rate of 2 K/hr down to room temperature.

In the same system, Effenberger [216] characterized $Cu_3Na_5(PO_4)_3(HPO_4)$. In these experiments, the author used 2 g of a stoichiometric mixture of $NaNO_3$ and Na_2HPO_4 added with 1 g of metal copper. After two days of heating at 493 K, the experimenter obtained single crystals of $Cu_3Na_5(PO_4)_3(HPO_4)$ mixed with Cu_2O, $Cu_2(OH)PO_4$, and Cu, but, if 1–2 cm^3 of HNO_3 are added to the starting components, a pure crop of $Cu_3Na_5(PO_4)_3(HPO_4)$ is obtained.

5.1.10.5. Flux processes

These processes although extensively used for the preparation of conden-sed phosphates could not be used for monophosphates because, in most of the usual phosphoric fluxes, the condensation occurs at relatively low tempera-tures. Fluxes of chlorides or fluorides were successfully used for the production of a few divalent-cation monophosphates. A mixture of zinc chloride and potas-sium chloride, corresponding to an eutectic composition in this system, was also used for the preparation of $ZnKPO_4$ single crystals. Na_2MoO_4 was used as flux agent by Elammari *et al.* [217] for the crystal growth of $ZnNaPO_4$. Never-theless, the use of a phosphoric flux in the single crystal production of $ZnCsPO_4$ was performed successfully by Blum [218]. After a previous study of a part of the system $Cs_4P_2O_7-Zn_3(PO_4)_2-ZnO$, the author concluded that cesium diphos-phate, $Cs_4P_2O_7$, was probably the most appropriate flux. He could obtain crystals of $ZnCsPO_4$ up to 7 mm long by the slow cooling (0.5 K/hr), between 1393 and 1073 K, of a melt of $ZnCsPO_4$ and $Cs_4P_2O_7$ in the molar ratio 1/1.

5.1.10.6. Degradation of condensed phosphates

As we wrote at the beginning of this chapter, monophosphates are the final term of the hydrolytic degradation of all condensed phosphates. For obvious reasons, this process seems illogical and is consequently not commonly used for the synthesis of monophosphates. In some occasions, it proved to be fruitful.

For instance, Averbuch-Pouchot and Durif [219] characterized a new form of rubidium dihydrogenmonophosphate, RbH_2PO_4, that they obtained by the hydrolytic cleavage of rubidium cyclotetraphosphate, $Rb_4P_4O_{12}$:

$$Rb_4P_4O_{12} + 4H_2O \longrightarrow 4RbH_2PO_4$$

An aqueous solution of rubidium cyclotetraphosphate is heated at 333 K until it is converted into a gel. This gel is then kept at room temperature in a water-saturated atmosphere. After several days, crystals of RbH_2PO_4 appear in the gel. The yield of such a process is very poor.

In spite of its present purely academic interest, the degradation of condensed phosphates into monophosphates will be widely used for the production of the PO_4 radicals necessary for plant nutrition through the use of non-water soluble fertilizers mainly made of condensed phosphates in the future.

5.1.10.7. The use of boron monophosphate

Some very stable and refractory monophosphates are used for the elimination of nuclear wastes. The dissolution of some radioactive refuses, for a further use of a precipitation process, may be delicate, thus a purely solid-state process using mainly boron monophosphate was optimized. These reactions were more especially investigated for the elimination of iron, a major component of the nuclear wastes. Vasovic *et al.* [220] report the following reactions:

$$Fe_2O_3 + 2BPO_4 \longrightarrow 2FePO_4 + B_2O_3$$

$$2[FeCl_3.6H_2O] + 2BPO_4 \longrightarrow 2FePO_4 + B_2O_3 + 6HCl + 9H_2O$$

These two reactions are said to have good yields when performed at 1173 K, for 4–6 hours. In the two cases, the quartz form of $FePO_4$ is obtained. The same authors also describe another type of reaction:

$$BPO_4 + 3KCl \xrightarrow{1073\,K} BCl_3 + K_3PO_4$$

A change of valence is sometimes observed during these reactions. For instance, Bamberger [221] reports:

$$Ce^{IV}O_2 + BPO_4 \longrightarrow Ce^{III}PO_4 + 1/2B_2O_3 + 1/4O_2$$

The reaction was investigated up to 1273 K in argon or nitrogen atmosphere. Under these conditions, the oxygen evolution starts at 823–873 K. In the case of terbium oxide, the same phenomenon occurs:

$$1/7Tb_7O_{12} + BPO_4 \longrightarrow TbPO_4 + 1/2B_2O_3 + 3/28O_2$$

The same author observed the formation of mixed valency monophosphates in the elimination of iron:

$$Fe_2O_3 + mBPO_4 \longrightarrow Fe_x^{II} Fe_y^{III} (PO_4)_m + m/2B_2O_3 + (x-1)/4O_2$$

The same process has been applied to fluorides [221]:

$$AlF_3 + BPO_4 \xrightarrow{1223K} BF_3 + AlPO_4$$
$$NdF_3 + BPO_4 \longrightarrow BF_3 + NdPO_4$$

5.1.10.8. Organic-cation monophosphates

No specific processes are necessary for their synthesis. Most of the organic-cation monophosphates were prepared by a reaction involving the direct action of monophosphoric acid on the corresponding amine. Schematically the reaction is:

$$R-NH_2 + H_3PO_4 \longrightarrow [R-NH_3]^+. (H_2PO_4)^-$$

or

$$2R-NH_2 + H_3PO_4 \longrightarrow [R-NH_3]_2^+. (HPO_4)^{2-}$$

The reaction is performed at room temperature by mixing diluted aqueous solutions of the two components in a stoichiometric ratio and the resulting solution evaporated at room temperature.

References

1. U. Schülke and R. Kayser, *Z. anorg. allg. Chem.*, **531**, (1985), 167–176.

2. E. Thilo and H. Grunze, *Z. anorg. allg. Chem.*, **281**, (1955), 263–283.

3. I. Grunze, K. Dostal, and E. Thilo, *Z. anorg. allg. Chem.*, **302**, (1959), 221–229.

4. J. M. Verdier, *Thesis*, (Univ. of Paris, France, 1967).

5. M. T. Averbuch-Pouchot and A. Durif, *Acta Crystallogr.*, **C41**, (1985), 665–667.

6. M. T. Averbuch-Pouchot and A. Durif, *Acta Crystallogr.*, **C41**, (1985), 1555–1556.

7. D. M. Wiench and M. Jansen, *Z. anorg. allg. Chem.*, **501**, (1983), 95–101.

8. M. T. Averbuch-Pouchot and A. Durif, *J. Solid State Chem.*, **46**, (1983), 193–196.

9. A. Larbot and J. Durand, *Acta Crystallogr.*, **C39**, (1983), 12–15.

10. T. Lis, *Acta Crystallogr.*, **C50**, (1994), 484–487.

11. E. Tillmanns and W. H. Baur, *Acta Crystallogr.*, **B27**, (1971), 2124–2132.

12. M. Catti, G. Ferraris, and G. Ivaldi, *Acta Crystallogr.*, **B34**, (1978), 369–373.

13. W. H. Baur and A. A. Khan, *Acta Crystallogr.*, **B26**, (1970), 1584–1596.

14. M. Bagieu-Beucher, M. T. Averbuch-Pouchot, and M. Rzaigui, *Acta Crystallogr.*, **C47**, (1991), 1364–1366.

15. J. F. Sarver, *Trans. Brit. Ceram. Soc.*, **65**, (1966), 191–198.

16. F. L. Katnack and F. A. Hummel, *J. Electrochem. Soc.*, **105**, (1958), 125–133.

17. J. J. Brown and F. A. Hummel, *J. Electrochem. Soc.*, **111**, (1964), 1052–1057.

18. L. Merker and H. Wondratschek, *Z. anorg. allg. Chem.*, **306**, (1960), 25–29.

19. J. Berak, *Roczniki Chem.*, **32**, (1958), 17–22.

20. J. H. Welch and W. Gutt, *J. Chem. Soc.*, (1961), 4442–4444.

21. W. Fix, H. Heymann, and R. Heinke, *J. Am. Ceram. Soc.*, **52**, (1969), 346–347.

22. E. R. Kreidler and F. A. Hummel, *Inorg. Chem.*, **6**, (1967), 884–891.

23. R. A. Mc Cauley and F. A. Hummel, *Trans. Brit. Ceram. Soc.*, **67**, (1968), 619–628.

24. N. C. Webb, *Acta Crystallogr.*, **21**, (1966), 942–948.

25. B. M. Nirsha, T. V. Khomutova, A. A. Fakeev, V. M. Agre, B. V. Zhadanov, N. P. Kozlova, and V. A. Olikova, *Russ. J. Inorg. Chem.*, **26**, (1981), 799–802.

26. V. A. Kopilevich, *Russ. J. Inorg. Chem.*, **35**, (1990), 464–468.

27. R. C. Ropp, M. A. Aia, C. W. W. Hoffman, T. J. Veleker, and R. W. Mooney, *Anal. Chem.*, **31**, (1959), 1163–1166.

28. R. C. Ropp and M. A. Aia, *Anal. Chem.*, **34**, (1962), 1188–1191.

29. E. Thilo and I Grunze, *Z. anorg. allg. Chem.*, **290**, (1957), 209–222.

30. L. N. Schegrov, N. M. Antraptseva, and I. G. Ponomareva, *Izv. Akad. Nauk SSSR, Neorg. Mater.*, **25**, (1989), 308–312.

31. E. Thilo and I. Grunze, *Z. anorg. allg. Chem.*, **290**, (1957), 209–222.

32. J. C. Elliot, *Structure and Chemistry of the Apatites and Other Calcium Orthophosphates*, (Elsevier, Amsterdam, 1994).

33. M. V. Goloshchapov and T. N. Filatova, *Russ. J. Inorg. Chem.*, **14**, (1969), 424–426.

34. J. E. Salmon and H. Terrey, *J. Chem. Soc.*, (1950), 2813–2824.

35. J. Komrska and V. Satava, *Silikati*, **13**, (1969), 135–141.

36. Y. Cudennec, A. Lecerf, A. Riou, and Y. Gerault, *C. R. Acad. Sci., Sér. 2*, **301**, (1985), 93–98.

37. C. Calvo, *Canad. J. Chem.*, **43**, (1965), 436–445.

38. J. S. Stephens and C. Calvo, *Canad. J. Chem.*, **45**, (1967), 2303–2312.

39. C. Calvo, J. Phys. *Chem. Solids*, **24**, (1963), 141–149.

40. A. Riou, Y. Cudennec, and Y. Gerault, *Rev. Chim. Minér.*, **23**, (1986), 810–818.

41. R. J. Hill and J. B. Jones, *Am. Miner.*, **61**, (1976), 987–995.

42. F. Liebau, *Acta Crystallogr.*, **18**, (1965), 352–354.

43. R. S. Gamidov, V. P. Golovatschev, C. S. Mamedov, and N. V. Belov, *Dokl. Akad. Nauk, SSSR*, **150**, (1963), 381–384.

44. G. Y. Chao, *Z. Kristallogr.*, **130**, (1969), 261–266.

45. P. A. Sandomirskii, G. P. Klientova, M. A. Simonov, and N. V. Belov, *Dokl. Akad. Nauk SSSR*, **236**, (1977), 597–600.

46. A. Riou, Y. Cudennec, and Y. Gerault, *Acta Crystallogr.*, **C43**, (1987), 194–197.

47. M. T. Averbuch-Pouchot, *J. Appl. Crystallogr.*, **7**, (1974), 511–512.

48. L. Fred, F. L. Katnack, and F. A. Hummel, *J. Electrochem. Soc.*, **105**, (1958), 125–133.

49. E. Thilo and I. Schulz, *Z. anorg. allg. Chem.*, **269**, (1951), 201–208.

50. J. Saison, *C. R. Acad. Sci.*, **250**, (1960), 2374–2376.

51. M. A. de Schulten, *Bull. Soc. fr. Minér.*, **27**, (1904), 100–103.

52. A. Boudjada, A. Durif, and J. C. Guitel, *Acta Crystallogr.*, **B36**, (1980), 133–135.

53. R. Herak, B. Prelesnik, and M. Curik, *Z. Kristallogr.*, **164**, (1983), 25–30.

54. I. D. Brown and R. D. Shannon, *Acta Crystallogr.*, **A29**, (1973), 266–282.

55. A. G. Nord and P. Kirkegaard, *Chemica Scripta*, **15**, (1980), 27–39.

56. L. J. Spencer, *Mineral. Mag.*, **15**, (1908), 1–38.

57. A. Whittaker, *Acta Crystallogr.*, **B31**, (1975), 2026–2035.

58. W. Demel, *Ber. dtsch Chem. Ges.*, **12**, (1879), 1171–1174.

59. M. T. Averbuch-Pouchot and A. Durif, *Bull. Soc. fr. Minér. Cristallogr.*, **95**, (1972), 511–512.

60. M. T. Averbuch-Pouchot, A. Durif, J. C. Guitel, I. Tordjman, and M. Laügt, *Bull. Soc. fr. Minér. Cristallogr.*, **96**, (1973), 278–280.

61. L. M. Dikareva, G. G. Sadikov, M. A. Porai–Koshits, I. B. Baranovskii, S. S. Abdullaev, and R. N. Schelokov, *Russ. J. Inorg. Chem.*, **27**, (1982), 236–240.

62. W. A. Dollase, *Acta Crystallogr.*, **B25**, (1969), 2298–2302.

63. S. Jaulmes and C. Durif, *C. R. Acad. Sci.*, *Sér. C*, **262**, (1966), 1530–1533.

64. R. Masse and A. Durif, *J. Solid State Chem.*, **73**, (1988), 468–472.

65. G. Wallez, *Thesis*, (Univ. P. et M. Curie, Paris, France, 1993).

66. E. L. Rakotomahina-Ralaisoa, *Thesis*, (Univ. of Grenoble, France, 1972).

67. D. Blum, A. Durif, and M. T. Averbuch-Pouchot, *Ferroelectrics*, **69**, (1986), 283–292.

68. M. T. Averbuch-Pouchot, *Mat. Res. Bull.*, **4**, (1969), 851–858.

69. M. Andratschke, *Z. anorg. allg. Chem.*, **620**, (1994), 361–365.

70. L. Elammari, *Thesis*, (Univ. of Rabat, Morocco, 1989).

71. O. V. Yakubovich, M. A. Simonov, and O. K. Mel'nikov, *Kristallografiya*, **35**, (1990), 42–46.

72. N. I. Golovastikov, *Sov. Phys. Crystallogr.*, **6**, (1962), 733–739.

73. G. Engel, *Neues Jahrb. Mineral.*, **127**, (1976), 197–211.

74. G. Giuseppetti and C. Tadini, *Tschermak's Mineral. Petrogr. Mitt.*, **20**, (1973), 1–12.

75. M. T. Averbuch-Pouchot, *Thesis*, (Univ. of Grenoble, France, 1974).

76. M. Andratschke and A. Feltz, *Z. anorg. allg. Chem.*, **582**, (1990), 179–189.

77. L. Elammari, J. Durand, L. Cot, and B. Elouadi, *Z. Kristallogr.*, **180**, (1987), 137–140.

78. O. V. Yakubovich and O. K. Mel'nikov, *Kristallografiya*, **34**, (1989), 62–66.

79. E. Schultz, *Z. Kristallogr.*, **132**, (1970), 450–451.

80. M. T. Averbuch-Pouchot and A. Durif, *Mat. Res. Bull.*, **8**, (1973), 353–356.

81. M. Andratschke and P. Hartmann, *Z. Chem.*, **30**, (1990), 415–416.

82. M. T. Averbuch-Pouchot and A. Durif, *Mat. Res. Bull.*, **3**, (1968), 719–722.

83.   W. L. Bragg and G. B. Brown, *Z. Kristallogr.*, **63**, (1926), 538–556.

84.   R. E. Newnham and M. J. Redman, *J. Am. Ceram. Soc.*, **48**, (1965), 547.

85.   F. Kubel, *Z. Kristallogr.*, **209**, (1994), 755.

86.   D. Destenay, *Mem. Soc. Roy. Sci. Liège*, **10**, (1950), 5–28.

87.   F. Hanic, M. Handlovic, K. Burdova, and J. Majling, *J. Cryst. Spect. Res.*, **12**, (1982), 99–127.

88.   S. Geller and J. L. Durand, *Acta Crystallogr.*, **13**, (1960), 325–331.

89.   R. E. Newnham, R. P. Santoro, and M. J. Redman, *J. Phys. Chem. Solids*, **26**, (1965), 445–447.

90.   M. Mercier and J. Gareyte, *Solid State Comm.*, **5**, (1967), 139–142.

91.   A. Byström, *Ark. Kem. Miner. Geol.*, **17**, (1943), 1–4.

92.   P. B. Moore, *Am. Miner.*, **57**, (1972), 1333–1344.

93.   L. Elammari, B. Elouadi, and W. Depmeier, *Acta Crystallogr.*, **C44**, (1988), 1357–1359.

94.   Y. A. Ivanov, M. A. Simonov, and N. V. Belov, *Sov. Phys. Crystallogr.*, **19**, (1974), 96–97.

95.   M. T. Averbuch-Pouchot and A. Durif, *Mat. Res. Bull.*, **8**, (1973), 1–8.

96.   J. Moring and E. Kostiner, *J. Solid State Chem.*, **61**, (1986), 379–383.

97.   Y. Le Page and G. Donnay, *Canad. Miner.*, **15**, (1977), 518–521.

98.   R. W. G. Wyckoff, *Crystal Structures*, Vol. 3, (John Wiley and Sons, New York, 1965).

99.   C. Palache, H. Berman, and C. Frondel, *Dana's System of Mineralogy*, Vol. 2, (John Wiley and Sons, New York, 1963).

100. M. A. Bredig, *J. Phys. Chem.*, **46**, (1942), 747–764.

101. C. Calvo and R. Faggiani, *Can. J. Chem.*, **53**, (1975), 1849–1853.

102. A. W. Kolsi, M. Quarton, and W. Freundlich, *J. Solid State Chem.*, **36**, (1981), 107–111.

103. C. W. Struck and J. G. White, *Acta Crystallogr.*, **15**, (1962), 290–291.

104. R. Masse and A. Durif, *J. Solid State Chem.*, **71**, (1987), 574–576.

105. M. Ben Amara, M. Vlasse, G. Le Flem, and P. Hagenmuller, *Acta Crystallogr.*, **C39**, (1983), 1483–1485

106. R. Klement and R. Uffelmann, *Naturwiss.*, **29**, (1941), 300–301.

107. R. Klement and P. Kresse, *Z. anorg. allg. Chem.*, **310**, (1961), 53–65.

108. J. A. McGinnety, *Acta Crystallogr.*, **B28**, (1972), 2845–2852.

109. K. Okada and J. Ossaka, *Acta Crystallogr.*, **B36**, (1980), 919–921.

110. M. Miyake, H. Morikawa, and S. I. Iwai, *Acta Crystallogr.*, **B36**, (1980), 532–536.

111. L. Elammari and B. Elouadi, *Acta Crystallogr.*, **C45**, (1989), 1864–1867.

112. A. Elfakir, J. P. Souron, F. Robert, and M. Quarton, *C. R. Acad. Sci.*, *Sér. 2*, **309**, (1989), 199–203.

113. L. H. Brixner and C. M. Foris, *Mat. Res. Bull.*, **10**, (1975), 31–34.

114. M. Bagieu-Beucher and R. Masse, *Z. Kristallogr.*, **188**, (1989), 5–10.

115. L. Elammari, B. Elouadi, and W. Depmeier, *J. Solid State Chem.*, **76**, (1988), 266–269.

116. M. Quarton and A. W. Kolsi, *Acta Crystallogr.*, **C39**, (1983), 664–667.

117. A. Kawahara, T. Kageyama, I. Watanabe, and J. Yamakawa, *Acta Crystallogr.*, **C49**, (1993), 1275–1277.

118. M. Hata and F. Marumo, *Acta Crystallogr.*, **B38**, (1982), 239–241.

119. I. Tordjman, J. C. Guitel, A. Durif, M. T. Averbuch-Pouchot, and R. Masse, *Mat. Res. Bull.*, **13**, (1978), 983–988.

120. M. Quarton and M. T. Oumba, *Mat. Res. Bull.*, **18**, (1983), 967–974.

121. J. L. Schoemaker, E. Kostiner, and J. B. Anderson, *Z. Kristallogr.*, **152**, (1980), 317–332.

122. H. Effenberger, *Z. Kristallogr.*, **168**, (1984), 113–119.

123. V. I. Lyutin, A. G. Tutov, V. V. Llyukhin, and N. V. Belov, *Sov. Phys. Dokl.*, **18**, (1973), 21–24.

124. C. Goria, *Atti Cl. Scienze Fisiche*, **92**, (1958), 96–104.

125. G. E. R. Schulze, *Z. physik. Chem.*, **B24**, (1934), 215–240.

126. J. D. Mackenzie, W. L. Roth, and R. H. Wentorf, *Acta Crystallogr.*, **12**, (1959), 79.

127. F. Dachille and L. S. D. Glasser, *Acta Crystallogr.*, **12**, (1959), 820–821.

128. C. Frondel, *Dana's System of Mineralogy*, Vol. 3, (Wiley and Sons, New York, 1962).

129. H. Strunz, *Z. Kristallogr.*, **103A**, (1941), 228–229.

130. K. Kosten and H. Arnorld, *Z. Kristallogr.*, **152**, (1980), 119–133.

131. A. Goiffon, G. Bayle, R. Astier, J. C. Jumas, M. Maurin, and E. Philippot, *Rev. Chim. Minér.*, **20**, (1983), 338–350.

132. E. C. Shafer, M. W. Shafer, and R. Roy, *Z. Kristallogr.*, **107**, (1956), 263–275.

133. R. C. L. Mooney, *Acta Crystallogr.*, **9**, (1956), 113–117.

134. R. Debnath and J. Chaudhuri, *J. Solid State Chem.*, **46**, (1983), 193–196.

135. R. Glaum, R. Gruehn, and M. Möller, *Z. anorg. allg. Chem.*, **543**, (1986), 111–116

136. P. Rittner and R. Glaum, *Z. Kristallogr.*, **209**, (1994), 162–169.

137. J. P. Attfield, P. D. Battle, and A. K. Cheetham, *J. Solid State Chem.*, **57**, (1985), 357–361.

138. J. Tudo and D. Carton, *C. R. Acad. Sci.*, **289**, (1979), 219–221.

139. N. Kinomura, M. Shimada, M. Koizumi, and S. Kume, *Mat. Res. Bull.*, **11**, (1976), 457–460.

140. A. Leclaire, A. Benmoussa, M. M. Borel, A. Grandin, and B. Raveau, *Eur. J. Solid State Inorg. Chem.*, **28**, (1991), 1323–1333.

141. W. O. Mulligan, D. F. Mullica, G. W. Beall, and L. A. Boatner, *Inorg. Chim. Acta*, **60**, (1982), 39–43.

142. K. K. Palkina, *Izv. Akad. Nauk SSSR, Neorg. Mater.*, **18**, (1982), 1413–436.

143. E. Patscheke, H. Fuess, and G. Will, *Chem. Phys. Letters*, **2**, (1968), 47–50.

144. W. O. Mulligan, D. F. Mullica, G. W. Beall, and L. A. Boatner, *Acta Crystallogr.*, **C39**, (1983), 23–24.

145. L. Vegard, *Phil. Mag.*, **1**, (1926), 1151–1193.

146. L. Vegard, *Z. Kristallogr.*, **67**, (1928), 482–485.

147. M. Kizilyalli and A. J. E. Welch, *J. Appl. Crystallogr.*, **9**, (1976), 413–414.

148. G. W. Beall, L. A. Boatner, D. F. Mullica, and W. O. Milligan, *J. Inorg. Nucl. Chem.*, **43**, (1981), 101–105.

149. F. Weigel, V. Scherer, and H. Henschel, *J. Am. Ceram. Soc.*, **48**, (1965), 383–384

150. R. C. L. Mooney-Slater, *Z. Kristallogr.*, **117**, (1962), 371–385.

151. R. Masse and A. Durif, *C. R. Acad. Sci.*, *Sér. 2*, **300**, (1985), 849–851.

152. R. C. L. Mooney, *Acta Crystallogr.*, **3**, (1950), 337–340.

153. K. K. Palkina, *Izv. Akad. Nauk SSSR, Neorg. Mater.*, **18**, (1982), 1413–436.

154. A. Hezel and S. D. Ross, *J. Inorg. Nucl. Chem.*, **29**, (1967), 2085–2089.

155. A. Clearfield, R. H. Blessing, and J. A. Stynes, *J. Inorg. Nucl. Chem.*, **30**,

(1968), 2249–2258.

156. A. Clearfield and G. D. Smith, *Inorg. Chem.*, **8**, (1969), 431–436.

157. G. Alberti, P. Cardini-Galli, U. Costantino, and E. Torracca, *J. Inorg. Nucl. Chem.*, **29**, (1967), 571–578.

158. A. N. Christensen, E. K. Andersen, I. G. K. Andersen, G. Alberti, M. Nielsen, and M. S. Lehman, *Acta Chem. Scand.*, **44**, (1990), 865–872.

159. S. Yamanaka and M. Tanaka, *J. Inorg. Nucl. Chem.*, **41**, (1979), 45–48.

160. B. Matkovic and M. Sljukic, *Croatica Chem. Acta*, **37**, (1965), 115–116.

161. B. Matkovic, M. Sljukic, and B. Prodic, *Croatica Chem. Acta*, **38**, (1966), 69–70.

162. B. Matkovic, B. Prodic, M. Sljukic, and S. W. Peterson, *Croatica Chem. Acta*, **40**, (1968), 147–161.

163. M. Sljukic, B. Matkovic, B. Prodic, and D. Anderson, *Z. Kristallogr.*, **130**, (1969), 148–161.

164. M. Topic and B. Prodic, *J. Appl. Crystallogr.*, **2**, (1969), 230–232.

165. B. Matkovic, B. Kojic-Prodic, M. Sljukic, M. Topic, R. D. Willet, and F. Pullen, *Inorg. Chimica*, **4**, (1970), 571–576.

166. M. Topic, B. Kojic-Prodic, and S. Popovic, *Czech. J. Phys.*, **B20**, (1970), 1003–1006.

167. L. O. Hagman and P. Kierkgaard, *Acta Chem. Scand.*, **22**, (1968), 1822–1832.

168. A. Clearfield, B. D. Roberts, and M. A. Subramanian, *Mat. Res. Bull.*, **19**, (1984), 219–226.

169. M. Sugantha, U. V. Varadaraju, and G. V. Subba Rao, *J. Solid State Chem.*, **111**, (1994), 33–39.

170. A. I. Kryukova, *Russ. J. Inorg. Chem.*, **36**, (1991), 1108–1111.

171. B. Matkovic, B. Prodic, and M. Sljukic, *Bull. Soc. Chim. France*, Special issue, (1968), 1777–1779

172. S. Kamoun, A. Jouini, M. Kamoun, and A. Daoud, *Acta Crystallogr.*, **C45**, (1989), 481–482.

173. L. Giardia, F. Garbassi, and M. Calcaterra, *Acta Crystallogr.*, **B29**, (1973), 1826–1829.

174. M. T. Averbuch-Pouchot, A. Durif, and J. C. Guitel, *Acta Crystallogr.*, **C43**, (1987), 1896–1898.

175. M. T. Averbuch-Pouchot and A. Durif, *Acta Crystallogr.*, **C43**, (1987),

1894–1896.

176. M. T. Averbuch-Pouchot, A. Durif, and J. C. Guitel, *Acta Crystallogr.*, **C44**, (1988), 1968–1972.

177. M. T. Averbuch-Pouchot, A. Durif, and J. C. Guitel, *Acta Crystallogr.*, **C44**, (1988), 890–892.

178. M. T. Averbuch-Pouchot, A. Durif, and J. C. Guitel, *Acta Crystallogr.*, **C44**, (1988), 99–102.

179. M. Bagieu-Beucher, A. Durif, and J. C. Guitel, *Acta Crystallogr.*, **C45**, (1989), 421–423.

180. M. Bagieu-Beucher and R. Masse, *Acta Crystallogr.*, **C47**, (1991), 1642–1645.

181. M. T. Averbuch-Pouchot, *Acta Crystallogr.*, **C49**, (1993), 813–815.

182 M. P. Attfield, R. E. Morris, and A. K. Cheetham, *Acta Crystallogr.*, **C50**, (1994), 981–984.

183. J. Loub, *Acta Crystallogr.*, **B47**, (1991), 468–473.

184. M. T. Averbuch-Pouchot, *C. R. Acad. Sci., Sér. 2*, **318**, (1994), 191–196.

185. D. E. C. Corbridge, *Acta Crystallogr.*, **9**, (1956), 991–994.

186. D. R. Powell, S. K. Smith, T. C. Farrar, and F. K. Ross, *Acta Crystallogr.*, **C50**, (1994), 342–346.

187. W. Jingxian, S. Genbo, H. Qingzhen, and L. Jinkui, *J. Struct. Chem.*, **7**, (1988), 148–151.

188. W. Lange, *Chem. Ber.*, 62, (1929), 786–792.

189. W. Lange, *Chem. Ber.*, 62, (1929), 793–801.

190. M. Amadori, *Atti Accad. Lincei*, **21II**, (1913), 688–692.

191. A. F. Bernt, *Acta Crystallogr.*, **B30**, (1974), 529–530.

192. J. Durand, W. Granier, and L. Cot, *C. R. Acad. Sci.*, **277**, (1973), 101–103.

193. J. Durand, L. Cot, and J. L. Galigné, *Acta Crystallogr.*, **B30**, (1974), 1565–1569.

194. M. T. Robinson, *J. Phys. Chem*, **62**, (1958), 925–928.

195. A. Perloff, *Acta Crystallogr.*, **B28**, (1972), 2183–2191.

196. A. F. Bernt and J. M. Sylvester, *Acta Crystallogr.*, **B28**, (1972), 2191–2193.

197. J. Durand, W. Granier, L. Cot, and C. Avinens, *C. R. Acad. Sci.*, **277**, (1973), 13–14.

198. J. L. Galigné, J. Durand, and L. Cot, Acta Crystallogr., **B30**, (1974), 697–701.

199. J. Durand, L. Cot, and J. L. Galigné, Acta Crystallogr., **B34**, (1978), 388–391.

200. J. Durand, W. Granier, L. Cot, and J. L. Galigné, Acta Crystallogr., **B31**, (1975), 1533–1535.

201. A. F. Shihada, B. K. Hassan, and A. T. Mohammed, Z. anorg. allg. Chem., **466**, (1980), 139–144.

202. V. B. Tul'chinski, I. G. Ryss, and V. I. Zubov, Russ. J. Inorg. Chem., **11**, (1966), 1446–1447.

203. J. Weidlein, Z. anorg. allg. Chem., **358**, (1968), 13–20.

204. R. W. Harrison, R. C. Thomson, and J. Trotter, J. Chem. Soc. (A), (1966), 1775–1780.

205. J. Trotter and S. H. Whitlow, J. Chem. Soc. (A), (1967), 1383–1386.

206. H. Müller and K. Dehnicke, Z. anorg. allg. Chem., **350**, (1967), 231–236.

207. J. Weidlein, Z. anorg. allg. Chem., **358**, (1968), 13–20.

208   K. K. Palkina, S. I. Maksimova, N. T. Chibiskova, T. A. Tripol'skaya, G. U. Vol'f, and T. B. Kuvshinova, Izv. Akad. Nauk SSSR., Neorg. Mater., **27**, (1991), 1028–1031.

209. K. K. Palkina, S. I. Maksimova, N. T. Chibiskova, T. B. Kuvshinova, and A. N. Volodina, Izv. Akad. Nauk SSSR., Neorg. Mater., **20**, (1984), 1557–1560.

210. M. Mathew and L. W. Schroeder, Acta Crystallogr., **B35**, (1979), 11–13.

211. J. R. Lehr, E. H. Brown, A. W. Frazier, J. P. Smith, and R. P. Thrasher, Tenn. Val. Auth. Chem. Eng. Bull., No 6, (1967).

212. A. Durif, Bull. Soc. fr. Minéral. Cristallogr., **94**, (1971), 536–537.

213. E. Banks, R. Chianelli, and K. Korenstein, Inorg. Chem., 14, (1975), 1634–1639.

214. H. K. Henisch, Crystal Growth in Gels, (Pennsylvannia State Univ. Press, Univ. Park and London, 1970).

215. A. Kawahara, T. Kageyama, I. Watanabe, and J. Yamakawa, Acta Crystallogr., **C49**, (1993), 1275–1277.

216. H. Effenberger, Z. Kristallogr., **172**, (1985), 97–104.

217. L. Elammari, J. Durand, L. Cot, and B. Elouadi, Z. Kristallogr., **180**, (1987), 137–140.

218. D. Blum, Thesis, (Univ. of Grenoble, France, 1986).

219. M. T. Averbuch-Pouchot and A. Durif, *Acta Crystallogr.*, **C41**, (1985), 665–667.

220. D. D. Vasovic and D. R. Stojakovic, *J. Am. Ceram. Soc.*, **77**, (1994), 1372–1374.

221. C. E. Bamberger, *J. Am. Ceram. Soc.*, **65**, (1982), 107–108.

5.2. OLIGOPHOSPHATES

5.2.1. Introduction

With oligophosphates we enter into the domain of condensed phosphates. This field of phosphate chemistry was recently covered by a detailed compilation performed by Durif [1]. Thus, in the following sections devoted to condensed phosphate chemistry, we will report briefly the state of the art, except for parts of this chemistry in which major developments occurred during the last few years, for instance in the case of cyclophosphates with the recent discovery of the P_9O_{27} ring as well as the rapid improvements in the chemistry of large rings as $P_{10}O_{30}$ and $P_{12}O_{36}$.

Oligophosphates include anions corresponding to the general formula $[P_nO_{(3n+1)}]^{(n+2)-}$ (see Chapter 4). A good number of diphosphates (n = 2) and triphosphates (n = 3) are presently well characterized, but compounds with n = 4 and 5 are rather rare. By paper chromatography analysis, the existence of oligophosphates with n > 5 was clearly established by Griffith and Buxton [2]. However, such compounds have not been produced in a good crystalline state and no structural data are available yet.

5.2.2. Diphosphates

Diphosphates which are frequently called pyrophosphates are the most numerous among the oligophosphates. In 1816, Berzelius was the first to characterize such a salt. Ever since they have been the object of hundreds of investigations. Nevertheless, the areas of ignorance in this field remain important. For instance, $Na_4P_2O_7$, which was chemically well characterized since the beginning of the last century is known to have six crystalline modifications, but only a structural model was proposed for one of them. Another typical example is the crystal structure of $Li_4P_2O_7$ which was not determined before 1994.

Among condensed phosphate chemistry, acidic anions are generally very rare. On the contrary, the existence of acidic anions is a common fact for the first two terms of the oligophosphates, diphosphates and triphosphates. In diphosphate chemistry, the existence of $[HP_2O_7]^{3-}$, $[H_2P_2O_7]^{2-}$, and $[H_3P_2O_7]^-$ groups are well established for a long time. The geometrical behavior of such groups in crystalline materials is very particular. They are always interconnected by very strong hydrogen bonds which build infinite entities as chains, ribbons or planes, commonly but inappropriately called "macroanions". In Chapter 6, we report the various geometrical configurations of these "macroanions" in phosphate chemistry.

5.2.2.1. General properties of diphosphates

Very little is known about the anhydrous $M_4^I P_2O_7$ neutral monovalent-cation diphosphates. The potassium salt is very hygroscopic and the ammonium derivative loses ammonia in air and at room temperature, and transforms into $(NH_4)_3HP_2O_7$ [3]. The lithium, silver, and thallium derivatives appear as stable compounds in normal conditions.

Neutral $M_2^{II} P_2O_7$ divalent-cation diphosphates and their various hydrates were extensively investigated at a structural point of view, but very little is known of their chemical and physical properties. They are stable in normal conditions.

Thermal Behavior – $M_4^I P_2O_7$ diphosphates are generally stable up to their melting temperatures. Some of these melting temperatures are known: 1158 K for the lithium salt [4–5], 1261 K for the sodium derivative [6], and 843 [7] or 916 K [8] for the silver diphosphate. Some are polymorphic: $Li_4P_2O_7$ undergoes a transition at 903 K [3–4], $Ag_4P_2O_7$ at 623–647 K [7–8], $K_4P_2O_7$ at 773 K, $Tl_4P_2O_7$ at 618 K [9], and $Na_4P_2O_7$ has six crystalline forms [6], with transitions at 676, 785, 791, 816, and 827 K.

By firing, all $M_2^I H_2P_2O_7$ diphosphates decompose into long-chain polyphosphates:

$$M_2^I H_2P_2O_7 \longrightarrow 2M^I PO_3 + H_2O$$

$M_2^{II} P_2O_7$ neutral divalent-cation diphosphates are also generally stable till their melting point. We report some of these temperatures in Table 5.2.1.

Things are different for $M^{II} H_2P_2O_7$ divalent-cation dihydrogendiphosphates whose thermal behavior is extremely dependent on the nature of the associated cation. For M^{II} = Mg, Mn, Fe, Co, Ni, Cu, and Zn, the thermal evolution leads to the formation of cyclotetraphosphates:

$$2M^{II} H_2P_2O_7 \longrightarrow M_2^{II} P_4O_{12} + 2H_2O$$

whereas, with M^{II} = Ca, Sr, Ba, and Pb, this evolution leads to long-chain poly-phosphates:

$$M^{II}H_2P_2O_7 \longrightarrow M^{II}(PO_3)_2 + H_2O$$

The thermal decomposition of some complex dihydrogendiphosphates produce very unexpected compounds. Thus, Chudinova *et al.* [10] obtained a cyclooctaphophate, $Ga_2K_2P_8O_{24}$, by firing $GaK(H_2P_2O_7)_2$:

$$2GaK(H_2P_2O_7)_2 \xrightarrow{\text{603 K}} Ga_2K_2P_8O_{24} + 4H_2O$$

Avaliani *et al.* [11] characterized a cyclotetraphosphate, $InNaP_4O_{12}$, during the thermal evolution of $InNa(H_2P_2O_7)_2$:

$$InNa(H_2P_2O_7)_2 \xrightarrow{\text{588–678 K}} InNaP_4O_{12} \xrightarrow{\text{973–1023 K}} In(PO_3)_3 + \text{melt}$$

Table 5.2.1
Melting temperatures of some $M_2^{II}P_2O_7$ diphosphates.

Formula	mp (K)	Ref.	Formula	mp (K)	Ref.
$Mg_2P_2O_7$	1655	[12]	$Mn_2P_2O_7$	1474	[15]
$Co_2P_2O_7$	1513	[13]	$Ca_2P_2O_7$	1626	[16]
$Ni_2P_2O_7$	1668	[13]	$Sr_2P_2O_7$	1648	[17]
$Zn_2P_2O_7$	1293	[14]	$Ba_2P_2O_7$	1703	[18]
$Cd_2P_2O_7$	1393	[14]	$Pb_2P_2O_7$	1098	[19]

We will illustrate this wide category of phosphates with some examples taken from monovalent, divalent and trivalent-monovalent cation derivatives.

5.2.2.2. Monovalent-cation diphosphates

Very little is known of the *anhydrous neutral salts* $M_4^IP_2O_7$. The classical anhydrous sodium salt, $Na_4P_2O_7$, characterized since the beginning of the last century by several chemists, Berzelius, Clark, and Graham and whose existence was the basic argument for the elaboration of the first classification of condensed phosphates by Graham, is still poorly investigated. This salt is said to have six crystalline modifications [6], but nothing is known of the atomic arrangements of these various forms outside of a structural model proposed by Leung and Calvo [20]. On the contrary, the decahydrate, $Na_4P_2O_7.10H_2O$, is a well investigated compound and is commercially available.

The crystal structure of the lithium salt, $Li_4P_2O_7$, was not determined by Yabukovich and Mel'nikov [21] before 1994.

In spite of a careful study performed by Yamada and Koizumi [7], the silver salt, $Ag_4P_2O_7$, is poorly known, especially since its unit-cell parameters are still a matter of controversy.

The *dihydrogendiphosphates*, $M_2^IH_2P_2O_7$, and their hydrates are better known because they are easier to prepare in a good crystalline state. Their main crystallographic features are reported in Table 5.2.2.

Table 5.2.2
Main crystallographic features of monovalent cation dihydrogendiphosphates.

Formula	a α	b β	c (Å) γ (°)	S. G.	Z	Ref.
$Na_2H_2P_2O_7$	27.49	12.35	6.856	*Fddd*	16	[22]
$Ag_2H_2P_2O_7$	27.78(2)	12.385(6)	7.026(4)	*Fddd*	4	[23]
$Na_2H_2P_2O_7.6H_2O$	14.099(6)	6.959(4) 117.69(4)	13.455(8)	*C2/c*	4	[24]
$K_2H_2P_2O_7$	7.001(1) 101.26(3)	8.918(2) 106.20(2)	6.576(1) 107.73(3)	$P\bar{1}$	2	[25]
$K_2H_2P_2O_7.1/2H_2O$	17.96(1)	6.958(5) 120.9(1)	14.24(1)	*C2/c*	8	[26–27]
$(NH_4)_2H_2P_2O_7$	9.058(7)	11.199(8) 108.40(1)	7.764(6)	$P2_1/a$	4	[28–29]
$Rb_2H_2P_2O_7.1/2H_2O$	19.55(4)	10.534(3)	7.784(3)	$Pna2_1$	8	[30–31]
$Cs_2H_2P_2O_7$	7.967(3)	9.055(3) 90.00(1)	11.404(3)	*Cc* or *C2/c*	4	[30, 32]
$Tl_2H_2P_2O_7$	17.896(8)	7.094(2) 120.09(5)	14.698(6)	*C2/c*	4	[33]

We will illustrate this section with two recent structural descriptions: those of $Li_4P_2O_7$ and $Ag_2H_2P_2O_7$. For the latter material which is an important reagent for the synthesis of dihydrogendiphosphates, we add some comments dealing with its preparation and its use.

The $Li_4P_2O_7$ *atomic arrangement* – This salt is monoclinic, $P2_1/n$, with $Z = 4$ and the following unit-cell parameters:

$$a = 5.190(2), \quad b = 13.902(3), \quad c = 7.901(3) \text{ Å}, \quad \beta = 89.97(3)°$$

Its atomic arrangement, recently determined by Yakubovich and Mel'nikov [21], is a packing of LiO_4 tetrahedra and P_2O_7 groups. As shown by Figure 5.2.1, a projection along the **a** direction, the phosphoric entities, all located around the planes $z = 1/4$ and 3/4, are interconnected by layers of lithium atoms spreading around the planes $z = 0$ and $1/2$. Inside a lithium layer, all the lithium atoms have fourfold tetrahedral coordination. By edge and corner sharing, the LiO_4 tetrahedra are assembled to build a two-dimensional framework perpen-

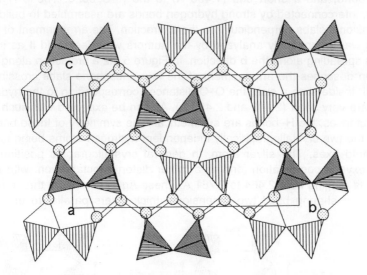

Figure 5.2.1. Projection, along the **a** direction, of the atomic arrangement in $Li_4P_2O_7$. The P_2O_7 groups are shown by polyhedral representation and the grey circles represent the lithium atoms. Solid lines correspond to the Li–O bonds.

dicular to the **c** direction. The phosphoric group has no internal symmetry. Within the four crystallographically independent LiO_4 tetrahedra, the Li–O distances range between 1.842 and 2.095 Å.

Structure and properties of $Ag_2H_2P_2O_7$ – This salt was often observed during the decyclization of various silver cyclophosphates. For instance, in the case of silver cyclotriphosphate monohydrate, the ring-opening scheme is given by:

$$2\ Ag_3P_3O_9.H_2O + H_2O \longrightarrow 3Ag_2H_2P_2O_7$$

It was probably due to the lack of a crystallization process that no crystal data were reported for this salt before 1992, when Averbuch-Pouchot and Durif [23] described a procedure for the production of single crystals and performed the structure determination. This salt is orthorhombic, *Fddd*, with $Z = 16$ and the following unit-cell dimensions:

$$a = 27.78(2), \quad b = 12.385(6), \quad c = 7.026(4)\ \text{Å}$$

In this compound, the phosphoric group has a twofold internal symmetry, the central oxygen atom being located on a twofold axis. Inside the phosphoric tetrahedron, three types of P–O distances are observed, a long one (1.598 Å) corresponding to the P–O–P bond, two intermediate (1.540 and 1.525 Å) to the P–OH bond, and a short one (1.490 Å) to the P–O bond. These $H_2P_2O_7$ entities, interconnected by strong hydrogen bonds are assembled to build thick bidimensional slabs perpendicular to the **a** direction. The arrangement of these entities was erroneously analyzed by the authors who described it as infinite ribbons spreading along the **b** direction. In Figure 5.2.2 a projection along the **c** direction describes the arrangement of the four phosphoric slabs crossing the unit cell. Inside such a slab, the O–O distances corresponding to the hydrogen bonds are very short (2.442 and 2.453 Å). As can be expected with such short O–O distances, the H–bonds are symmetrical. The symmetry of these bonds is binary, the two crystallographically independent hydrogen atoms being located on twofold axes. The silver atom, in general crystallographic position, has sixfold oxygen coordination forming a rather distorted octahedron, with Ag–O distances ranging from 2.464 to 2.781 Å. These AgO_6 octahedra through edge and corner sharing build two-dimensional thick layers parallel to the (b, c)

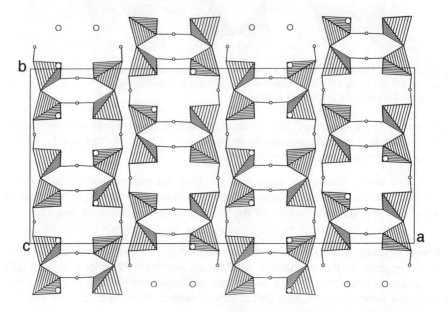

Figure 5.2.2. Projection, along the **c** direction, of the atomic arrangement in $Ag_2H_2P_2O_7$. The phosphoric groups are shown by polyhedral representation. The smaller circles are the hydrogen atoms, the larger ones the silver atoms.

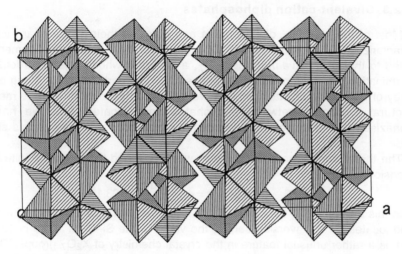

Figure 5.2.3. Projection, along the **c** direction, of the layers of silver polyhedra in $Ag_2H_2P_2O_7$.

plane. The general appearence of this layer arrangement is shown in Figure 5.2.3, a projection along the **c** axis, whereas the arrangement of the polyhedra inside one layer is provided by Figure 5.2.4, a projection along the **a** axis.

Since its characterization, this salt is extensively used for the preparation of dihydrogendiphosphates through an exchange reaction:

$$Ag_2H_2P_2O_7 + 2M^ICl \longrightarrow M^I_2H_2P_2O_7 + 2AgCl$$

The corresponding sodium salt, $Na_2H_2P_2O_7$, has been chemically well characterized for a long time and is isotypic.

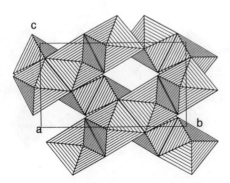

Figure 5.2.4. Projection, along the **a** direction, of a layer of silver polyhedra in $Ag_2H_2P_2O_7$.

5.2.2.3. Divalent-cation diphosphates

Most of the $M_2^{II}P_2O_7$ diphosphates and many of their hydrates are well characterized to date. Among them, a good number have atomic arrangements related to that of *thortveitite*, $Sc_2Si_2O_7$, a disilicate whose crystal structure determination was performed by Zachariasen [34] as early as 1930. Along our survey of phosphate crystal chemistry we have very few opportunities to meet structural analogies between phosphates and silicates, thus we have emphasized this example and described the atomic arrangement of this silicate.

The thortveitite is monoclinic, $C2/m$, with $Z = 2$ and the following unit-cell dimensions:

$$a = 6.56, \quad b = 8.58, \quad c = 4.74 \text{ Å}, \quad \beta = 103.13°$$

In this arrangement, the central oxygen atom of the Si_2O_7 disilicate group, being located on an inversion center, the value of the Si–O–Si angle is 180° which is a rather unusual feature in the crystal chemistry of X_2O_7 groups. The disilicate group has a high internal symmetry since the silicon atom and one oxygen atom are located in a mirror plane. In Figure 5.2.5 a projection along the **c** direction shows how the disilicate groups form layers around the planes $y = 0$ and $1/2$, whereas scandium atoms which are all located in the planes $x = 0$ and $1/2$, form corrugated layers between the anionic planes. The scandium atoms, located on twofold axis, have an octahedral oxygen coordination. The

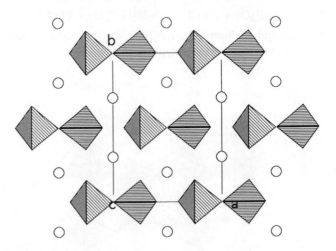

Figure 5.2.5. Projection, along the **c** axis, of the atomic arrangement in thortveitite. The Si_2O_7 groups are shown by polyhedral representation and the Sc atoms represented by open circles.

Table 5.2.3
Main crystallographic data for some $M_2^{II}P_2O_7$ diphosphates.

Formula	a α	b β	c (Å) γ (°)	S. G.	Z	Ref.
β-$Mg_2P_2O_7$	6.494(7)	8.28(1) 103.8(1)	4.522(5)	$C2/m$	2	[35–36]
β-$Zn_2P_2O_7$	6.61(1)	8.29(1) 105.4(2)	4.51(1)	$C2/m$	2	[37]
β-$Ni_2P_2O_7$ at 853 K	6.501	8.239 104.14	4.480	$C2/m$	2	[38]
β-$Cu_2P_2O_7$	6.827(8)	8.118(10) 108.85(10)	4.576(6)	$C2/m$	2	[39]
$Mn_2P_2O_7$	6.633	8.584 102.67	4.546	$C2/m$	2	[40]
$Fe_2P_2O_7$	6.649(2) 90.04(3)	8.484(2) 103.89(3)	4.488(1) 92.82(3)	$P\bar{1}$	2	[41]

unusual situation of the disilicate group around an inversion center provoked a re-investigation of this salt by Cruickshank *et al.* [42]. Using better cystal data, these authors performed refinements with the $C2$ and Cm space groups, but concluded finally that the Zachariasen's results were probably the right ones. In Table 5.2.3, we report the main crystal data for some $M_2^{II}P_2O_7$ diphosphates which are isotypic or closely related to thortveitite.

Most of the $M_2^{II}P_2O_7$ diphosphates are polymorphic. We report some examples of their transition temperatures. The Mg, Cu, Zn, Sr, Ba, and Pb derivatives are dimorphic, with transformation temperatures at 343, 373, 401, 1048, 1073 and 943 K, respectively. The nickel salt has four crystalline forms and calcium has three.

Good surveys of the crystal chemistry of anhydrous divalent-cation diphosphates were published by Nord and Kierkegaard [43] and Brown and Calvo [44].

5.2.2.4. Trivalent-monovalent cation diphosphates

Monovalent-trivalent cation diphosphates appear with various stoichiometries: $M^{III}M^{I}P_2O_7 \cdot xH_2O$, $M^{III}M^{I}(H_2P_2O_7)_2$, $M_2^{III}M^{I}H_3(P_2O_7)_2$ or $M_2^{III}M_6^{I}(P_2O_7)_3$. Among them, the anhydrous $M^{III}M^{I}P_2O_7$ compounds are the most numerous; more than sixty-five are presently known. We have therefore selected this class of compounds to illustrate this section. It is common to read that these compounds belong to two structure types, known as $FeNaP_2O_7$ (I) and $FeNaP_2O_7$ (II) types. In fact, five structure types were investigated in this family. The first structural type is common to $FeLiP_2O_7$ and $InLiP_2O_7$, and the great majority of the other ones are isotypic either with $FeNaP_2O_7$ (I) or with $FeNaP_2O_7$ (II). The fourth structure type includes compounds with $M^I = K$ and $M^{III} = Er, Y, Ho,$ and Dy, whereas α-$TiNaP_2O_7$ is the only example of the last structure type till now.

We illustrate the crystal chemistry of these $M^{III}M^IP_2O_7$ salts with the description of these five structural types. They all have common features and include a three-dimensional framework built by independent $M^{III}O_6$ octahedra, sharing their corners with the diphosphate groups to create channels in which are located the monovalent cations. The main differences rest with the connections established between the $M^{III}O_6$ octahedra and the neighbouring phosphoric groups. Each $M^{III}O_6$ octahedron shares its six corners with six different P_2O_7 groups in YKP_2O_7 and α-$TiNaP_2O_7$, whereas it also shares six in $InLiP_2O_7$, $AlKP_2O_7$ and $FeNaP_2O_7$ (II), but only with five phosphoric groups this time.

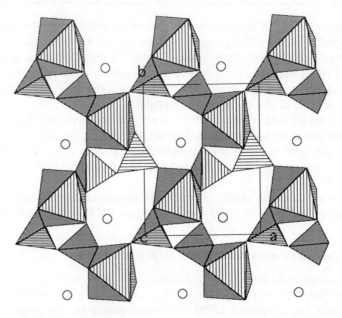

Figure 5.2.6. Projection of the structure of $InLiP_2O_7$ along the **c** axis. A polyhedral representation is used for the InO_6 octahedra and the P_2O_7 groups. Open circles represent the lithium atoms.

The InLiP$_2$O$_7$ atomic arrangement – This diphosphate is monoclinic, *P2$_1$*, with *Z* = 2 and the following unit-cell dimensions:

$$a = 7.084(2), \quad b = 8.436(2), \quad c = 4.908(3) \text{ Å}, \quad \beta = 110.75(2)°$$

Tranqui *et al.* [45] confirmed that this compound is isotypic with that of the iron-lithium salt previously reported by Genkina *et al.* [46]. The projection of this structure along the **c** direction (Figure 5.2.6) shows the large channels containing the lithium atoms. This projection can give the impression that the InO$_6$ octahedron has two common edges with two PO$_4$ tetrahedra belonging to two different P$_2$O$_7$ groups. Probably for this reason, the structure was erroneously described by Tranqui *et al.* [45]. In fact, owing to the short period along the **c** direction, the two tetrahedra are superimposed in projection and each of them shares one corner with the same InO$_6$ group. An examination of this arrangement shows that each InO$_6$ octahedron shares its six corners with five adjacent P$_2$O$_7$ groups as shown by Figure 5.2.7, a perspective view of the surrounding of an isolated InO$_6$ group.

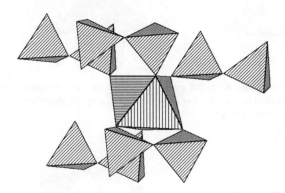

Figure 5.2.7. Perspective view of the surrounding of an isolated InO$_6$ octahedron.

The AlKP$_2$O$_7$ atomic arrangement – This salt, belonging to the structural type commonly called FeNaP$_2$O$_7$ (I), is monoclinic, *P2$_1$/c*, with *Z* = 4 and the following unit-cell dimensions:

$$a = 7.308(8), \quad b = 9.662(6), \quad c = 8.025(4) \text{ Å}, \quad \beta = 106.69(7)°$$

The structure determination was first performed by Ng and Calvo [47]. The representation of this arrangement, in projection along the **c** direction and given in Figure 5.2.8, shows clearly the large channels containing the potassium atoms. The three-dimensional network of AlO$_6$ and P$_2$O$_7$ is built by two centrosymmetrical layers, spreading perpendicular to the **c** direction around the

Figure 5.2.8. Projection, along the **c** direction, of the atomic arrangement in $AlKP_2O_7$. A polyhedral representation is used for the AlO_6 octahedra and the phosphoric groups. Open circles represent the potassium atoms.

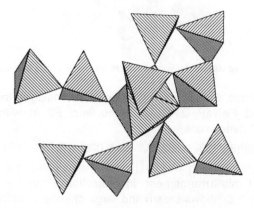

Figure 5.2.9. Perspective view of an AlO_6 octahedron and its five neighboring phosphoric groups.

planes $z = 1/4$ and $3/4$. Inside this framework, each AlO_6 octahedron shares its six corners with five adjacent phosphoric groups. A perspective view of an AlO_6 octahedron and of its five P_2O_7 neighbors is given in Figure 5.2.9, while Figure

Figure 5.2.10. Projection, along the **c** direction, of an isolated layer of AlO_6 octahedra and phosphoric groups. Potassium atoms are omitted.

5.2.10 represents, in projection along **c**, an isolated layer of the $AlO_6-P_2O_7$ framework. In the latter representation, as well as in Figure 5.2.8, the apparently unshared corners of the AlO_6 octahedron are, in fact, shared with PO_4 groups belonging to phosphoric groups of the adjacent layers.

The FeNaP$_2$O$_7$ (II) atomic arrangement – The crystal structure determination of this third structural type was performed by Gabelica-Robert *et al.* [48]. FeNaP$_2$O$_7$ (II) is also monoclinic, $P2_1/c$, with $Z = 4$ and the following unit-cell dimensions:

$$a = 7.324(1), \quad b = 7.905(1), \quad c = 9.575(1) \text{ Å}, \quad \beta = 111.86(1)°$$

In Figure 5.2.11 a projection along the **a** direction shows that layers of phosphoric groups are sited around the planes $z = 0$ and $1/2$, alternating with parallel layers of FeO_6 octahedra centered by the planes $z = 1/4$ and $3/4$. This structure was originally described by Gabelica-Robert *et al.* [48] as a *cage*

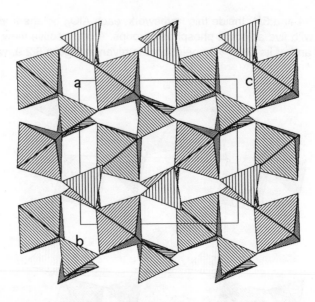

Figure 5.2.11. Projection, along the **a** direction, of the atomic arrangement in FeNaP$_2$O$_7$ (II), showing the alternating layers of phosphoric groups and FeO$_6$ octahedra. Na atoms are omitted.

structure whose main building unit is a FeP$_2$O$_{11}$ entity made by a P$_2$O$_7$ group sharing two of its oxygen atoms with a FeO$_6$ octahedron. This way of describing the arrangement was further used for the descriptions of other similar compounds frequently. The sodium atoms are located in channels parallel to

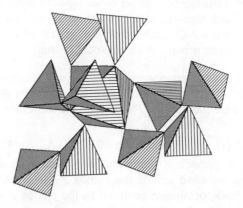

Figure 5.2.12. Perspective view of the surrounding of an iron atom in FeNaP$_2$O$_7$ (II).

the [110] direction. In the FeO_6–P_2O_7 framework, each FeO_6 octahedron shares its six corners with five different phosphoric entities. A perspective view of the neighbouring of an iron atom is represented in Figure 5.2.12.

Crystal structure determinations of numerous diphosphates isotypic with $FeNaP_2O_7(II)$ or $AlKP_2O_7$ are repeatedly published even today.

The YKP_2O_7 atomic arrangement – This atomic arrangement was recently determined by Hamady *et al.* [49]. YKP_2O_7 is orthorhombic, *Cmcm*, with $Z = 4$ and the following unit-cell dimensions:

$$a = 5.716(1), \quad b = 9.216(1), \quad c = 12.244 \text{ Å}$$

A projection of this structure along the **a** direction is drawn in Figure 5.2.13. This arrangement can also be described as a layer arrangement. Layers of YO_6 octahedra, centered by the planes $z = 0$ and $1/2$, and layers of phosphoric anions, centered by the planes $z = 1/4$ and $3/4$, alternate perpendicularly to the **c** direction. In this stacking, each YO_6 octahedron shares its six corners with six different P_2O_7 groups. This situation, not clearly visible in Figure 5.2.13, is illustrated more clearly by Figure 5.2.14, the perspective view of an isolated YO_6 octahedron and of its six P_2O_7 neighbors. The yttrium atom which is located on a centrosymmetrical position *(2/m)* is surrounded by a very regular octahedron of oxygen atoms, with Y–O distances ranging from 2.227 to 2.262 Å.

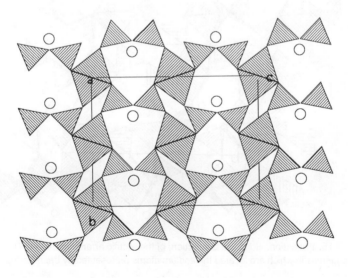

Figure 5.2.13. Projection of the structure of YKP_2O_7 along the **a** axis. Both YO_6 octahedra and P_2O_7 groups are shown by polyhedral representation, the open cirles represent the K atoms.

   The P_2O_7 group has an unusual internal symmetry. The central oxygen atom is located on a position of symmetry mm. The phosphorus atoms lie in one of these mirrors as well as one external oxygen atom of each PO_4 tetrahedron.

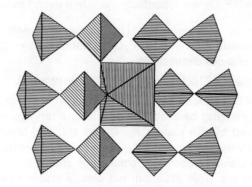

Figure 5.2.14. Perspective view of an YO_6 octahedron and of its six P_2O_7 neighbors.

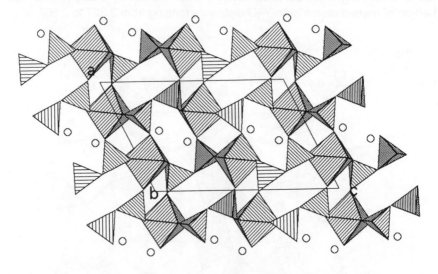

Figure 5.2.15. Projection, along the **b** direction, of the atomic arrangement in α-$TiNaP_2O_7$ showing the channels in which are located the sodium atoms represented by open circles.

The α-$TiNaP_2O_7$ atomic arrangement – This diphosphate, first characterized by Leclaire *et al.* [50], is monoclinic, $P2_1/c$, with $Z = 4$ and the following unit-cell dimensions:

$$a = 8.697(1), \quad b = 5.239(7), \quad c = 13.293(3) \text{ Å}, \quad \beta = 116.54(1)°$$

The six corners of the TiO_6 octahedron are shared, as in YKP_2O_7, with six different phosphoric groups. As in all the other types of $M^{III}M^{I}P_2O_7$ diphosphates, the three-dimensional framework built by the TiO_6 octahedra and the P_2O_7 groups generates channels in which are located the monovalent cations. As illustrated by Figure 5.2.15, these channels are parallel to the **b** direction and half of them are empty. The pseudo-hexagonal surrounding of a TiO_6 group is represented in Figure 5.2.16. Leclaire *et al.* [50] noted some similarities between this arrangement and that of β-cristobalite.

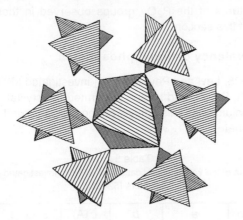

Figure 5.2.16. Projection, along the **c** direction, of an isolated TiO_6 octahedron sharing its six corners with six different P_2O_7 groups

General comments on the $M^{III}M^{I}P_2O_7$ derivatives – In these five structural types, the monovalent cations form files all located in relatively wide channels created by the $M^{III}O_6$–P_2O_7 frameworks. The situation of these arrays in the channels differs in these arrangements. In $InLiP_2O_7$ and YKP_2O_7, the Li and K atoms form linear eccentric arrays in the channels, whereas, in the three other structural types, the monovalent cations form zigzag chains centering the tunnels. Moreover, one must note that in α-$TiNaP_2O_7$ one channel out of two is empty.

In order to avoid repetitive sentences on the descriptions of these five structural types, we summarize in Table 5.2.4. some geometrical features concerning the oxygen coordinations of the associated cations in these five structural types.

Table 5.2.4

Ranges of the metal-oxygen distances observed in the $M^{III}O_6$ octahedra and $M^{I}O_n$ polyhedra in the $M^{III}M^{I}P_2O_7$ diphosphates.

Formula	Range of M^{III}–O (Å)	Range of M^{I}–O (Å)	n	Ref.
$InLiP_2O_7$	2.085–2.179	1.957–2.118	4	[45]
$AlKP_2O_7$	1.841–1.925	2.739–3.185	10	[47]
$FeNaP_2O_7$	1.938–2.053	2.384–2.662	6	[48]
YKP_2O_7	2.227–2.262	2.961–3.200	10	[49]
$TiNaP_2O_7$	1.923–2.079	2.429–2.936	9	[50]

The main features of the P_2O_7 groups observed in these examples are given at the end of this section.

5.2.2.5. Higher-valency cation diphosphates

As early as 1935, Levi and Peyronel [51] investigated $M^{IV}P_2O_7$ diphosphates for M^{IV} = Si, Ti, Sn, Zr, and Hf and performed the crystal structure determination of the zirconium salt. This study was the first structural investigation of a condensed phosphate.

Table 5.2.5

Crystallographic data of the various modifications of SiP_2O_7, investigated by Liebau and co-workers [52–56].

Form	a α	b β	c (Å) $\gamma(°)$	S. G.	Z
(AI)	22.428(6)			$Pa3$	108
(AII)	22.4	22.4	14.9	tetr.	72
(AIII)	4.73	6.33 90.1	14.71	$P2_1/c$	4
(AIV)	4.73	12.02 91.3	7.62	$P2_1/n$	4
(BI)	8.18	8.18	11.85	$P6_3...$	6

Among this series of compounds, including now in addition to those first investigated by Levi and Peyronel [51], GeP_2O_7, CeP_2O_7, PtP_2O_7, MoP_2O_7, ReP_2O_7, ThP_2O_7, and UP_2O_7, the silicium derivative was the most investigated. Through a series of very careful studies of the SiO_2–P_2O_5 system by Liebau and co-workers [52–56], seven forms of SiP_2O_7 were characterized. Five of them were investigated by the same authors who discussed the structural relations between them. A determination of the atomic arrangement of the cubic form was also performed by Tillmanns *et al.* [57]. Table 5.2.5 reports the main crystal data for various forms of SiP_2O_7.

The $GeO_2-P_2O_5-H_2O$ system was also carefully investigated and the phase-equilibrium diagram constructed by Mal'shikov *et al.* [58]. GeP_2O_7 is observed as a congruent-melting compound (mp = 1513 K). Two forms of this salt are reported by the authors: a monoclinic one which is stable between 823 K and 1323 K and a cubic one which is stable from 1323 K to its melting point. In addition, they observed that an hydrated form corresponding to the formula $GeP_2O_7.1.6H_2O$ can be obtained below 823 K. This third salt is hexagonal.

Averbuch-Pouchot and Durif [59–60] prepared and determined the crystal structures of several pentavalent-bivalent cation diphosphates, $Nb_2Mg(P_2O_7)_2$, $Nb_2Co(P_2O_7)_2$, and $Ta_2Cd(P_2O_7)_2$. The first two are isotypic. These authors discussed the close structural relationships between these atomic arrangements and that of the cubic $M^{IV}P_2O_7$ series.

$Sb^{III}Sb^{V}(P_2O_7)_2$ described by Verbaere *et al.* [61] also has an atomic arrangement closely related to that of the cubic $M^{IV}P_2O_7$ series.

5.2.2.6. Organic-cation diphosphates

A relatively small number of organic-cation diphosphates have been well characterized to date. The pioneers in this field were Adams and Ramdas [62–64] who reported three guanidinium derivatives, $[C(NH_2)_3]_3HP_2O_7$, $[C(NH_2)_3]_4P_2O_7.H_2O_2.3/2H_2O$ and $[C(NH_2)_3]_4P_2O_7.H_2O$ as early as 1977.

More recently, three ethylenediammonium diphosphates were described: $[NH_3-(CH_2)_2-NH_3]_3(HP_2O_7)_2.2H_2O$ and $[NH_3-(CH_2)_2-NH_3]_2P_2O_7$ by Kamoun *et al.* [65–66] and $[NH_3-(CH_2)_2-NH_3]_2H_2P_2O_7$ by Averbuch-Pouchot and Durif [67]. Ethanolammonium dihydrogendiphosphate, $[NH_3-(CH_2)_2-OH]_2H_2P_2O_7$, was also characterized by the latter authors [68].

5.2.2.7. Chemical preparations of diphosphates

As for the chemical preparations of monophosphates, many starting materials are commercially available for the synthesis of diphosphates. Sodium, potassium, and ammonium diphosphates, sodium dihydrogendiphosphate and diphosphoric acid are common chemicals. It is necessary nevertheless to report that some types of preparations are more or less specific for this category of phosphates.

The use of diphosphoric acid – Many alkali diphosphates were obtained by neutralizing diphosphoric acid with the appropriate amounts of the corresponding alkali carbonates. In spite of its apparent simplicity, the reaction must be optimized carefully. For instance to prepare $CsH_3P_2O_7.H_2O$, Larbot *et al.* [69] observed that a ratio 1/1 of $H_4P_2O_7$ and Cs_2CO_3 must be used in the initial solution. When this process is used, the reaction tank must be kept at a low temperature to avoid the decondensation of the acid.

Thermal methods – Many diphosphates can be prepared by the classical thermal methods. For instance, most of divalent-cation diphosphates can be elaborated as polycrystalline specimens according to the following type of reaction:

$$2M^{II}CO_3 + 2(NH_4)_2HPO_4 \longrightarrow M^{II}_2P_2O_7 + 4NH_3 + 2CO_2 + 3H_2O$$

The temperature and the time of heating are mainly dependent on the nature of the divalent cation involved in the reaction.

Flux processes – The various steps for the condensation of monophosphoric acid were carefully investigated in function of temperature and it is well-known that the first step, corresponding to the formation of diphosphoric acid $H_4P_2O_7$, occurs at a relatively low temperature (T < 473 K). This property was often used to prepare diphosphates by flux methods using monophosphoric acid as a starting reagent. For instance, Durif and Averbuch-Pouchot [70] prepared single crystals of $MnHP_2O_7$ by heating a mixture of H_3PO_4 and $MnCO_3$ containing a large excess of phosphoric acid for 18 hours at 473 K.

When using phosphoric acid as a flux, one must take note that the state of polymerization of this acid, at a given temperature, is very dependent on the nature of the added cations.

Sodium polyphosphate, $NaPO_3$, has sometimes been used as a flux agent. Thus, Klement [71] obtained $ZnNa_2P_2O_7$, $Na_2MgP_2O_7$, $CaNa_2P_2O_7$, $Sr_2P_2O_7$, and $Ba_2P_2O_7$ by the action of the corresponding divalent oxides on a large excess of molten $NaPO_3$.

$$2NaPO_3 + MgO \longrightarrow MgNa_2P_2O_7$$

Exchange reactions – As we have said before, since its recent characterization and the description of a convenient procedure for its preparation [72], $Ag_2H_2P_2O_7$ was successfully used for the preparations of several dihydrogendiphosphates through an exchange process derived from the Boullé's method. An example of such a reaction is:

$$Ag_2H_2P_2O_7 + 2CsCl \longrightarrow Cs_2H_2P_2O_7 + 2AgCl$$

Several dihydrogendiphosphates, whose preparations were difficult to master by using more classical processes, were easily synthesized by this method. This process was mainly used for the production of organic derivatives.

In some investigations, $Ag_4P_2O_7$ was also used as a reagent in similar exchange reactions.

Dehydration-condensation of mono- or dihydrogenmonophosphates – Many diphosphates can be prepared as polycrystalline samples by the dehydration-condensation of mono- or dihydrogenmonophosphates. Thus, $Na_4P_2O_7$ and $Na_2H_2P_2O_7$ can be prepared by heating Na_2HPO_4 or NaH_2PO_4:

$$2Na_2HPO_4 \longrightarrow Na_4P_2O_7 + H_2O$$

$$2NaH_2PO_4 \longrightarrow Na_2H_2P_2O_7 + H_2O$$

Divalent-cation diphosphates can also be prepared by the same process, for instance, $Ca_2P_2O_7$:

$$2CaHPO_4.2H_2O \xrightarrow[-4H_2O]{393\ K} 2CaHPO_4 \xrightarrow[-H_2O]{723\ K} \gamma\text{-}Ca_2P_2O_7 \xrightarrow{1023\ K} \beta\text{-}Ca_2P_2O_7$$

Some divalent-cation dihydrogendiphosphates can also be synthesized by a very similar process, but using the dihydrogenmonophosphate as a starting material instead. Thus, $SrH_2P_2O_7$ can be obtained from $Sr(H_2PO_4)_2$:

$$Sr(H_2PO_4)_2 \longrightarrow SrH_2P_2O_7 + H_2O\ [73]$$

In this particular case, the reaction must be run at 463–483 K which is extremely low. One must also not forget that in reactions of this type, a prolonged heating or a slight increase of the reaction temperature leads to a more condensed phosphate:

$$SrH_2P_2O_7 \xrightarrow{593–603\ K} Sr(PO_3)_2 + H_2O$$

In a similar way, $MgH_2P_2O_7$ was prepared by firing $Mg(H_2PO_4)_2.2H_2O$:

$$Mg(H_2PO_4)_2.2H_2O \xrightarrow{368–403\ K} Mg(H_2PO_4)_2 + 2H_2O \xrightarrow{488–523\ K} MgH_2P_2O_7\ [74]$$

The corresponding manganese diphosphate, $MnH_2P_2O_7$, can be obtained from $Mn(H_2PO_4)_2.2H_2O$ through an almost identical scheme [74] as well as $CdH_2P_2O_7$:

$$Cd(H_2PO_4)_2.2H_2O \xrightarrow{373\ K} Cd(H_2PO_4)_2 \xrightarrow{453\ K} CdH_2P_2O_7 \xrightarrow{473\ K} Cd_2P_4O_{12}$$

The use of BPO_4 – The use of BPO_4 as an exchange agent that was reported in Section 5.1.10.7., devoted to chemical preparations of monophosphates, is also suitable for the elaboration of some diphosphates. Vasovic and Stojakovic [75] reported the following reactions:

$$2Mn_2O_3 + 4BPO_4 \longrightarrow 2Mn_2P_2O_7 + 2B_2O_3 + O_2$$

$$2MnO_2 + 2BPO_4 \longrightarrow Mn_2P_2O_7 + B_2O_3 + O_2$$

Yields are good when these reactions are run at 1073–1103 K for 4 hours.

A similar reaction was used by Bamberger [76] to elaborate uranium and neptunium diphosphates:

$$M^{IV}O_2 + 2BPO_4 \longrightarrow M^{IV}P_2O_7 + B_2O_3$$

Degradation of condensed phosphates – During the degradation of condensed phosphoric anions, diphosphates are always observed before the final step leading to monophosphates, but generally they apear as components of a complex phase containing various oligophosphates. With the exception of some accidental reactions, this process is not appropriate for the production of diphosphates.

The use of exchange resins – The existence of diphosphoric acid as a common commercial chemical limits this process to specific reactions in which one needs a pure and freshly prepared acid. Resins are sometimes used to transform a neutral diphosphate into an acidic one.

Preparations of the organic derivatives – No specific process was necessary to develop the preparation of organic-cation diphosphates. Most of them were prepared using two main processes:
  – the direct action of diphosphoric acid on the corresponding amine, according to the following scheme:

$$2(R\text{–}NH_2) + H_4P_2O_7 \longrightarrow (R\text{–}NH_3)_2H_2P_2O_7$$

  – an exchange reaction involving the use of $Ag_2H_2P_2O_7$. In this case the amine chloride is used as a starting material and the reaction is:

$$2(R\text{–}NH_3Cl) + Ag_2H_2P_2O_7 \longrightarrow (R\text{–}NH_3)_2H_2P_2O_7 + 2AgCl$$

In both cases, the reaction must be run at room temperature or below in order to avoid the hydrolysis of the P_2O_7 group.

5.2.2.8. The diphosphate anion

The diphosphate group which is built by two corner-sharing PO_4 tetrahedra is the simplest condensed anion. Using tens of accurate X-ray structural determinations, we can examine what are the main geometrical features of the P_2O_7 group.

In most structural studies of diphosphates, many characteristics of the geometrical configuration of this anion are discussed and, among them, mainly the P–O–P angle and the relative orientation of the two constituting tetrahedra. For the latter aspect, a terminology was developed in order to distinguish between "eclipsed" (cis) or "staggered" (trans) conformation. In fact, outside very rare examples, involving very restrictive symmetry elements, a P_2O_7 group is never purely staggered or purely eclipsed. Thus, in our opinion, such a group can be clearly described by the P–P distance, the P–O–P angle, and its internal symmetry when one or several of its components are located on symmetry elements.

P_2O_7 groups may adopt various internal symmetries: 2, m, $\bar{1}$ are the most common, but, recently, a diphospate group with mm internal symmetry was reported [49]. In the present survey, we will classify the examples according to these internal symmetries. The $\bar{1}$ *internal symmetry*, corresponding to a central oxygen atom located on an inversion center, leads to a P–O–P angle of 180°. In the case of *mirror symmetry*, two situations are observed. This symmetry element is frequently perpendicular to the P–P direction and consequently includes the central oxygen atom. In some rarer cases, five components of the P_2O_7 group are located in the mirror plane, the two phosphorus atoms, the central oxygen atom, and two external oxygen atoms. This latter configuration was observed in only two compounds, $Ta_2Cd(P_2O_7)_3$ and in a mixed-anion phosphate, $CaNb_2O(P_4O_{13})(P_2O_7)$ to date. In the case of *twofold internal symmetry*, the central oxygen atom is located on a binary axis. This internal symmetry is not very common and was recently observed in $Cs_2H_2P_2O_7$, $Ag_2H_2P_2O_7$, and ethylenediammonium dihydrogendiphosphate. In the case of *mm symmetry*, the first mirror plane which is perpendicular to the P–P direction, contains the central oxygen atom, whereas the second mirror contains the two phosphorus atoms and one external oxygen atom of each tetrahedron. Most of the P_2O_7 groups have no internal symmetry, but 2 and m symmetries are commonly observed nevertheless.

The $\bar{1}$ internal symmetry, corresponding to a central oxygen atom located on an inversion center and leading so to a P–O–P angle of 180°, was the subject of many controversies. One must admit that in all the previous studies leading to the conclusion of the existence of such a symmetry in a P_2O_7 group, the central oxygen atom has a high thermal factor indicating either a disordered location for this oxygen atom or an error of symmetry or of unit-cell assignment.

Inside a PO_4 tetrahedron belonging to a diphosphate group, the values observed for O–O and P–O distances, and O–P–O angles never differ significantly from those that we will observe repeatedly along our survey of condensed phosphoric anions. Two main types of P–O distances must be distinguished: the P–O(L) distances corresponding to the P–O–P bond and the P–O(E) distances corresponding to the unshared or external oxygen atoms. When the diphosphate group is an acidic one, a third type of P–O distance, corresponding to the P–OH bonds is to be considered. This P–OH distance has a value that is generally intermediate between P–O(L) and P–O(E):

$$P-O(L) > P-OH > P-O(E)$$

In spite of the apparently high distortion inside the two constituting tetrahedra, the averaged P–O distance inside a tetrahedron never differs significantly from

1.540 Å, an average value calculated for several tens of accurate crystal structures. The O–P–O average angle is always very close to 109.1°, a value calculated under the same conditions

Table 5.2.6

Some geometrical features of P_2O_7 groups with no internal symmetry.

Formula	P–P (Å)	P–O–P (°)	Ref.
$Na_4P_2O_7$	2.936	127.5	[77]
$(NH_4)_2H_2P_2O_7$	2.980	133.7	[78]
$Co_2P_2O_7$	2.998	142.6	[79]
β-$Ca_2P_2O_7$	2.955	130.5	[80]
	2.991	137.8	
$FeNaP_2O_7$ (II)	2.937	133.0	[81]
$FeKP_2O_7$	2.843	124.3	[81]
$Nb_2Co(P_2O_7)_3$	2.954	138.3	[59]
$Nb_2Mg(P_2O_7)_3$	2.961	140.0	[59]
$Ta_2Cd(P_2O_7)_3$	2.962	139.8	[60]
$CsMoO_2HP_2O_7$	2.862	126.6	[82]
$(NH_4)_2MoO_2P_2O_7$	2.911	131.3	[83]
$(eda)_2H_2P_2O_7$	2.893	128.1	[67]
$(etha)_2H_2P_2O_7$	2.960	132.9	[68]
$Cu_2(eda)_3(HP_2O_7)_2.3H_2O$	2.902	128.6	[84]
$Te(OH)_6.Cs_2H_2P_2O_7$	2.907	129.5	[85]
$Te(OH)_6.K_3HP_2O_7.H_2O$	2.971	132.6	[86]
$Te(OH)_6.2(NH_4)_2H_2P_2O_7$	2.903	128.5	[87]

eda = ethylenediammonium or $[NH_3–(CH_2)_2–NH_3]^{2+}$
etha = ethanolammonium or $[NH_3–(CH_2)_2–OH]^+$

Moreover, for acidic diphosphate groups, the P–O–H angles are always observed within a range of 110–120°.

We report in Tables 5.2.6 and 5.2.7 some numerical values observed for the P–P distances and the P–O–P angles selected in recent and accurate structure determinations. We classified these data in function of the internal symmetries of the diphosphate groups.

The P–P distances, ranging from 2.843 to 3.116 Å in the reported examples, are within a range that we shall find, with a very small number of exceptions, for all the other types of condensed phosphoric anions that we will examine along this book. It must be noticed that P–P distances longer than 3 Å are rare, but frequently observed in the case of centrosymmetric groups what can be simply explained since an evident correlation exists between the P–P value and the P–O–P angle. Aside from this case, it seems that no correlation can be established between the internal symmetry of a diphosphate group and the P–P distance.

Table 5.2.7

Some geometrical features of P_2O_7 groups with various internal symmetries.

Formula	P–P (Å)	P–O–P (°)	Ref.
P_2O_7 groups with twofold symmetry			
(eda)$_2$P$_2$O$_7$	2.968	141.9	[66]
Ag$_2$H$_2$P$_2$O$_7$	2.910	131.2	[72]
Cs$_2$H$_2$P$_2$O$_7$	2.963	132.7	[88]
P_2O_7 groups with mirror symmetry			
Rb$_2$H$_2$P$_2$O$_7$.1/2H$_2$O	2.964	133.5	[89]
	3.009	137.7	
Fe$_3$(P$_2$O$_7$)$_2$	2.870	128.1	[90]
	3.012	142.2	
Ta$_2$Cd(P$_2$O$_7$)$_3$	2.936	135.8	[60]
CaNb$_2$O(P$_4$O$_{13}$)(P$_2$O$_7$)	3.005	145.6	[91]
Centrosymmetrical P_2O_7 groups			
Ni$_2$P$_2$O$_7$	3.116	180.0	[92]
Nb$_2$Co(P$_2$O$_7$)$_3$	3.064	180.0	[59]
Nb$_2$Mg(P$_2$O$_7$)$_3$	3.054	180.0	[59]
P_2O_7 groups with mm symmetry			
YKP$_2$O$_7$	2.946	129.6	[49]

– eda = ethylenediammonium or [NH$_3$–(CH$_2$)$_2$–NH$_3$]$^{2+}$
– when the atomic arrangement includes several crystallographically independent diphosphate groups with different internal symmetries the same chemical formula can appear in several parts of the table.

For *the P–O–P angles*, if one excepts the centrosymmetrical diphosphate groups in which this angle is 180°, the values commonly observed vary between 127 and 146° which is a range of values that we shall find repeatedly along the examination of more condensed phosphoric anions.

As can be observed in most classes of condensed phosphates, no valuable correlation could be established between the geometrical configuration of the diphosphate anions and the nature of the associated cations up to now.

References

1.   A. Durif, *Crystal Chemistry of Condensed Phosphates*, (Plenum Press, New York, 1995).

2.   E. J. Grifith and R. L. Buxton, *J. Am. Chem. Soc.*, **89**, (1967), 2884–2890.

3.   A. W. Frazier, J. P. Smith, and J. R. Lehr, *J. Agr. Food Chem.*, **13**, (1965), 316–322.

4.   J. Nakano, T. Yamada, and S. Miyazawa, *J. Am. Ceram. Soc.*, **62**, (1979), 465–467.

5.   T. Y. Tien and F. A. Hummel, *J. Am. Ceram. Soc.*, **44**, (1961), 206–208.

6.   V. B. Lazarev, I. D. Sokolova, G. A. Sharpataya, I. S. Shaplygin, and I. B. Markina, *Thermochim. Acta,* **86**, (1985), 243–249.

7.   T. Yamada and H. Koizumi, *J. Crystal Growth*, **64**, (1983), 558–562.

8.   R. K. Osterheld and T. J. Mozer, *J. Inorg. Nucl. Chem.*, **35**, (1973), 3463–3465.

9.   K. Dostal, V. Kocman, and V. Ehrenbergrova, *Z. anorg. allg. Chem.*, **367**, (1969), 80–91.

10. N. N. Chudinova, M. A. Avaliani, L. S. Guzeeva, and I. V. Tananaev, *Izv. Akad. Nauk SSSR, Neorg. Mater.*, **14**, (1978), 2054–2057.

11. M. A. Avaliani, N. N. Chudinova, and I. V. Tananaev, *Izv. Akad. Nauk SSSR, Neorg. Mater.*, **20**, (1984), 282–286.

12. J. Berak, *Roczniki Chem.*, **32**, (1958), 17–22.

13. J. F. Sarver, *Trans. Brit. Ceram. Soc.*, **65**, (1966), 191–198.

14. J. J. Brown and F. A. Hummel, *J. Electrochem. Soc.*, **111**, (1964), 1052–1057.

15. Z. A. Konstant and A. I. Dimante, *Izv. Akad. Nauk SSSR, Neorg. Mater.*, **13**, (1977), 99–103.

16. E. P. Egan and Z. T. Wakefield, *J. Am. Chem. Soc.*, **79**, (1957), 558–561.

17. E. R. Kreidler and F. A. Hummel, *Inorg. Chem.*, **6**, (1967), 884–891.

18. R. A. Mc Cauley and F. A. Hummel, *Trans. Brit. Ceramic Soc.*, **67**, (1968), 619–628.

19. L. Merker and H. Wondratschek, *Z. anorg. allg. Chem.*, **306**, (1960), 25–29.

20. K. Y. Leung and C. Calvo, *Canad. J. Chem.*, **50**, (1972), 2519–2526.

21. O. V. Yakubovich and O. K. Mel'nikov, *Kristallografiya*, **39**, (1994), 815–820.

22. E. Ingerson and G. W. Morey, *Am. Min.*, **28**, (1943), 488–498.

23. M. T. Averbuch-Pouchot and A. Durif, *Eur. J. Solid State Inorg. Chem.*, **29**, (1992), 993–999.

24. R. L. Collin and M. Willis, *Acta Crystallogr.*, **B27**, (1971), 291–302.

25. A. Larbot, J. Durand, A. Norbert, and L. Cot, *Acta Crystallogr.*, **C39**, (1983), 6–8.

26. D. S. Emmerson and D. E. C. Corbridge, *Phosphorus*, **2**, (1972), 159–160.

27. Y. Dumas, J. L. Galigné, and J. Falgueirettes, *Acta Crystallogr.*, **B29**, (1973), 2913–2918.

28. A. W. Frazier, J. P. Smith, and J. R. Lehr, *J. Agr. Food Chem.*, **13**, (1965), 316–322.

29. M. T. Averbuch-Pouchot and A. Durif, *Eur. J. Solid State Inorg. Chem.*, **29**, (1992), 191–198.

30. A. Larbot, A. Norbert, and L. Cot, *C. R. Acad. Sci.*, Sér. C, **289**, (1979), 185–187.

31. M. T. Averbuch-Pouchot and A. Durif, *C. R. Acad. Sci.*, Sér. 2, **316**, (1993), 469–476.

32. M. T. Averbuch-Pouchot and A. Durif, *C. R. Acad. Sci.*, Sér. 2, **316**, (1993), 41–46.

33. M. T. Averbuch-Pouchot and A. Durif, to be published

34. W. H. Zachariasen, *Z. Kristallogr.*, **73**, (1930), 1–6.

35. K. Lukaszewicz, *Roczniki Chem.*, **35**, (1961), 31–35.

36. C. Calvo, *Canad. J. Chem.*, **43**, (1965), 1139–1146.

37. C. Calvo, *Canad. J. Chem.*, **43**, (1965), 1147–1153.

38. A. Pietraszko and K. Lukaszewicz, *Bull. Acad. Polon. Sci. Ser. Sci. Chim.*, **16**, (1968), 183–187.

39. B. E. Robertson and C. Calvo, *Can. J. Chem.*, **46**, (1968), 605–612.

40. T. Stefanidis and A. G. Nord, *Acta Crystallogr.*, **C40**, (1984), 1995–1999.

41. J. T. Hoggins, J. S. Swinnea, and H. Steinfink, *J. Solid State Chem.*, **50**,

(1983), 278–283.

42.   D. W. J. Cruickshank, H. Lynton, and G. A. Barclay, *Acta Crystallogr.*, **15**, (1962), 491–498.

43   A. G. Nord and P. Kierkegaard, *Chemica Scripta*, **15**, (1980), 27–39.

44   I. D. Brown and C. Calvo, *J. Solid State Chem.*, **1**, (1970), 173–179.

45.   D. Tranqui, S. Hamdoune, and Y. Le Page, *Acta Crystallogr.*, **C43**, (1987), 201–202.

46.   E. A. Genkina, B. A. Maksimov, V. A. Timofeeva, A. B. Bykov, and O. K. Mel'Nikov, *Dokl. Akad. Nauk SSSR*, **284**, (1985), 864–867.

47.   H. N. Ng and C. Calvo, *Canad. J. Chem.*, **51**, (1973), 2613–2620.

48.   F. Gabelica-Robert, M. Goreaud, Ph. Labbé, and B. Raveau, *J. Solid State Chem.*, **45**, (1982), 389–395.

49.   A. Hamady, M. F. Zid, and T. Jouini, *J. Solid State Chem.*, **113**, (1994), 120–124.

50.   A. Leclaire, A. Benmoussa, M. M. Borel, A. Grandin, and B. Raveau, *J. Solid State Chem.*, **77**, (1988), 299–305.

51.   G. R. Levi and G. Peyronel, *Z. Kristallogr.*, **92**, (1935), 190–209.

52.   F. Liebau, G. Bissert, and N. Koppen, *Z. anorg. allg. Chem.*, **359**, (1968), 113–134.

53.   F. Liebau and G. Bissert, *Bull. Soc. Chim. Fr.*, (1968), 1742–1744.

54.   G. Bissert and F. Liebau, *Acta Crystallogr.*, **B26**, (1970), 233–240.

55.   F. Liebau and K. H. Hesse, *Naturwiss.*, **56**, (1969), 634–635.

56.   K. F. Hesse, *Acta Crystallogr.*, **B35**, (1979), 724–725.

57.   E. Tillmanns, W. Gebert, and W. H. Baur, *J. Solid State Chem.*, **7**, (1973), 69–84.

58.   A. E. Mal'shikov, O. V. Egorova, and I. A. Bondar, *Russ. J. Inorg. Chem.*, **33**, (1988), 722–726.

59.   M. T. Averbuch-Pouchot and A. Durif, *Z. Kristallogr.*, **180**, (1987), 195–202.

60.   M. T. Averbuch-Pouchot and A. Durif, *Acta Crystallogr.*, **C43**, (1987), 1861–1863.

61.   A. Verbaere, S. Oyetola, D. Guyomard, and Y. Piffard, *J. Solid State Chem.*, **75**, (1988), 217–224.

62.   J. M. Adams and V. Ramdas, *Acta Crystallogr.*, **B32**, (1976), 3224–3227.

63.   J. M. Adams and V. Ramdas, *Acta Crystallogr.*, **B33**, (1977), 3654–3657.

64. J. M. Adams and V. Ramdas, *Acta Crystallogr.*, **B34**, (1978), 2150–2156.

65. S. Kamoun, A. Jouini, and A. Daoud, *C. R. Acad. Sci.*, Sér. 2, **308**, (1989), 923–925.

66. S. Kamoun, A. Jouini, and A. Daoud, *J. Solid State Chem.*, **99**, (1992), 18–28.

67. M. T. Averbuch-Pouchot and A. Durif, *C. R. Acad. Sci.*, Sér. 2, **316**, (1993), 187–192.

68. M. T. Averbuch-Pouchot and A. Durif, *Eur. J. Solid State Inorg. Chem.*, **29**, (1992), 411–418.

69. A. Larbot, J. Durand, and A. Norbert, *Rev. Chim. Minér.*, **17**, (1980), 548–554.

70. A. Durif and M. T. Averbuch-Pouchot, *Acta Crystallogr.*, **B38**, (1982), 2883–2885.

71. R. Klement, *Chem. Ber.*, **93**, (1960), 2314–2316.

72. M. T. Averbuch-Pouchot and A. Durif, *Eur. J. Solid State Inorg. Chem.*, **29**, (1992), 993–999.

73. R. C. Ropp and M. A. Aia, *Anal. Chem.*, **34**, (1962), 1288–1291.

74. L. N. Schegrov, N. M. Antraptseva, and I. G. Ponomareva, *Izv. Akad. Nauk SSSR, Neorg. Mater.*, **25**, (1989), 308–312.

75. D. D. Vasovic and D. R. Stojakovic, *J. Am. Ceram. Soc.*, **77**, (1994), 1372–1374.

76. C. E. Bamberger, *J. Am. Ceram. Soc.*, **65**, (1982), 107–108.

77. K. Y. Leung and C. Calvo, *Canad. J. Chem.*, **50**, (1972), 2519–2526.

78. M. T. Averbuch-Pouchot and A. Durif, *Eur. J. Solid State Inorg. Chem.*, **29**, (1992), 191–198.

79. N. Krishnamachari and C. Calvo, *Acta Crystallogr.*, **B28**, (1972), 2883–2885.

80. N. C. Webb, *Acta Crystallogr.*, **21**, (1966), 942–948.

81. J. P. Gamondes, F. d'Yvoire, and A. Boullé, *C. R. Acad. Sci.*, **272**, (1971), 49–52.

82. M. T. Averbuch-Pouchot, *J. Solid State Chem.*, **79**, (1989), 296–299.

83. M. T. Averbuch-Pouchot, *Acta Crystallogr.*, **C44**, (1988), 2046–2048.

84. M. T. Averbuch-Pouchot and A. Durif, *C. R. Acad. Sci.*, Sér. 2, **316**, (1993), 187–192.

85. M. T. Averbuch-Pouchot and A. Durif, *Eur. J. Solid State Inorg. Chem.*,

30, (1993), 1153–1162.

86.   M. T. Averbuch-Pouchot and A. Durif, *Acta Crystallogr.*, **B39**, (1983), 27–28.

87.   M. T. Averbuch-Pouchot and A. Durif, *Acta Crystallogr.*, **C48**, (1992), 973–975.

88.   M. T. Averbuch-Pouchot and A. Durif, *C. R. Acad. Sci.*, Sér. 2, **316**, (1993), 41–46.

89.   M. T. Averbuch-Pouchot and A. Durif, *C. R. Acad. Sci.*, Sér. 2, **316**, (1993), 469–476.

90.   M. Ijjaali, G. Venturini, R. Gerardin, B. Malaman, and C. Gleitzer, *Eur. J. Solid State Inorg. Chem.*, **28**, (1991), 983–998.

91.   M. T. Averbuch-Pouchot, *Z. anorg. allg. Chem.*, **545**, (1987), 118–124.

92.   R. Masse, J. C. Guitel, and A. Durif, *Mat. Res. Bull.*, **14**, (1979), 337–341.

5.2.3. Triphosphates

Although the triphosphate anion was suspected by some chemists during the last century, it is not before the early fifties that its existence was clearly established thanks to the elaboration by Morey and co-workers [1–3] of two basic phase-equilibrium diagrams, $NaPO_3$–$Na_4P_2O_7$ and KPO_3–$K_4P_2O_7$. These two diagrams, represented in Figures 5.2.17 and 5.2.18, show clearly the existence of $Na_5P_3O_{10}$ and $K_5P_3O_{10}$. The structural investigations of two forms of $Na_5P_3O_{10}$ by Corbridge [4] and Davies and Corbridge [5] confirmed the suspected geometry of the triphosphate anion soon after. More than one hundred triphosphates have been characterized to date.

5.2.3.1. General properties of triphosphates

With the exception of the sodium salt which is extensively investigated, very little is known of the basic physical properties of triphosphates. However, for some of them which have been characterized during the elaboration of various phase-equilibrium diagrams, *the melting or decompostion temperatures* are well established. We report, in Table 5.2.8, some of these investigations and the corresponding data. One can notice that all the reported triphosphates melt incongruently.

There are no general rules for the *thermal behavior* of triphosphates. From a relatively large number of experiments, one can simply note that this behavior seems to be mainly dependent on the nature of the associated cations like in all

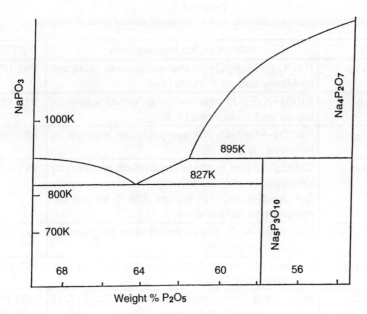

Figure 5.2.17. The NaPO₃–Na₄P₂O₇ phase-equilibrium diagram.

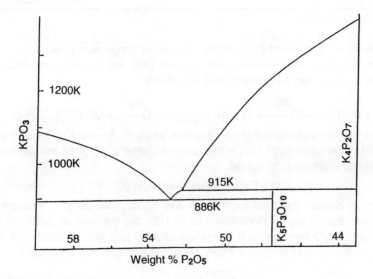

Figure 5.2.18. The KPO₃–K₄P₂O₇ phase-equilibrium diagram.

Table 5.2.8
Various investigations performed for the characterization of some triphosphates.

Formula	Nature of the investigation	mp (K)
$Na_5P_3O_{10}$	$NaPO_3–Na_4P_2O_7$ phase-equilibrium diagram by Morey and co-workers [1–3]	895 (d)
$K_5P_3O_{10}$	$KPO_3–K_4P_2O_7$ phase-equilibrium diagram by Morey and co-workers [1–3]	915 (d)
$Rb_5P_3O_{10}$	$RbPO_3–Rb_4P_2O_7$ phase-equilibrium diagram by Krivovyazov *et al.* [6]	889 (d)
$Cs_5P_3O_{10}$	$CsPO_3–Cs_4P_2O_7$ phase-equilibrium diagram by Krivovyazov *et al.* [6] Sotnikova-Yuzhik [7] reports 878 K for the decomposition temperature.	791 (d)
$Zn_3K_4(P_3O_{10})_2$	$Zn(PO_3)_2–K_2O$ phase-equilibrium diagram by Krivovyazov *et al.* [8]	876 (d)
$ZnK_3P_3O_{10}$	$KPO_3–ZnO$ phase-equilibrium diagram by Krivovyazov *et al.* [8]	931 (d)
$AlCs_2P_3O_{10}$ $GaCs_2P_3O_{10}$ $CrCs_2P_3O_{10}$ $FeCs_2P_3O_{10}$	Solid state investigation of the $M^{III}Cs_2P_3O_{10}$ compounds by Lyutsko *et al.* [9]	1053 (d) 1008 (d) 1310 (d) 1253 (d)

d: incongruent melting

other classes of condensed phosphates. We report some exemples.

As investigated by Selevich and Lyustko [10], the thermal evolution of $MnH_2P_3O_{10}.2H_2O$ can be schematized as follows:

$$MnH_2P_3O_{10}.2H_2O \xrightarrow{363–523\,K} MnH_2P_3O_{10} + 2H_2O \xrightarrow{653–713\,K} Mn(PO_3)_3 + H_2O$$

leading to Mn(III) polyphosphate as the final product. A very similar path was observed by Mel'nikov *et al.* [11] in the case of $GaH_2P_3O_{10}$, then the C-form of gallium polyphosphate is obtained:

$$GaH_2P_3O_{10} \longrightarrow Ga(PO_3)_3\ (C)$$

In some cases, the final product of the decomposition is a mixture of phosphates. According to Rodicheva *et al.* [12], the thermal evolution of the $Ln_5(P_3O_{10})_3.xH_2O$ salts leads to a mixture of poly- and monophosphates. For instance, in the case of lanthanum they observed the following reaction:

$$La_5(P_3O_{10})_3.16H_2O \xrightarrow{1223\,K} 2La(PO_3)_3 + 3LaPO_4 + 16H_2O$$

For mixed-cation triphosphates, the decomposition schemes as can be expected are very various and in some cases unexpected.

According to Chudinova et al. [13] and Hilmer et al. [14–15], $EuKHP_3O_{10}$ has a still more complex path of decomposition leading to a mixture of three phosphates at 833 K:

$$2EuKHP_3O_{10} \longrightarrow EuK(PO_3)_4 + EuPO_4 + KPO_3 + H_2O$$

A mixed diphosphate, $GaKP_2O_7$, is the final product in the evolution of $GaKHP_3O_{10}$ investigated by Chudinova et al. [16]:

$$GaKHP_3O_{10} \xrightarrow{823\,K} GaKP_2O_7 + 1/2P_2O_5 + 1/2H_2O$$

but, according to Guzeeva et al. [17], a mixed cyclotriphosphate, $MnKP_3O_9$, appears when $MnKHP_3O_{10}$ decomposes:

$$2Mn^{III}KHP_3O_{10} \xrightarrow{823–873\,K} 2Mn^{II}KP_3O_9 + H_2O + 1/2O_2$$

$Mg_2NH_4P_3O_{10}.7H_2O$ and $Mn_2NH_4P_3O_{10}.6H_2O$ which were investigated by Ol'shevskaya et al. [18], transform into the corresponding cyclotetraphosphates $M_2P_4O_{12}$ when heated above 773 K, whereas the calcium derivative, $Ca(NH_4)_3P_3O_{10}.2H_2O$, leads to the polyphosphate $Ca(PO_3)_2$ at the same temperature.

At the same temperature, the two zinc salts, $Zn_2NH_4P_3O_{10}.7H_2O$ and $Zn_3(NH_4)_4(P_3O_{10})_2.9H_2O$ behave differently:

$$Zn_2NH_4P_3O_{10}.7H_2O \longrightarrow Zn_2P_4O_{12}$$

$$Zn_3(NH_4)_4(P_3O_{10})_2.9H_2O \longrightarrow Zn(PO_3)_2$$

The most stable compound in the Ga_2O_3–NH_4–P_2O_5–H_2O system, investigated by Chudinova et al. [19], is $GaNH_4HP_3O_{10}$. This triphosphate transforms into gallium cyclohexaphosphate, $Ga_2P_6O_{18}$, when heated at 723 K.

$$2GaNH_4HP_3O_{10} \xrightarrow{723\,K} Ga_2P_6O_{18} + 2H_2O + 2NH_3$$

The formation of gallium–cesium cyclododecaphosphate together with a mixed gallium–cesium diphosphate was observed by Grunze et al. [20] during the thermal decomposition of $GaCsHP_3O_{10}$:

$$6GaCsHP_3O_{10} \longrightarrow Ga_3Cs_3P_{12}O_{36} + 3GaCsP_2O_7 + 3H_2O$$

Being trimorphic, the temperature of decomposition of $GaCsHP_3O_{10}$ is a function of the form used for the experiment and varies from 823 to 853 K.

In some cases, the rate of heating has a strong influence on the nature of the transformation. As observed by Guzeeva and Tananaev [21], $MnCsHP_3O_{10}$ is stable till 923 K and then decomposes according to these two different schemes:

$$2MnCsHP_3O_{10} \longrightarrow 2MnCsP_2O_7 + P_2O_5 + H_2O$$

$$6MnCsHP_3O_{10} \longrightarrow 3MnCsP_2O_7 + Mn_3Cs_3P_{12}O_{36} + 3H_2O$$

The first scheme of decomposition is observed for rapid heating, the second for slow heating rates.

As in many decomposition processes, intermediate steps can occur. For instance, Prodan et al. [22] observed the following transformation during the thermal evolution of $Na(NH_4)_4P_3O_{10}.4H_2O$:

$$Na(NH_4)_4P_3O_{10}.4H_2O \xrightarrow{\;323\,K\;} Na(NH_4)_3HP_3O_{10}.H_2O + NH_3 + 3H_2O$$

With relatively few exceptions, triphosphates are difficult to crystallize in aqueous media. This probably explains why most of the well characterized triphosphates were discovered in various systems investigated by flux methods at relatively high temperatures.

As for diphosphates, the existence of acidic triphosphate anions, $[HP_3O_{10}]^{4-}$, $[H_2P_3O_{10}]^{3-}$, leads to the formation of infinite anionic networks which we will describe in Chapter 6.

5.2.3.2. Alkali and monovalent-cation triphosphates

Many monovalent-cation triphosphates are described in chemical literature, but the atomic arrangements of only five of them are clearly established to date: the two crystalline forms of $Na_5P_3O_{10}$, its hexahydrate, $Na_5P_3O_{10}.6H_2O$, $K_3H_2P_3O_{10}.H_2O$ and $Na(NH_4)_4P_3O_{10}.4H_2O$.

Owing to their important applications in the detergents industry, the sodium derivatives were extensively studied. Their chemical preparations as well as their chemical and physical properties were reported in hundreds of publications. The crystallographic investigations of these salts started very early. Bonneman and Bassiere [23] performed the first X-ray investigation of the hexahydrate as early as 1938. Dymon and King [24] proposed a method to obtain single crystals of the two anhydrous forms and performed their first crystallographic investigations. Later, Corbridge [4] and Davies and Corbridge [5] determined the crystal structures of these two forms. The crystal structure of the hexahydrate was determined by Dyroff [25] and later re-examined by Wiench et al. [26].

The reversible transition from phase (II) to phase (I) of $Na_5P_3O_{10}$ occurs at about 688 K and it is easier than the reverse process. Form (I) is more water soluble than form (II). Corbridge [4–5] tried to explain this difference by the presence of a fourfold coordinated sodium atom in form (I), which represents a relatively unstable local arrangement.

Several crystalline potassium triphosphates have been described, but, among them, only $K_3H_2P_3O_{10} \cdot H_2O$ was the subject of a structural characterization performed by Lyutsko and Johansson [27] who also report a detailed procedure of crystal growth. This compound has a typical layer arrangement discussed in Chapter 6 devoted to the networks built by acidic anions.

Table 5.2.9
Summary of various investigations on mononovalent-cation triphosphates.

Characterized salt	Nature of the investigation
$Li_5P_3O_{10} \cdot 5H_2O$	Obtained by Sotnikova-Yuzhik *et al.* [30] by alkaline hydrolytic decyclization of lithium cyclotriphosphate in a LiOH solution at 313–353 K. The scheme is: $$Li_3P_3O_9 + 2 \, LiOH \longrightarrow Li_5P_3O_{10} + H_2O$$
$Ag_5P_3O_{10}$ $Ag_5P_3O_{10} \cdot 2H_2O$	Chemical preparation and properties investigated by Lee [31] and Lamotte and Merlin [32]. The anhydrous salt has two crystalline forms.
$K_2H_3P_3O_{10} \cdot 2H_2O$ $K_5P_3O_{10} \cdot 4H_2O$ $K_3H_2P_3O_{10} \cdot H_2O$	Various physico-chemical investigations by Shashkova *et al.* [33, 34], Lyutsko *et al.* [35], Prodan and Shashova [36], Prodan and Bulavknina [37], Prodan *et al.* [38]
$(NH_4)_3H_2P_3O_{10}$ $(NH_4)_3H_2P_3O_{10} \cdot 2H_2O$ $(NH_4)_9H(P_3O_{10})_2 \cdot 2H_2O$ $(NH_4)_4HP_3O_{10}$ $(NH_4)_5P_3O_{10} \cdot H_2O$ $(NH_4)_5P_3O_{10} \cdot 2H_2O$ $(NH_4)_5P_3O_{10} \cdot 2H_2O_2 \cdot H_2O$	Investigation of the NH_3–$H_5P_3O_{10}$–H_2O system at 273 and 298 K by Farr *et al.* [39] and Frazier *et al.* [40] Crystal data for $(NH_4)_5P_3O_{10} \cdot xH_2O$ are reported by Waerstad and Mac Clellan [41]. $(NH_4)_5P_3O_{10} \cdot 2H_2O_2 \cdot H_2O$, has been isolated by Sotnikova-Yuzhik and Prodan [42].
$Rb_5P_3O_{10}$ $Cs_5P_3O_{10}$	Investigation of the $RbPO_3$–$Rb_4P_2O_7$ and $CsPO_3$–$Cs_4P_2O_7$ phase-equilibrium diagrams by Krivoviazov *et al.* [6]
$Na_xK_{(5-x)}P_3O_{10} \cdot 2H_2O$ $2 < x < 3$	Investigation of the $Na_5P_3O_{10}$–$K_5P_3O_{10}$–H_2O system by Bulavkina *et al.* [43] and Griffith and Buxton [44]
$Na(NH_4)_4P_3O_{10} \cdot H_2O$ $Na(NH_4)_3HP_3O_{10} \cdot H_2O$	Various physico-chemical investigations by Prodan *et al.* [22, 28]
$K_2(NH_4)_3P_3O_{10} \cdot 2H_2O$ $K_2(NH_4)_2HP_3O_{10} \cdot nH_2O$ $K_2(NH_4)_3P_3O_{10} \cdot nH_2O$ $(2 < n < 3)$	Prepared by Shashkova *et al.* [45] by action of gaseous ammonia on $K_2H_3P_3O_{10} \cdot 2H_2O$

The $Na_5P_3O_{10}-(NH_4)_5P_3O_{10}-H_2O$ system was investigated by Prodan *et al.* [28]. One of the intermediate compounds, $Na(NH_4)_4P_3O_{10}.4H_2O$, was studied with more details. Its thermal behavior which was also investigated by Prodan *et al.* [22] was reported at the beginning of this section. Later, Averbuch-Pouchot and Durif [29] produced single crystals of $Na(NH_4)_4P_3O_{10}.4H_2O$ and reported a detailed determination of its atomic arrangement.

A number of other monovalent-cation triphosphates were chemically studied along various investigations that we have summarized in Table 5.2.9.

We report below the description of the atomic arrangements of the two crystalline forms of $Na_5P_3O_{10}$.

The $Na_5P_3O_{10}$ (I) atomic arrangement [4] – $Na_5P_3O_{10}$ (I) is monoclinic, *C2/c*, with $Z = 4$ and the following unit cell:

$$a = 9.61(3), \quad b = 5.34(2), \quad c = 19.73(5) \text{ Å}, \quad \beta = 112.0(5)°$$

This atomic arrangement projected along the **b** axis is represented in Figure 5.2.19. The triphosphate anion has a twofold internal symmetry, its central phosphorus atom is located on a twofold axis. The anion axis is approximately parallel to the [$\overline{2}$01] direction. Two of the three crystallographically independent sodium atoms have a sixfold oxygen coordination building distorted octahedra, the last a fourfold. One of the sixfold coordinated sodium atom is located on an inversion center. The Na–O distances range between 2.26 and 2.41 Å in the

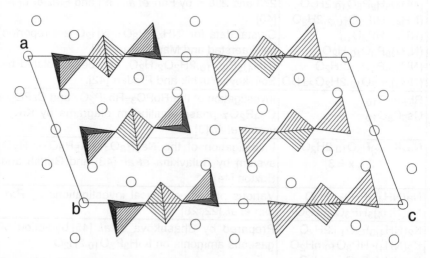

Figure 5.2.19. Projection, along the **b** axis, of the structure of $Na_5P_3O_{10}$ (I). The phosphoric groups are shown by polyhedral representation and the open circles figure the sodium atoms.

NaO_6 polyhedra and between 2.22 and 2.59 Å in the NaO_4 polyedron.

The $Na_5P_3O_{10}$ (II) atomic arrangement [5] – $Na_5P_3O_{10}$ (II) is monoclinic, *C2/c*, with $Z = 4$ and the following unit cell:

$$a = 16.00(5), \quad b = 5.24(2), \quad c = 11.25(3) \text{ Å}, \quad \beta = 93.0(5)°$$

As in form (I), the central phosphorus atom of the triphosphate anion is located on a twofold axis. Thus, this anion adopts this symmetry. Figure 5.2.20 represents a projection of this structure along the **b** direction. The three crystallographically independent sodium atoms are surrounded by oxygen atoms in distorted octahedral arrangements, which are linked by edge and corner sharing to form a complex three-dimensional network. A relatively large variation of the Na–O distance is observed inside these NaO_6 octahedra since this distance ranges from 2.25 to 2.76 Å.

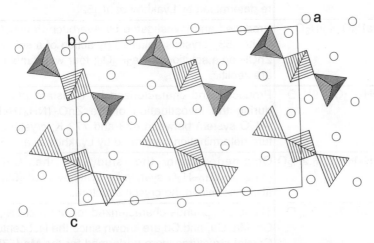

Figure 5.2.20. Projection, along the **b** axis, of the atomic arrangement in $Na_5P_3O_{10}$ (II). The drawing conventions are similar to those used in Figure 5.2.19.

5.2.3.3. Divalent and monovalent-divalent cation triphosphates

A few divalent-cation triphosphates are known presently, whereas more divalent-monovalent cation triphosphates are clearly characterized. However, only three divalent-cation triphosphates were accurately characterized: two zinc salts, $Zn_2HP_3O_{10}.6H_2O$ and $Zn_5(P_3O_{10})_2.17H_2O$, and one lead salt, $Pb_2HP_3O_{10}$.

The two zinc salts were prepared and characterized by Averbuch-Pouchot and co-workers [46–48]. Among the few examples of triphosphates, these two salts are prepared by the hydrolytic decyclization of P_3O_9 ring anions. Their atomic arrangements show that there are four zeolitic water molecules in the hexahydrate and seven in the heptadecahydrate.

Table 5.2.10

Various investigations on divalent-monovalent cation triphosphates.

Formula	Nature of the investigation
$Be_2NH_4P_3O_{10}$	Characterized during an investigation of the BeO–$(NH_4)_2O$–P_2O_5 system by a flux method by Bagieu-Beucher *et al.* [50]; structure determined by Averbuch-Pouchot *et al.* [51]
$Zn_2LiP_3O_{10}.8H_2O$	Prepared by evaporation of a 0.1 M equimolecular aqueous solution of $Li_5P_3O_{10}$ and $Zn(NO_3)_2$; structure determined by Lyakhov *et al.* [52]
$Zn_2NaP_3O_{10}.9H_2O$	Produced and investigated by Averbuch-Pouchot and Guitel [53]; crystals obtained by action of a solution of $ZnCl_2$ on a solution of $Na_3P_3O_9$; four water molecules are zeolitic.
$Zn_2NH_4P_3O_{10}.9H_2O$	Produced and characterized by Konstant *et al.* [54] during the investigation of the ZnO–$(NH_4)H_2PO_4$–H_2O system between 473 and 773 K; crystal structure determination performed by Lyakhov *et al.* [55]
$Ca(NH_4)_3P_3O_{10}.2H_2O$	Obtained during the study of the $CaCl_2$–$(NH_4)_5P_3O_{10}$–H_2O system by Lyakhov *et al.* [56] who also determined the crystal structure.
$M^{II}Na_3P_3O_{10}.12H_2O$	These compounds characterized for M^{II} = Zn, Ni, Mg, Co, Mn, Cu, and Cd are known since the last century. Crystal structures were performed for the Mn [57], Cu [58–59] and Cd salts [60].

$Pb_2HP_3O_{10}$ was synthesized by Worzala and Jost [49] by firing $Pb_2P_4O_{12}.2H_2O$ at 423 K. The crystals that they obtained contain H_3PO_4 which can be explained by the following scheme:

$$Pb_2P_4O_{12}.2H_2O \longrightarrow Pb_2HP_3O_{10} + H_3PO_4$$

As mentioned above, the divalent-monovalent cation triphosphates are more numerous. We list in Table 5.2.10 some data concerning the restricted number of those whose atomic arrangements were determined. Multiple other ones were chemically characterized during the investigation of various systems in aqueous solution or by flux methods, but most of them are not yet clearly identified.

We illustrate this class of triphosphates with the description of the atomic arrangement of $Be_2NH_4P_3O_{10}$.

The $Be_2NH_4P_3O_{10}$ atomic arrangement – $Be_2NH_4P_3O_{10}$ [50–51] is monoclinic, *C2/c*, with *Z* = 4 and has the following unit-cell dimensions:

$$a = 12.202(8), \quad b = 8.645(3), \quad c = 8.949(3) \text{ Å}, \quad \beta = 117.41(5)°$$

The present arrangement can be described as a three-dimensional network of P_3O_{10} groups and BeO_4 tetrahedra with the ammonium groups inserted in channels parallel to the **c** axis. Figure 5.2.21 gives a projection of the structure along the **b** direction. P_3O_{10} groups are all centered around planes *z* = 1/4 and 3/4 and are interconnected by BeO_4 tetrahedra located in planes *z* = 0 and 1/2. The central phosphorus atom of the P_3O_{10} group is located on a twofold axis, hence the anion adopts a binary symmetry which is rather common in this class of anions. The other representation, a projection along the **c** axis, given in Figure 5.2.22 shows how the arrangement of this three-dimensional network creates large channels parallel to the **c** direction in which are located the ammonium groups. Each BeO_4 tetrahedron shares its four corners with four adjacent PO_4 tetrahedra, belonging to three different P_3O_{10} groups. The Be–P distances are rather short, ranging from 2.811 to 2.884 Å. Inside a BeO_4

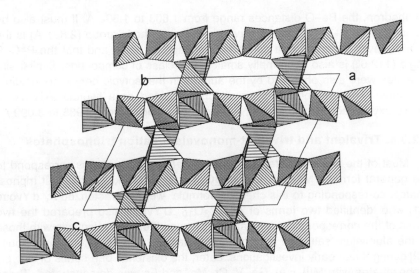

Figure 5.2.21. Projection, along the **b** direction, of the atomic arrangement in $Be_2NH_4P_3O_{10}$. BeO_4 and PO_4 tetrahedra are shown by polyhedral representation.

Figure 5.2.22. Projection of the atomic arrangement in $Be_2NH_4P_3O_{10}$ along the **c** axis. Open circles represent the NH_4 groups. PO_4 and BeO_4 tetrahedra are shown by polyhedral representation.

tetrahedron, the Be–O distances range from 1.603 to 1.662 Å. It must also be noted that the P–P distance observed in the phosphoric group (2.821 Å) is the shortest ever observed in a condensed phosphate anion and that the P–O–P angle (112.6°) is also abnormally small in this class of compounds. Similar singularities were also observed by the authors in the isotypic beryllium–rubidium derivative. The ammonium group which is located on a twofold axis has a sixfold oxygen coordination, with N–O distances ranging from 2.935 to 3.099 Å.

5.2.3.4. Trivalent and trivalent-monovalent cation triphosphates

Most of the well characterized *trivalent-cation triphosphates* correspond to the general formula $M^{III}H_2P_3O_{10}.nH_2O$. As far as we know, the first triphosphates, corresponding to this chemical formula, were characterized by d'Yvoire [61] who identified two forms of $AlH_2P_3O_{10}$. D'Yvoire also prepared the two forms of the corresponding iron salt and showed them to be isotypic with those of the aluminium salt. The chromium salt was identified by Rémy and Boullé [62] during these early investigations. Later, the existence of a series of isotypic triphosphates with M^{III} = Al, Ga, V, Cr, Mn, and Fe was demonstrated. These compounds do not exhibit ion-exchange properties in aqueous solutions, but only in molten salts at moderately high temperatures. For instance, in the case of cesium:

$$M^{III}H_2P_3O_{10} + 2CsNO_3 \longrightarrow M^{III}Cs_2P_3O_{10} + 2HNO_3$$

If other alkali nitrates are used, the exchange occurs but the reaction path is different:

$$M^{III}H_2P_3O_{10} + 2M^INO_3 \longrightarrow M^{III}M^IP_2O_7 + M^IPO_3 + 2HNO_3$$

This type of exchange reactions were extensively investigated by Lyutsko and co-workers (see Lyutsko *et al.* [63] and references herein).

ScH$_2$P$_3$O$_{10}$ and YbH$_2$P$_3$O$_{10}$ were characterized by Kanepe and Konstant [64] and Palkina *et al.* [65] respectively.

A number of hydrates were also reported: MnH$_2$P$_3$O$_{10}$.2H$_2$O by Guzeeva *et al.* [66], GaH$_2$P$_3$O$_{10}$.2H$_2$O by Mel'nikov *et al.* [67], and FeH$_2$P$_3$O$_{10}$.H$_2$O by Averbuch-Pouchot and Guitel [68]. The formation of the gallium salt by thermal condensation of Ga(H$_2$PO$_4$)$_3$ is interesting to note:

$$Ga(H_2PO_4)_3 \xrightarrow{413\,K} GaH_2P_3O_{10}.2H_2O \underset{\text{reversible}}{\overset{413-503\,K}{\longleftrightarrow}} GaH_2P_3O_{10} + 2H_2O$$

Various indium, yttrium and rare-earth triphosphates were reported in chemical literature. All these compounds correspond to the general formula M$_5^{III}$(P$_3$O$_{10}$)$_3$.nH$_2$O. In most of cases, these derivatives are poorly crystallized and their hydration state uncertain, but always high. No crystallographic investigation was performed on these salts.

Most of the well characterized *trivalent-monovalent cation triphosphates* have the general formula M^{III}M^{I}HP$_3$O$_{10}$ and many of them are dimorphous. D'Yvoire [69] was the first to describe a preparation of AlNH$_4$HP$_3$O$_{10}$ by firing Al(NH$_4$)$_2$P$_3$O$_{10}$.H$_2$O at 473 K:

$$Al(NH_4)_2P_3O_{10}.H_2O \longrightarrow AlNH_4HP_3O_{10} + NH_3 + H_2O$$

Later, the description of a crystal-growth method and the detailed structural investigation of this salt were reported by Averbuch-Pouchot *et al.* [70]. They also reported the chemical preparation and crystal data for the isotypic potassium salt, AlKHP$_3$O$_{10}$. Numerous M^{III}M^{I}HP$_3$O$_{10}$ have been well characterized to date. The nature of some of these investigations are gathered in Table 5.2.11.

Lanthanide- and bismuth-monovalent cation triphosphates with a similar chemical formula were also investigated. Nine LnM^{I}HP$_3$O$_{10}$ compounds, with M^I = K, Rb, Cs and Ln = Nd, Er, Yb, Tb, Dy, were prepared by Vinogradova and Chudinova [86]. EuKHP$_3$O$_{10}$ which was first prepared by Chudinova *et al.* [87], was later investigated by Hilmer *et al.* [88–89]. NdNH$_4$HP$_3$O$_{10}$ and CeNH$_4$HP$_3$O$_{10}$ were characterized by Chudinova *et al.* [90] and Vaivada and Konstant [91] respectively. The crystal growth procedure and the atomic arrangement of $BiNH_4HP_3O_{10}$ were reported by Averbuch-Pouchot and Bagieu-Beucher [92].

Table 5.2.11
Some investigations which lead to the characterization of $M^{III}M^{I}HP_3O_{10}$ triphosphates.

Formula	Nature of the investigation
$ScNaHP_3O_{10}$	Investigation of the Sc_2O_3–Na_2O–P_2O_5–H_2O system between 523 and 588 K by Avaliani [71]
$VM^{I}HP_3O_{10}$	$VNH_4HP_3O_{10}$ was described by Teterevkov and Mikhailovskaya [72] and later re-examined by Krasnikov *et al.* [73], whereas $VCsHP_3O_{10}$ was described by Klinkert and Jansen [74].
$CrM^{I}HP_3O_{10}$	The ammonium salt was characterized by Vaivada *et al.* [75]. The thermal evolution of this salt was examined by Lyustko and Pap [76]. $CrKHP_3O_{10}$, $CrRbHP_3O_{10}$ and $CrCsHP_3O_{10}$ were prepared by Grunze and Chudinova [77].
$MnM^{I}HP_3O_{10}$	$MnKHP_3O_{10}$ was obtained by Guzeeva *et al.* [78] and $MnRbHP_3O_{10}$ by Guzeeva and Tananaev [79] during the investigation of the MnO_2–Rb_2O–P_2O_5–H_2O system between 413 and 623 K. This compound, the most stable in the system, is produced over the whole temperature range. $MnCsHP_3O_{10}$, decomposing at 923 K, is the most stable compound in the MnO_2–Cs_2O–P_2O_5–H_2O system investigated by Guzeeva and Tananaev [21].
$(Fe, Al)M^{I}HP_3O_{10}$	Grunze and Grunze [80] characterized eight of them for M^{I} = K, NH_4, Rb, and Cs, M^{III} = Al and Fe. The structure of the iron–ammonium salt was determined by Krasnikov *et al.* [81].
$GaM^{I}HP_3O_{10}$	$GaM^{I}HP_3O_{10}$ were characterized for M^{I} = K, NH_4, and Cs by Chudinova *et al.* [19, 82–84]. The ammonium salt was obtained by firing at 623 K a mixture of Ga_2O_3 and $(NH_4)_2HPO_4$, the two other ones during investigations of the corresponding Ga_2O_3–M_2O–P_2O_5 systems.
$InM^{I}HP_3O_{10}$	The sodium, rubidium and cesium derivatives were obtained by Avaliani *et al.* [85] during an investigation of the In_2O_3–M_2O–P_2O_5 systems.

Many other trivalent-monovalent cation triphosphates with various formulæ were also reported in chemical literature during the last 30 years, but they have not been investigated from a structural point of view to date and for some of the reported compounds the nature of the anion remains to be confirmed.

5.2.3.5. Chemical preparation of triphosphates

As mentioned at the beginning of this section, in most cases triphosphates are difficult to obtain in a good crystalline state by the classical reactions of aqueous chemistry. This assumption seems to be well confirmed by the fact that majority of the presently known triphosphates were characterized during experiments involving the use of flux methods.

$Na_5P_3O_{10}$ preparation – The most common starting material for the preparation of triphosphates is the sodium salt which has been commercially available in a very good state of purity for a long time. The first preparation of a pure crystalline $Na_5P_3O_{10}$ was performed by Schwartz [93] by melting together $Na_4P_2O_7$ and $NaPO_3$ in the stoichiometric ratio:

$$Na_4P_2O_7 + NaPO_3 \longrightarrow Na_5P_3O_{10}$$

but several other processes involving the commonly available sodium monophosphates can be used.

Ion-exchange resins – Most of the alkali triphosphates can be prepared from the sodium salt by the use of ion-exchange resins. In the first step, the triphosphoric acid $H_5P_3O_{10}$ is produced by passing an aqueous solution of the sodium salt through the appropriate resin, then the acid is immediately neutralized by the required hydroxides or carbonates. During the last step, the reactor must be cooled in order to avoid the hydrolytic scission of the triphosphate anion. Various systems involving $(NH_4)_5P_3O_{10}$ or $K_5P_3O_{10}$ as components were performed by using alkali triphosphates prepared by this process.

Flux methods – A number of triphosphates were obtained during investigations run with a large excess of phosphoric acid at relatively low temperatures. In most cases, the optimization of the initial component ratios and the temperature leads to a reproducible procedure for their preparations.

Classical methods of aqueous chemistry – Several mixed-cation triphosphates have been successfully prepared as well crystallized samples by using the classical methods of aqueous chemistry. The $M^{II}Na_3P_3O_{10}.12H_2O$ (M^{II} = Mg, Ni, Co, Cu...) series of triphosphates can be prepared simply by the following reaction:

$$Na_5P_3O_{10} + M^{II}SO_4 \longrightarrow M^{II}Na_3P_3O_{10} + Na_2SO_4$$

Decyclization of cyclotriphosphate-ring anions – Along our survey of triphosphates, we reported several examples of the production of P_3O_{10} anions by hydrolytic decyclization of cyclotriphosphate anions according to the following scheme:

$$H_3P_3O_9 + H_2O \longrightarrow H_5P_3O_{10}$$

In spite of its apparent simplicity and elegance, this procedure of synthesis, well reproducible in some cases, is most of time accidental and will very probably remain academic.

5.2.3.6. The triphosphate anion

No more than twenty atomic arrangements of triphosphates have been determined with reasonable accuracy up to now. We report, in Table 5.2.12, the main features describing the geometry of fourteen P_3O_{10} groups observed in crystal structures determined with an acceptable accuracy. We limit this description to what we consider as essential to depict a P_3O_{10} group: the P–P distances, the P–O–P and P–P–P angles.

If one examines the internal symmetry of the P_3O_{10} groups, one observes that about half of them have a twofold symmetry, the central phosphorus atom is located on a binary axis. The remaining ones have no internal symmetry or a pseudo-binary one. Figure 5.2.23 describes some examples of such phosphoric groups.

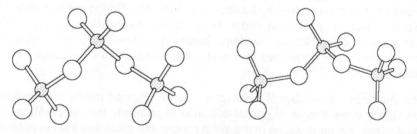

Figure 5.2.23. Some examples of $[P_3O_{10}]^{5-}$ anions: on the left-hand side the P_3O_{10} group observed in $Be_2NH_4P_3O_{10}$, on the right-hand side that observed in $FeH_2P_3O_{10}$. The first one has binary symmetry, the second one has no symmetry.

For the *P–P distances and P–O–P angles*, the values reported here are within the limits generally observed in condensed phosphate crystal chemistry, with the exception of those reported for $Be_2NH_4P_3O_{10}$ where the P–P distance

(2.821 Å) and the P–O–P angle (112.6°) are the smallest ever encountered in a condensed phosphate anion. The situation is similar in the isotypic compound, $Be_2RbP_3O_{10}$ in which one observes: P–P = 2.818 Å and P–O–P = 122.4°.

On the contrary, the great dispersion of *the P–P–P angles* varying from 84.5 to 151.3° may appear unusual but a similar range of values is commonly observed in the crystal chemistry of the long-chain polyphosphates and in large ring-anions like cyclohexaphosphates.

Table 5.2.12
Main geometrical features in the P_3O_{10} groups in fourteen examples of triphosphates.

Formula	P–P (Å)	P–P–P (°)	P–O–P (°)	Sym.	Ref.
$Na_5P_3O_{10}$ (I)	2.87	151.5	121.8	Twofold	[4]
$Na_5P_3O_{10}$ (II)	2.87	156.8	121.5	Twofold	[5]
$Na_5P_3O_{10}.6H_2O$	2.881 2.887	151.3	124.4 123.4	None	[25–26]
$Na(NH_4)_4P_3O_{10}.4H_2O$	2.944	103.9	130.6	Twofold	[29]
$K_3H_2P_3O_{10}.H_2O$	2.978 2.991	98.1	139.0 138.4	None	[27]
$Be_2NH_4P_3O_{10}$	2.821	105.4	112.6	Twofold	[50–51]
$Be_2RbP_3O_{10}$	2.818	105.3	122.4	Twofold	[94]
$Zn_2HP_3O_{10}.6H_2O$	2.913 2.915	98.0	129.0 130.2	None	[46]
$Zn_2NaP_3O_{10}.9H_2O$	2.912 2.906	98.0	130.0 128.9	None	[53]
$Zn_2NH_4P_3O_{10}.7H_2O$	2.916 2.920	98.4	128.9 130.0	None	[55]
$FeH_2P_3O_{10}$	2.932 2.933	124.0	133.5 134.1	None	[27]
$AlNH_4HP_3O_{10}$	2.919	127.7	132.2	Twofold	[70]
$FeNH_4HP_3O_{10}$	2.897	112.2	130.5	Twofold	[81]
$BiNH_4HP_3O_{10}$	2.878 2.978	84.5	126.3 137.6	None	[92]

Inside the three constituting PO_4 tetrahedra of the triphosphoric groups, no special features are observed when these entities are compared with those measured in any other condensed phosphoric anion.

178   *Topics in Phosphate Chemistry*

References

1.   G. W. Morey and E. Ingerson, *Am. J. Sci.*, **242**, (1944), 1.

2   E. Ingerson and G. W. Morey, *Am. Mineralogist*, **28**, (1943), 48.

3   G. W. Morey, *J. Am. Chem., Soc.*, **76**, (1954), 4724.

4.   D. E. C. Corbridge, *Acta Crystallogr.*, **13**, (1960), 263–269.

5.   D. R. Davies and D. E. C. Corbridge, *Acta Crystallogr.*, **11**, (1958), 315–319.

6.   E. L. Krivoviazov, V. P. Volkova, and N. K. Vostresenskaya, *Izv. Akad. Nauk SSSR, Neorg. Mater.*, **9**, 761–765 (1970).

7.   V. A. Sotnikova-Yuzhik, *Izv. Akad. Nauk SSSR, Neorg. Mater.*, **27**, (1991), 1011–1013.

8.   E. L. Krivoviazov, N. K. Voskresenskaya, and K. K. Palkina, *Izv. Akad. Nauk SSSR, Neorg. Mater.*, **6**, (1969), 1057–1061.

9.   V. A. Lyutsko, O. G. Pap, and N. M. Ksenofontova, *Izv. Akad. Nauk SSSR, Neorg. Mater.*, **22**, (1986), 1773–1777.

10. A. F. Selevich and V. D. Lyutsko, *Russ. J. Inorg. Chem.*, **29**, (1984), 364–369.

11. P. P. Mel'Nikov, V. A. Efremov, A. K. Stepanov, T. S. Romanova, and L. N. Komissarova, *Russ. J. Inorg. Chem.*, **21**, (1976), 26–28.

12. G. V. Rodicheva, I. V. Tananaev, and N. M. Romanova, *Izv. Akad. Nauk SSSR, Neorg. Mater.*, **17**, (1981), 126–130.

13. N. N. Chudinova, N. V. Vinogradova, G. M. Balagina, and K. K. Palkina, *Izv. Akad. Nauk SSSR, Neorg. Mater.*, **13**, (1977), 1494–1499.

14. W. Hilmer, N. N. Chudinova, K. H. Jost, and N. V. Vinogradova, *Izv. Akad. Nauk SSSR, Neorg. Mater.*, **15**, (1979), 332–334.

15. W. Hilmer, N. N. Chudinova, K. H. Jost, and N. V. Vinogradova, *Izv. Akad. Nauk SSSR, Neorg. Mater.*, **15**, (1979), 1123–1125.

16. N. N. Chudinova, M. A. Avaliani, L. S. Guzeeva, and I. V. Tananaev, *Izv. Akad. Nauk SSSR, Neorg. Mater.*, **14**, (1978), 2054–2060.

17. L. S. Guzeeva, A. V. Lavrov, and I. V. Tananaev, *Izv. Akad. Nauk SSSR, Neorg. Mater.*, **18**, (1982), 1850–1855.

18. O. P. Ol'shevskaya, T. N. Galkova, and E. A. Prodan, *Russ. J. Inorg. Chem.*, **27**, (1982), 947–949.

19. N. N. Chudinova, I. Grunze, and L. S. Guzeeva, *Izv. Akad. Nauk SSSR, Neorg. Mater.*, **23**, (1987), 616–621.

20.   I. Grunze, K. K. Palkina, N. N. Chudinova, L. S. Guzeeva, M. A. Avaliani, and S. I. Maksimova, *Izv. Akad. Nauk SSSR, Neorg. Mater.*, **23**, (1987), 610–615.

21.   L. S. Guzeeva and I. V. Tananaev, *Izv. Akad. Nauk SSSR, Neorg. Mater.*, **24**, (1988), 651–654.

22.   E. A. Prodan, B. M. Galogadja, P. N. Petruskaia, and B. H. Kordjev, *Dokl. Akad. Nauk SSSR*, **25**, (1981), 163–165.

23.   P. Bonneman and M. Bassiere, *C. R. Acad. Sci.*, **206**, (1938), 1379–1380.

24.   J. J. Dymon and A. J. King, *Acta Crystallogr.*, **4**, (1951), 378–379.

25.   D. R. Dyroff, *Thesis*, (Calif. Inst. Technol., Pasadena, USA, 1965).

26.   D. M. Wiench, M. Jansen, and R. Hoppe, *Z. anorg. allg. Chem.*, **488**, (1982), 80–86.

27.   V. Lyutsko and G. Johansson, *Acta Chem. Scand.*, **A38**, (1988), 663–669.

28.   E. A. Prodan, L. I. Petrovskaya, and V. N. Korzhuev, *Russ. J. Inorg. Chem.*, **25**, (1980), 1013–1017.

29.   M. T. Averbuch-Pouchot and A. Durif, *Acta Crystallogr.*, **C41**, (1985), 1553–1555.

30.   V. A. Sotnikova-Yuzhik, G. V. Pelsyak, and E. A. Prodan, *Russ. J. Inorg. Chem.*, **32**, (1987), 1505–1507.

31.   J. D. Lee, *J. Chem. Soc. (A), Inorg. Phys. Theor.*, **12**, (1968), 2881–2882.

32.   A. Lamotte and J. C. Merlin, *Bull. Soc. Chim. Fr.*, (1968), 4311–4312.

33.   I. L. Shashkova, V. A. Lyutsko, and E. A. Prodan, *Izv. Akad. Nauk SSSR, Neorg. Mater.*, **23**, (1987), 986–991.

34.   I. L. Shashkova, E. A. Prodan, and V. A. Lyutsko, *Izv. Akad. Nauk SSSR, Neorg. Mater.*, **24**, (1988), 259–263.

35.   V. A. Lyutsko, I. L. Shashkova, and E. A. Prodan, *Zhur. Prik. Spektr.*, **3**, (1982), 415–420.

36.   E. A. Prodan and I. L. Shashkova, *Russ. J. Inorg. Chem.*, **28**, (1983), 180–183.

37.   E. A. Prodan and N. V. Bulavkina, *Izv. Akad. Nauk SSSR, Neorg. Mater.*, **17**, (1981), 1662–1667.

38.   E. A. Prodan, I. L. Shashkova, and L. A. Lesnikovich, *Russ. J. Inorg. Chem.*, **24**, (1979), 1454–1458.

39.   T. D. Farr, J. D. Fleming, and J. D. Hatfield, *J. Chem. Eng. Data*, **12**,

(1967), 141–142.

40. A. W. Frazier, J. P. Smith, and J. R. Lehr, *J. Agr. Food Chem.*, **13**, (1965), 316–322.

41. K. R. Waerstad and G. H. Mc Clellan, *J. Appl. Crystallogr.*, **7**, (1974), 404–405.

42. V. A. Sotnikova-Yushik and E. A. Prodan, *Russ. J. Inorg. Chem.*, **26**, (1981), 848–849.

43. N. V. Bulavkina, E. A. Prodan, and L. I. Petrovskaya, *Russ. J. Inorg. Chem.*, **29**, (1984), 1104–1106.

44. E. J. Griffith and R. L. Buxton, *J. Chem. Eng. Data,* **13**, (1968), 145–148.

45. I. L. Shashkova, E. A. Prodan, and V. A. Lyutsko, *Izv. Akad. Nauk SSSR, Neorg. Mater.*, **24**, (1988), 259–263.

46. M. T. Averbuch-Pouchot and J. C. Guitel, *Acta Crystallogr.*, **B32**, (1976), 1670–1673.

47. M. T. Averbuch-Pouchot and A. Durif, *J. Appl. Crystallogr.*, **8**, (1975), 564.

48. M. T. Averbuch-Pouchot, A. Durif, and J. C. Guitel, *Acta Crystallogr.*, **B31**, (1975), 2482–2486.

49. H. Worzala and K. H. Jost, *Z. anorg. allg. Chem.*, **445**, (1978), 36–46.

50. M. Bagieu-Beucher, A. Durif, and M. T. Averbuch-Pouchot, *J. Appl. Crystallogr.*, **9**, (1976), 52.

51. M. T. Averbuch-Pouchot, A. Durif, J. Coing-Boyat, and J. C. Guitel, *Acta Crystallogr.*, **B33**, (1977), 203–205.

52. A. S. Lyakhov, V. A. Lyutsko, L. I. Prodan, and K. K. Palkina, *Izv. Akad. Nauk SSSR, Neorg. Mater.*, **27**, (1991), 1014–1018.

53. M. T. Averbuch-Pouchot and J. C. Guitel, *Acta Crystallogr.*, **B33**, (1977), 1427–1431.

54. Z. A. Konstant, I. Sikach, and A. P. Dindune, *Izv. Akad. Nauk SSSR, Neorg. Mater.*, **20**, (1984), 1893–1897.

55. A. S. Lyakhov, V. A. Lyutsko, T. N. Galkova, and K. K. Palkina, *Russ. J. Inorg. Chem.*, **36**, (1991), 1715–1718.

56. A. S. Lyakhov, V. A. Lyutsko, T. N. Galkova, and K. K. Palkina, *Russ. J. Inorg. Chem.*, **38**, (1993), 1174–1178.

57. M. Herceg, *2ᵈ European Crystallogr. Meeting*, (Keszthely, Hungary, 1974).

58. A. Jouini and A. Durif, *C. R. Acad. Sci.*, Sér. 2, **297**, (1983), 573–575.

59. A. Jouini, M. Dabbabi, M. T. Averbuch-Pouchot, A. Durif, and J. C. Guitel,

Acta Crystallogr., **C40**, (1984), 728–730.

60. V. Luytsko and G. Johansson, *Acta Chem. Scand.*, **A38**, (1984), 415–417.

61. F. d' Yvoire, *Bull. Soc. Chim. Fr.*, (1962), 1224–1236.

62. P. Rémy and A. Boullé, *C. R. Acad. Sci.*, **258**, (1964), 927–929.

63. V. A. Lyutsko, A. S. Lyakhov, G. K. Tuchkovskii, and K. K. Palkina, *Russ. J. Inorg. Chem.*, **36**, (1981), 662–664.

64. Z. Ya. Kanepe and Z. A. Konstant, *Izv. Akad. Nauk SSSR, Neorg. Mater.*, **19**, (1983), 969–971.

65. K. K. Palkina, S. I. Maksimova, and V. G. Kuznetsov, *Izv. Akad. Nauk SSSR, Neorg. Mater.*, **15**, (1979), 2168–2170.

66. L. S. Guzeeva, A. V. Lavrov, and I. V. Tananaev, *Izv. Akad. Nauk SSSR, Neorg. Mater.*, **18**, (1982), 1850–1855.

67. P. P. Mel'nikov, V. A. Efremov, A. K. Stepanov, T. S. Romanova, and L. N. Komissarova, *Russ. J. Inorg. Chem.*, **21**, (1976), 26–28.

68. M. T. Averbuch-Pouchot and J. C. Guitel, *Acta Crystallogr.*, **B33**, (1977), 1613–1615.

69. F. d' Yvoire, *Bull. Soc. Chim. Fr.*, (1962), 1224–1236.

70. M. T. Averbuch-Pouchot, A. Durif, and J. C. Guitel, *Acta Crystallogr.*, **B33**, (1977), 1436–1438.

71. M. A. Avaliani, *Izv. Akad. Nauk SSSR, Neorg. Mater.*, **26**, (1990), 2647–2648.

72. A. I. Teterevkov and G. K. Mikhailovskaya, *Russ. J. Inorg. Chem.*, **25**, (1980), 781–782.

73. V. V. Krasnikov, Z. A. Konstant, and V. S. Fundamenskii, *Izv. Akad. Nauk SSSR, Neorg. Mater.*, **19**, (1983), 1373–1378.

74. B. Klinkert and M. Jansen, *Z. anorg. allg. Chem.*, **567**, (1988), 77–86.

75. M. A. Vaivada, Z. A. Konstant, and V. V. Krasnikov, *Izv. Akad. Nauk SSSR, Neorg. Mater.*, **21**, (1985), 1555–1559.

76. V. A. Luytsko and O. G. Pap, *Russ. J. Inorg. Chem.*, **34**, (1989), 665–668.

77. I. Grunze and N. N. Chudinova, *Izv. Akad. Nauk SSSR, Neorg. Mater.*, **24**, (1988), 988–993.

78. L. S. Guzeeva, A. V. Lavrov, and I. V. Tananaev, *Izv. Akad. Nauk SSSR, Neorg. Mater.*, **18**, (1982), 1850–1855.

79. L. S. Guzeeva and I. V. Tananaev, *Izv. Akad. Nauk SSSR, Neorg. Mater.*, **24**, (1988), 646–650.

80.   I. Grunze and H. Grunze, *Z. anorg. allg. Chem.*, **512**, (1984), 39–47.

81.   V. V. Krasnikov, Z. A. Konstant, and V. S. Fundamenskii, *Izv. Akad. Nauk SSSR, Neorg. Mater.*, **19**, (1983), 1373–1378.

82.   N. N. Chudinova, I. V. Tananaev, and M. A. Avaliani, *Izv. Akad. Nauk SSSR, Neorg. Mater.*, **13**, (1977), 2234–2235.

83.   N. N. Chudinova, M. A. Avaliani, L. S. Guzeeva, and I. V. Tananaev, *Izv. Akad. Nauk SSSR, Neorg. Mater.*, **14**, (1978), 2054–2060.

84.   N. N. Chudinova, I. Grunze, L. S. Guzeeva, and M. A. Avaliani, *Izv. Akad. Nauk SSSR, Neorg. Mater.*, **23**, (1987), 604–609.

85.   M. A. Avaliani, N. N. Chudinova, and I. V. Tananaev, *Izv. Akad. Nauk SSSR, Neorg. Mater.*, **15**, (1979), 1688–1689.

86.   N. V. Vinogradova and N. N. Chudinova, *Izv. Akad. Nauk SSSR, Neorg. Mater.*, **19**, (1983), 116–119.

87.   N. N. Chudinova, N. V. Vinogradova, G. M. Balagina, and K. K. Palkina, *Izv. Akad. Nauk SSSR, Neorg. Mater.*, **13**, (1977), 1494–1499.

88.   W. Hilmer, N. N. Chudinova, K. H. Jost, and N. V. Vinogradova, *Izv. Akad. Nauk SSSR, Neorg. Mater.*, **15**, (1979), 332–334.

89.   W. Hilmer, N. N. Chudinova, K. H. Jost, and N. V. Vinogradova, *Izv. Akad. Nauk SSSR, Neorg. Mater.*, **15**, (1979), 1123–1125.

90.   N. N. Chudinova, L. P. Shklover, L. I. Shkol'nikova, A. E. Balanevskaya, and G. M. Balagina, *Izv. Akad. Nauk SSSR, Neorg. Mater.*, **14**, (1978), 1324–1328.

91.   M. A. Vaivada and Z. A. Konstant, *Izv. Akad. Nauk SSSR, Neorg. Mater.*, **22**, (1986), 2026–2028.

92.   M. T. Averbuch-Pouchot and M. Bagieu-Beucher, *Z. anorg. allg. Chem.*, **552**, (1987), 171–180.

93.   F. Schwarz, *Z. anorg. Chem.*, **9**, (1895), 249–266.

94.   M. T. Averbuch-Pouchot and A. Durif, *C. R. Acad. Sci.*, Sér. 2, **316**, (1993), 609–614.

5.2.4. Tetraphosphates

5.2.4.1. Introduction

The existence of tetraphosphates was suspected a long time ago, but it is not before 1955 that Ostereld and Langguth [1] and Langguth *et al.* [2] showed during the elaboration of the $Pb_2P_2O_7$–$Pb(PO_3)_2$ phase-equilibrium diagram (Figure 5.2.24) the existence of $Pb_3P_4O_{13}$ which is an incongruent melting compound decomposing at 973 K. From various considerations based both on the chemical formula and on paper chromatography experiments they concluded that it was a tetraphosphate.

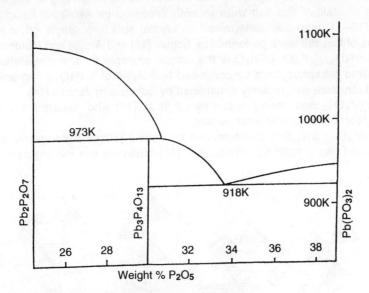

Figure 5.2.24. The $Pb_2P_2O_7$–$Pb(PO_3)_2$ phase-equilibrium diagram.

For the first structural proof of the existence of the P_4O_{13} group, it was not till 1976 that Averbuch-Pouchot and Durif [3] and Durif et al. [4] reported the chemical preparation and the crystal structure of $(NH_4)_2SiP_4O_{13}$.

Tetraphosphates are to date still relatively rare compounds. Chemical literature reports more than thirty characterizations, but only six atomic arrangements were determined. A P_4O_{13} group has also been observed in a mixed-anion phosphate $CaNb_2O(P_2O_7)(P_4O_{13})$, investigated recently by Averbuch-Pouchot [5].

5.2.4.2. The present state of tetraphosphate chemistry

For this survey of tetraphosphates we first report in details the six tetraphosphates well characterized at a structural point of view and then the other examples found in chemical literature in the form of a condensed table. One must be cautious with the true nature of these latter examples. In all cases, they are poorly crystallized derivatives which are not clearly characterized even now.

The two forms of $Ba_3P_4O_{13}$ were first reported by Mc Cauley and Hummel [6] and in [7]. Single crystals of the two forms were prepared by Millet *et al.* [8] who reported their unit-cell dimensions. A crystal structure of the low-temperature form, proposed by Gatehouse *et al.* [9], needs to be revised.

The existence of $Pb_3P_4O_{13}$ was reported at the beginning of this section. Single crystals of this salt were recently prepared by Averbuch-Pouchot and Durif [10–11] who also determined its crystal structure. Some other investigations of this salt were performed by Schulz [12] and Argyle and Hummel [13].

$[Co(NH_3)_6]_2P_4O_{13}.5H_2O$ is the unique example for the preparation of a crystalline tetraphosphate by controlled hydrolysis of a P_4O_{12} ring anion. Its crystal structure was recently determined by Schulz and Jansen [14].

$Cr_2P_4O_{13}$ was characterized by Lii *et al.* [15] who reported a complete description of the atomic arrangement.

$Bi_2P_4O_{13}$ was first characterized by Schulz [16] as a congruent melting compound (mp = 1083 K). Hilmer *et al.* [17] confirmed this melting point during

Figure 5.2.25. Projection of $(NH_4)_2SiP_4O_{13}$ along the **c** axis. The PO_4 tetrahedra and the SiO_6 octahedra are shown by polyhedral representation and open circles are the ammonium groups.

a general investigation of the condensed bismuth phosphates and later Bagieu-Beucher and Averbuch-Pouchot [18] synthesized single crystals and reported a detailed determination of its atomic arrangement.

The existence of $(NH_4)_2SiP_4O_{13}$ was reported in the introduction. The corresponding germanium salt, $(NH_4)_2GeP_4O_{13}$, was characterized by Averbuch-Pouchot [19] as an isotype. We will describe the crystal structure of the silicon derivative as an example of this class of compound.

The atomic arrangement of $(NH_4)_2SiP_4O_{13}$ – This salt is triclinic, $P\bar{1}$, with $Z = 2$ and the following unit-cell dimensions :

$$a = 15.14(1), \quad b = 7.684(5), \quad c = 4.861(5) \text{ Å}$$
$$\alpha = 97.86(1), \quad \beta = 96.74(1), \quad \gamma = 83.89(1)°$$

This atomic arrangement is a typical layer packing. Layers built by a two-dimensional network of SiO_6 octahedra and P_4O_{13} groups are located around planes $x = 1/4$ and $3/4$ and alternate with layers containing the ammonium groups and located in planes $x = 0$ and $1/2$. In Figure 5.2.25 a projection of this structure

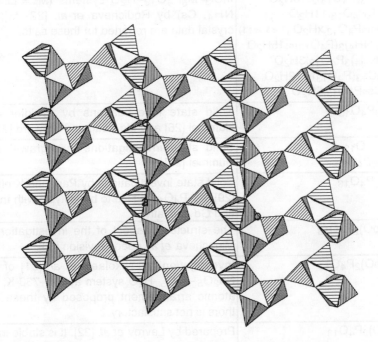

Figure 5.2.26. Projection of a SiP_4O_{13} layer along the **a** axis. The PO_4 tetrahedra and the SiO_6 octahedra are shown by polyhedral representation. The ammonium groups are omitted.

along the **c** axis shows clearly its layered arrangement. Figure 5.2.26 represents, in projection along the **a** axis, the details of the SiP_4O_{13} layer

Table 5.2.13

Summary of some investigations performed in tetraphosphate chemistry

Formula	Type of investigation or preparation
$(NH_4)_6P_4O_{13}.6H_2O$	Prepared by Griffith [20] from $Pb_3P_4O_{13}$ by an exchange reaction (see "chemical preparations"); not very stable, at 323 K is totally decomposed within 8 hours.
$(NH_4)_6P_4O_{13}.2H_2O$ $(NH_4)_4H_2P_4O_{13}$	Study by Waerstad and Mc Clellan [21] of the $NH_3–H_6P_4O_{13}–H_2O$ system at 273 K and investigation in dilute solutions at 298 K of the hydrolysis rate of $(NH_4)_6P_4O_{13}$
$In_2P_4O_{13}.10H_2O$ $In_3Li_3(P_4O_{13})_2.16H_2O$ $InLi_3P_4O_{13}.11H_2O$ $InNa_3P_4O_{13}.xH_2O$ (7 < x < 11) $In_3(NH_4)_3(P_4O_{13})_2.15H_2O$ $In(NH_4)_3P_4O_{13}.3H_2O$ $In_3Cs_3(P_4O_{13})_2.12H_2O$ $InCs_3P_4O_{13}.5H_2O$	Investigation at 273 K or 298 K (Na) of several $InCl_3–M_6^IP_4O_{13}–H_2O$ systems (M^I = Li, Na, NH_4, Cs) by Rodicheva et al. [22–25]; no crystal data are reported for these salts.
$Sr_3P_4O_{13}$	Solid state investigations by Kreidler and Hummel [26] and Mc Keag and Steward [27]
$Y_2P_4O_{13}$ $Gd_2P_4O_{13}$	Solid state investigations by Agrawal and Hummel [28]
$La_2P_4O_{13}$	Solid state investigations by Park and Kreidler [29]; $La_2P_4O_{13}$ is said to be isotypic with the Y and Gd derivatives.
$(MoO)_2P_4O_{13}$	The structural aspect of the investigation of Minacheva et al. needs revision [30].
$(NbO)_2P_4O_{13}$	Investigation by Nikolaev et al. [31] of the $Nb_2O_5–P_2O_5–H_2O$ system at 703–733 K; the atomic arrangement proposed by these authors is not satisfactory.
$(VO)_2P_4O_{13}$	Prepared by Lavrov et al. [32]. It is stable in air up to 873 K, but decomposes at 943 K: $(VO)_2P_4O_{13} \longrightarrow 2VO(PO_3)_2 + 1/2O_2$

located around the plane $x = 1/4$. Each SiO_6 octahedron shares its six oxygen atoms with four adjacent P_4O_{13} groups. This network creates large rectangular voids at the center of the unit cell. The SiO_6 octahedron is rather regular, with Si–O distances ranging from 1.762 to 1.788 Å and O–Si–O angles varying from 86.5 to 93.2°. The two independent ammonium groups who establish the cohesion between the SiP_4O_{13} planes have different coordinations. Within a range of 3.20 Å, the first has a fivefold oxygen coordination and the second a sixfold.

The other tetraphosphates known presently are reported in Table 5.2.13. with some brief comments, either on the nature of the investigations which lead to their characterization or on their properties.

5.2.4.3. Chemical preparations of tetraphosphates

Considering the rather restricted number of well characterized tetraphosphates, it appears clearly that to date no general procedure exists. The only starting materials which could possibly be used are $Pb_3P_4O_{13}$ for exchange reactions and a water soluble salt, $(NH_4)_6P_4O_{13}.6H_2O$ for classical reactions of aqueous chemistry, but this latter material is not very stable (Table 5.2.13). Griffith [20] was the first to describe a reproducible method for the preparation of convenient amounts of a water-soluble tetraphosphate, $(NH_4)_6P_4O_{13}.6H_2O$, through a very simple metathesis reaction using $Pb_3P_4O_{13}$ as the starting material:

$$Pb_3P_4O_{13} + 3(NH_4)_2S \xrightarrow{H_2O} (NH_4)_6P_4O_{13}.6H_2O + 3PbS$$

The same procedure was used by Waerstad and Mac Clellan [21] to synthesize $(NH_4)_6P_4O_{13}.2H_2O$ and $(NH_4)_4H_2P_4O_{13}$.

In spite of the possibilities offered by this procedure, all the recent developments in the chemistry of tetraphosphates came from various other methods like the elaborations of phase-equilibrium diagrams, the systematic investigations of various systems by flux methods, aqueous processes, or the hydrolytic decyclyzation of a P_4O_{12} ring. We report some of these preparations.

For a long time, the only suggested method for the preparation of water-soluble tetraphosphates was the alkaline *hydrolysis of cyclotetraphosphates* based on the following scheme:

$$H_4P_4O_{12} + H_2O \longrightarrow H_6P_4O_{13}$$

Thilo and Ratz [33] tried to prepare $Na_6P_4O_{13}$ by this process. They used the hydrolytic decyclization of sodium cyclotetraphosphate with a solution of sodium hydroxide according to the following scheme:

$$Na_4P_4O_{12} + 2NaOH \xrightarrow{H_2O} Na_6P_4O_{13}.xH_2O$$

The reaction was run at 313 K, controlled by paper-chromatographic analysis, and is completed after about 100 hours. The resulting tetraphosphate is then precipitated by acetone as an oily syrup which could not be crystallized. According to the authors, this salt decomposes rapidly:

$$Na_6P_4O_{13}.xH_2O \longrightarrow 2Na_3HP_2O_7.H_2O$$

Griffith [20] did not succeed in reproducing this salt and reported that all attempts to produce a crystalline potassium tetraphosphate by the same procedure proved to be fruitless. The only successful use of this process seems to be the preparation of $[Co(NH_3)_6]_2P_4O_{13}.5H_2O$ by Schultz and Jansen [14].

Flux processes were used for the preparation of most of the other ones:

– $(NH_4)_2SiP_4O_{13}$ by heating a mixture of 1 g of silica wool and 20 g of $(NH_4)_2HPO_4$ at 623 K for 12 hours

– $Cr_2P_4O_{13}$ by heating a mixture of MoO_3, Cr_2O_3, and P_2O_5 established with a $Cr_2O_3 / P_2O_5 = 3/5$ molar ratio at 1303 K in a sealed silica tube

– $Bi_2P_4O_{13}$ by heating a mixture of K_2CO_3 (1.5 g), Bi_2O_3 (1 g), and H_3PO_4 (5 cm^3) at 473 K for several hours and then at 513 K for two days

– $(VO)_2P_4O_{13}$ by reacting of a mixture of H_3PO_4 and V_2O_5 established with a starting ratio P/V of 6–10 at 493–573 K

For all the compounds prepared by this method, the true nature of the anion was not known before the structural investigations.

5.2.4.4. The tetraphosphate group

The main geometrical features of the six well-known P_4O_{13} groups to date are gathered in Table 5.2.14. As for triphosphates, we report only the values directly connected with the condensation phenomenon in this table: P–P distances, as well as P–O–P and P–P–P angles. With so few examples, any

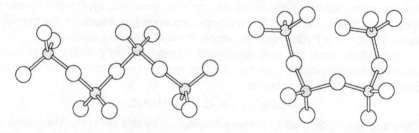

Figure 5.2.27. Representation of some P_4O_{13} groups: on the left-hand side the P_4O_{13} group observed in $Nb_2OCa(P_4O_{13})(P_2O_7)$, on the right-hand side that observed in $Cr_2P_4O_{13}$. The first one has binary symmetry, the second one has no symmetry.

discussion aimed at comparing their geometries or establishing correlations is fruitless. One can simply note that two P_4O_{13} groups have twofold internal symmetry, while the four remaining ones have no internal symmetry. With the exception of the P_4O_{13} group observed in the chromium salt which has a very accentuated horse-shoe conformation, most of these groups are zigzag chains of tetrahedra, quite comparable with a fragment of a long-chain polyphosphate. All that was mentioned before regarding the P–P distances and the P–P–P and P–O–P angles when discussing the triphosphate group remains valuable for the tetraphosphate group.

Figure 5.2.27 gives some examples of geometries observed among the P_4O_{13} groups.

Table 5.2.14

Main geometrical features in the six P_4O_{13} groups known to date.

Formula	P–P (Å)	P–P–P (°)	P–O–P (°)	Sym.	Ref.
$Pb_3P_4O_{13}$	2.895	84.6	131.3	none	[10–11]
	3.034	85.5	143.6		
	2.898		128.7		
$Cr_2P_4O_{13}$	2.979	108.6	136.8	none	[15]
	2.997	92.6	142.7		
	2.932		135.0		
$Bi_2P_4O_{13}$	2.974	95.8	140.9	2	[18]
	2.941		134.4		
$(NH_4)_2SiP_4O_{13}$	2.912	95.6	131.8	none	[3–4]
	2.931	108.6	133.7		
	2.856		125.8		
$[Co(NH_3)_6]_2P_4O_{13}.5H_2O$	2.958	121.0	134.4	none	[14]
	2.919	111.7	129.6		
	2.993		139.8		
$Nb_2OCa(P_4O_{13})(P_2O_7)$	2.848	96.3	126.3	2	[5]
	2.927		134.7		

References

1. R. K. Osterheld and R. P. Langguth, *J. Phys. Chem.*, **59**, (1955), 76–80.

2. R. O. Langguth, R. K. Osterheld, and E. Karl-Kroupa, *J. Phys. Chem.*, **60**, (1956), 1335–1336.

3.   M. T. Averbuch-Pouchot and A. Durif, *J. Solid State Chem.*, **18**, (1976), 391–393.

4.   A. Durif, M. T. Averbuch-Pouchot, and J. C. Guitel, *Acta Crystallogr.*, **B32**, (1976), 2957–2960.

5.   M. T. Averbuch-Pouchot, *Z. anorg. allg. Chem.*, **545**, (1987), 118–124.

6.   R. A. Mc Cauley and F. A. Hummel, *Trans. Br. Ceram. Soc.*, **67**, (1968), 619–625.

7.   *Powder Diffraction*, **1**, (1986), 80.

8.   J. M. Millet, H. S. Parker, and R. S. Roth, *J. Am. Ceram. Soc.*, **69C**, (1986), 103–105.

9.   B. M. Gatehouse, S. N. Platts, and R. S. Roth, *Acta Crystallogr.*, **C47**, (1991), 2285–2287.

10.   M. T. Averbuch-Pouchot and A. Durif, *C. R. Acad. Sci.*, Sér. 2, **303**, (1986), 543–545.

11.   M. T. Averbuch-Pouchot and A. Durif, *Acta Crystallogr.*, **C43**, (1987), 631–632.

12.   I. Schulz, *Z. anorg. allg. Chem.*, **257**, (1956), 106–112.

13.   J. F. Argyle and F. A. Hummel, *J. Am. Ceram. Soc.*, **43**, (1960), 542–547.

14.   F. Schulz and M. Jansen, *Z. anorg. allg. Chem.*, **543**, (1986), 152–160.

15.   K. H. Lii, Y. B. Chen, C. C. Su, and S. L. Wang, *J. Solid State Chem.*, **82**, (1989), 156–160.

16.   I. Schulz, *Z. anorg. allg. Chem.*, **257**, (1956), 106–112.

17.   N. Hilmer, N. N. Chudinova, and K. H. Jost, *Izv. Akad. Nauk SSSR, Neorg. Mater.*, **8**, (1978), 1507–1515.

18.   M. Bagieu-Beucher and M. T. Averbuch-Pouchot, *Z. Kristallogr.*, **180**, (1987), 165–170.

19.   M. T. Averbuch-Pouchot, *J. Appl. Crystallogr.*, **10**, (1977), 200.

20.   E. J. Griffith, *J. Inorg. Nucl. Chem.*, **26**, (1964), 1381–1383.

21.   K. R. Waerstad and G. H. Mc Clellan, *J. Appl. Crystallogr.*, **7**, (1974), 404–405.

22.   G. V. Rodicheva, E. N. Deichman, I. V. Tananaev, and Zh. K. Shaidarbekova, *Russ. J. Inorg. Chem.*, **22**, (1977), 1647–1650.

23.   G. V. Rodicheva, E. N. Deichman, I. V. Tananaev, and Zh. K. Shaidarbekova, *Russ. J. Inorg. Chem.*, **20**, (1975), 1316–1318.

24.   G. V. Rodicheva, E. N. Deichman, I. V. Tananaev, and Zh. K. Shaidarbe-

kova, *Russ.J. Inorg. Chem.*, **19**, (1974), 814–817.

25.   G. V. Rodicheva, E. N. Deichman, I. V. Tananaev, and Zh. K. Shaidarbekova, *Russ. J. Inorg. Chem.*, **19**, (1974), 1467–1470.

26.   E. R. Kreidler and F. A. Hummel, *Inorg. Chem.*, **6**, (1967), 884–891.

27.   A. H. McKeag and E. G. Steward, *British J. Appl. Phys.*, **4S**, (1955), 26.

28.   D. Agrawal and F. A. Hummel, *J. Electrochem. Soc.*, (1980), 1550–1554.

29.   H. D. Park and E. R. Kreidler, *J. Am. Ceram. Soc.*, **67**, (1984), 23–26.

30.   L. Kh. Minacheva, A. S. Antsyshkina, A. V. Lavrov, V. G. Sakhrova, V. P. Nikolaev, and M. A. Porai-Koshits, *Russ. J. Inorg. Chem.*, **24**, (1979), 51–53.

31.   V. P. Nikolaev, G. G. Sadikov, A. V. Lavrov, and M. A. Porai-Koshits, *Izv. Akad. Nauk SSSR, Neorg. Mater.*, **22**, (1986), 1364–1368.

32.   A. V. Lavrov, M. Ya. Voitenko, and L. A. Tezikova, *Izv. Akad. Nauk SSSR, Neorg. Mater.*, **14**, (1978), 2073–2077.

33.   E. Thilo and R. Ratz, *Z. anorg. allg. Chem.*, **260**, (1949), 255–266.

5.2.5. Pentaphosphates

The pentaphosphate anion, $[P_5O_{16}]^{7-}$, is the longest oligophosphate anion clearly characterized in this family till now. The corresponding phosphates are still very rare. To date, all the investigated pentaphosphates were characterized either during the elaboration of phase-equilibrium diagrams or during the investigations of various systems by flux methods. In all these cases, the true nature of the anion was recognized during the structural studies.

The geometry of a P_5O_{16} group was described for the first time by Smolin *et al.* [1] when they performed the structural determination of $Mg_2Na_3P_5O_{16}$, a compound discovered by Majling and Hanic [2] during the elaboration of a portion of the $MgO-Na_2O-P_2O_5$ phase-equilibrium diagram.

$V_2CsP_5O_{16}$ and $Fe_2CsP_5O_{16}$ are two isotypic compounds which were recently identified by Klinkert and Jansen [3] and the last example is provided by a complex compound, $Ta_2Rb_2H(PO_4)_2(P_5O_{16})$, containing both PO_4 and P_5O_{16} groups which was studied by Sadikov *et al.* [4].

To illustrate this class of oligophosphates we describe below the atomic arrangement of $Mg_2Na_3P_5O_{16}$.

The atomic arrangement of $Mg_2Na_3P_5O_{16}$ – This pentaphosphate is monoclinic, *P2/a*, with $Z = 2$ and the following unit-cell dimensions:

$$a = 18.617(5), \quad b = 6.844(3), \quad c = 5.174\ (3)\ \text{Å}, \quad \beta = 90.15(3)°$$

Figure 5.2.28 shows clearly the layer arrangement of this structure in which layers of P_5O_{16} groups alternate with layers of the associated cations. The central phosphorus atom of the phosphoric anion is located on a binary axis and this group adopts a twofold internal symmetry. The magnesium atom, located on a general crystallographic position, has a slightly distorted octahedral oxygen coordination. This MgO_6 octahedron shares its six oxygen atoms with five adjacent P_5O_{16} groups. One of the two crystallographically independent sodium atom, sited on the twofold axis, has sixfold oxygen coordination, the second one, in general position, has only five neighbors. Another way to describe this arrangement is to consider it as built by a three-dimensional network of P_5O_{16} groups and MgO_6 octahedra. This network is represented in Figure 5.2.29, a projection along the **c** axis, creates two kinds of channels parallel to the **c** axis in which are located the sodium atoms. This second representation also shows clearly the horse-shoe configuration of the phosphoric group and its twofold symmetry.

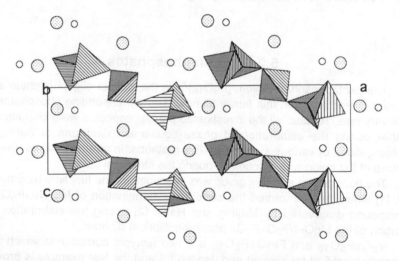

Figure 5.2.28. Projection of $Mg_2Na_3P_5O_{16}$ along the **b** axis. The P_5O_{16} groups are shown by polyhedral representation. The larger dotted circles are the Na atoms and the smaller open ones the Mg atoms.

General comments on pentaphosphates – It is evident that no general properties can be deduced from so few representatives. For the same reason, no general process of preparation exists. Thus, for this very limited class of phos-

Figure 5.2.29. Projection, along the **c** direction, of $Mg_2Na_3P_5O_{16}$. Both phosphoric anions and MgO_6 octahedra are shown by polyhedral representation. The dotted circles are the Na atoms.

phates, it is important to report in details the chemical preparations known presently for the four pentaphosphates.

Crystals of $Mg_2Na_3P_5O_{16}$, up to 2–3 mm long, were prepared by the slow cooling of a melt composed of 35 wt% $Mg_2P_2O_7$ and 65 wt% $NaPO_3$.

Crystals of $V_2CsP_5O_{16}$ and $Fe_2CsP_5O_{16}$ were produced by the slow cooling (10 K/h) of melts of Cs_2CO_3, H_3PO_4, and V_2O_3 (Fe_2O_3) heated up to 823–873 K for the vanadium salt and 873–923 K in the case of the iron salt. The vanadium compound can also be obtained by firing pure $VCsHP_3O_{10}$ at 823 K, according to the following scheme:

$$2VCsHP_3O_{10} \longrightarrow V_2CsP_5O_{16} + CsPO_3 + H_2O$$

$Ta_2Rb_2H(PO_4)_2(P_5O_{16})$ was identified during the investigation of the Ta_2O_5–Rb_2O–P_2O_5–H_2O system by flux methods in a temperature range of 573–653 K. The initial mixtures of the components were established with the following molar ratios: $P_2O_5/Rb_2O = 3–4$ and $P_2O_5/Ta_2O_5 = 16$.

In Figure 5.2.30, we report the representations of the three P_5O_{16} groups known presently. One can simply note that they all have a more or less accentuated horse-shoe conformation. One of them has no internal symmetry, the other has a twofold symmetry and the last a mirror symmetry. Table 5.2.15 gives the main geometrical features of these three P_5O_{16} groups.

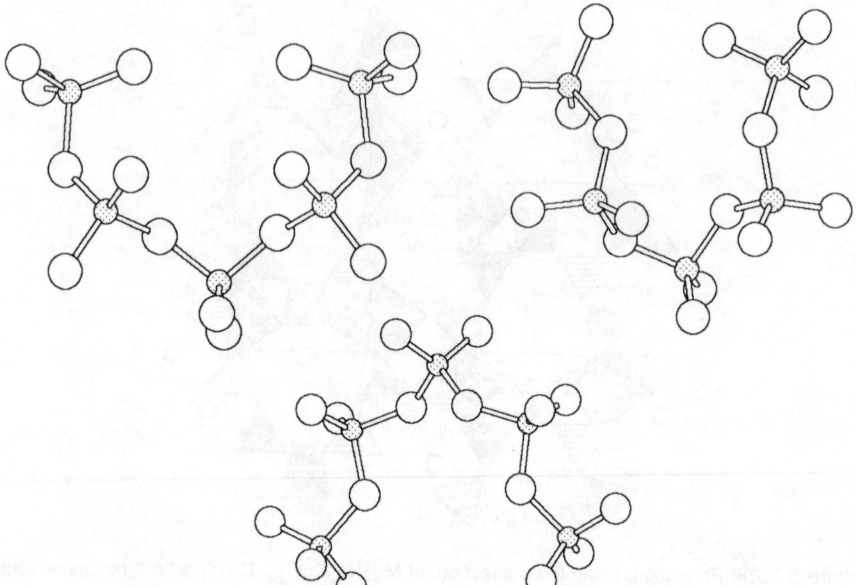

Figure 5.2.30. Polyhedral representations of the P_5O_{16} groups.
Upper (left): the pentaphosphate anion as observed in $Ta_2Rb_2H(PO_4)_2(P_5O_{16})$ (mirror symmetry)
Upper (right): the pentaphosphate anion as observed in $V_2CsP_5O_{16}$ (no symmetry)
Below: the pentaphosphate anion as observed in $Mg_2Na_3P_5O_{16}$ (twofold symmetry)

Table 5.2.15
Main geometrical features of the three P_5O_{16} groups.

Formula	P–P(Å)	P–P–P(°)	P–O–P(°)	Sym.	Ref.
$Mg_2Na_3P_5O_{16}$	2.949	120.0	136.7	2	[1–2]
	2.898	133.9	127.3		
$Ta_2Rb_2H(PO_4)_2(P_5O_{16})$	2.864	106.2	128.7	m	[4]
	2.879	122.2	129.7		
$V_2CsP_5O_{16}$	2.875	115.4	128.0	none	[3]
	2.908	87.3	129.6		
	2.817	113.1	122.4		
	2.857		126.3		

The case of trömelite – We cannot conclude this section devoted to oligophosphates without some mention about the unsolved trömelite problem. Trömelite is a calcium phosphate characterized by Hill *et al.* [5] during their investigation of the CaO–P$_2$O$_5$ system. It was originally identified as Ca$_7$(P$_5$O$_{16}$)$_2$, a formula corresponding to a possible pentaphosphate. This salt was further investigated by Van Wazer and Ohashi [6–7] who confirmed the existence of a pentaphosphate anion by paper-chromatography experiments. Later, Wieker *et al.* [8] produced crystalline trömelite and through a careful paper-chromatography analysis assigned the formula of an hexaphosphate, Ca$_4$P$_6$O$_{19}$, to this compound. These authors reported the following triclinic unit-cell dimensions for trömelite:

$$a = 9.40, \quad b = 13.39, \quad c = 7.07 \text{ Å}$$
$$\alpha = 109.5, \quad \beta = 87.9, \quad \gamma = 108.9°$$

Up to now, no atomic arrangement was determined for this compound and the nature of its anion remains uncertain today. In addition, the same authors claimed to have prepared [Co(NH$_3$)$_6$]$_2$(P$_6$O$_{19}$)$_3$.20H$_2$O starting from aqueous solutions of trömelite.

5.2.6. Substituted oligophosphates

Substituted oligophosphate anions are very rare. One can only report a few investigations dealing with some fluor-substituted triphosphates.

Feldman [9] obtained a series of *fluorotriphosphates* by the interaction of F$^-$ ions with cyclotriphosphate ions in aqueous solutions. An equilibrium reaction occurs and the P$_3$O$_9$ ring anion is opened to build a linear fluorotriphosphate anion, [P$_3$O$_9$F]$^{4-}$:

$$[P_3O_9]^{3-} + F^- \longrightarrow [P_3O_9F]^{4-}$$

This anion was isolated in the form of several crystalline salts:

(NH$_4$)$_4$P$_3$O$_9$F.H$_2$O, Ag$_4$P$_3$O$_9$F.H$_2$O, Ag$_3$KP$_3$O$_9$F, Ba$_2$P$_3$O$_9$F.5/2H$_2$O.

If the concentration of fluorine ions is high the [P$_3$O$_9$F]$^{4-}$ anion is degraded to [P$_2$O$_7$]$^{4-}$ and [PO$_2$F$_2$]$^-$. Feldman did the same observations in cyclotetraphosphate solutions with the formation of [P$_4$O$_{12}$F]$^{4-}$ anions, but did not produce any corresponding salts in a good crystalline state. Unfortunately, no structural investigations were reported for such derivatives.

References

1.   Yu. I. Smolin, Yu. F. Shepelev, A. I. Domanskii, and J. Majling, *Sov. Phys. Crystallogr.*, **23**, (1978), 715–717.

2.   J. Majling and F. Hanic, *J. Appl. Crystallogr.*, **12**, (1979), 244.

3.   B. Klinkert and M. Jansen, *Z. anorg. allg. Chem.*, **567**, (1988), 87–94.

4.   G. G. Sadikov, V. P. Nikolaev, A. V. Lavrov, and M. A. Porai-Koshits, *Dokl. Akad. Nauk SSSR*, **266**, (1982), 354–358.

5.   W. L. Hill, G. T. Faust, and D. S. Reynolds, *Am. J. Sci.*, **242**, (1944), 457–477.

6.   J. R. Van Wazer and S. Ohashi, *J. Am. Chem. Soc.*, **80**, (1958), 1010–1013

7.   S. Ohashi and J. R. Van Wazer, *J. Am. Chem. Soc.*, **81**, (1959), 830–832.

8.   W. Wieker, A. R. Grimmer, and E. Thilo, *Z. anorg. allg. Chem.*, **330**, (1964), 78–90.

9.   W. Feldman, *Z. anorg. allg. Chem.*, **338**, (1965), 235–244.

5.3. LONG-CHAIN POLYPHOSPHATES

As early as 1833, Graham prepared sodium polyphosphate by firing the monophosphate, NaH_2PO_4:

$$NaH_2PO_4 \longrightarrow NaPO_3 + H_2O$$

This same process was soon used to prepare KPO_3. These two compounds induced abundant but often confusing chemical literature. A good number of more or less water soluble varieties of $NaPO_3$ or KPO_3 were described. In most cases, distinction between these various forms was based on their solubility, behavior towards various chemical reagents, state of cristallinity, fibrous or not fibrous morphology. Various names were used to denote these forms which were known for a long time as Graham, Maddrell or Kurrol salts. During these early days, the polymeric nature of the anion was evidently not suspected and it is only recently that most of these forms were recognized as mixtures of various sodium or potassium polyphosphates with different chain lengths, the thermal history of the sample being the main parameter controlling the chain-length distribution. The development of modern techniques of investigation considerably reduced the number of these more or less hypothetical species. For instance, the existence of only two crystalline forms of sodium long-chain polyphosphate is clearly established today. Thus, in the following survey we shall not report an exhaustive examination of the tremendous chemical literature dealing with some of these polyphosphates, but examine what is only now firmly established for crystalline materials.

5.3.1. General Properties of Long-Chain Polyphosphates

The highly polymerized type of anion, found in what we denote as long-chain polyphosphates, has a formula approximating: $[PO_3]_n^{n-}$. There are some properties that are common to all long-chain polyphosphates. We have reported them briefly above.

5.3.1.1. Degree of polymerization

Long-chain polyphosphates were the subject of many physical investigations that we cannot report here. These studies which are mainly devoted to the estimation of their degree of polymerization can be summarize as follows:

a) the degree of polymerization is very high. The molecular weights estimated by various methods, sedimentation, dialysis, end-group titration, viscosity, pH titration, can vary from 500,000 to several million.

b) for a given compound the values obtained are mainly dependent on the thermal history of the specimen and of its stoichiometry. A very small departure from the theoretical O/P ratio, 3, can induce large variations in the degree of polymerization.

5.3.1.2. Stability and solubility

Polyphosphates are relatively stable compounds. Under normal conditions of temperature and hygrometry, they can be kept in a perfect state of crystallinity for many years.

As can be inferred from their estimated molecular weights, they are not water soluble.

5.3.1.3. Thermal behavior

All *monovalent and divalent-cation polyphosphates* are stable up to their melting points. Majling and Hanic [1] examined a number of phase-equilibrium diagrams involving them as components and noticed some dissipations in the melting temperatures reported in these various studies. In Table 5.3.1, we list some of these melting-temperature ranges. According to these authors, the dissipation of values must be attributed more to the sample purity than to the methods of temperature measurements.

The thermal behavior of *trivalent cation polyphosphates* is very different. In most cases, the $M^{III}(PO_3)_3$ polyphosphates transform into the corresponding monophosphates $M^{III}PO_4$ on prolongated heating:

$$M^{III}(PO_3)_3 \longrightarrow M^{III}PO_4 + P_2O_5 \quad [I]$$

For M^{III} = Al, Cr, Ga, V, Fe, Rh, Ti, Mo, Sc, and In, the reaction [I] occurs above 1273 K, but, as we have often observed in similar reactions, intermediate states

Table 5.3.1

Melting-temperature ranges observed for some monovalent and divalent-cation polyphosphates.

Formula	mp range (K)	Formula	mp range (K)
$LiPO_3$	911–939	$NaPO_3$	893–903
$AgPO_3$	754–763	KPO_3	1076–1086
$RbPO_3$	1053–1090	$CsPO_3$	993–1013
$TlPO_3$	707–736		
$Ca(PO_3)_2$	1237–1251	$Zn(PO_3)_2$	1126–1153
$Sr(PO_3)_2$	1255–1277	$Cd(PO_3)_2$	1150–1169
$Ba(PO_3)_2$	1123–1153	$Pb(PO_3)_2$	937–953

can occur during this process. Thus, for aluminum and scandium, the formation of the cyclotetraphosphates occurs before the final decomposition:

$$Al(PO_3)_3 \xrightarrow{\text{1283 K (2 hr)}} Al_4(P_4O_{12})_3$$

$$Sc(PO_3)_3 \xrightarrow{\text{1318 K (DTA)}} Sc_4(P_4O_{12})_3$$

The formation of diphosphates, $M_4^{III}(P_2O_7)_3$, is observed for M^{III} = Cr and In, whereas $Tl(PO_3)_3$ is stable up to its melting point (1063 K).

For yttrium, lanthanide, and bismuth polyphosphates, the scheme of decomposition given by [I] is also the rule and begins at various temperatures. For instance, Bukhalova *et al.* [2] found a temperature of 1203 K for the beginning of this decomposition in the case of $Ce(PO_3)_3$. The rate of P_2O_5 vaporization losses for $La(PO_3)_3$, leading to the monophosphate as in reaction [I], was measured by Park and Kreidler [3] at 1173 and 1473 K.

For *mixed-cation polyphosphates*, the path of the evolution is mainly dependent on the nature of the associated cations. For instance, in the case of $Ce(NH_4)_2(PO_3)_5$, the first step of the evolution corresponds to the formation of an ultraphosphate:

$$Ce(NH_4)_2(PO_3)_5 \xrightarrow{\text{823–943 K}} CeP_5O_{14} + 2NH_3 + H_2O$$

at higher temperatures, the final step leads to the monophosphate:

$$CeP_5O_{14} \longrightarrow CePO_4 + 2P_2O_5$$

For *acidic polyphosphates*, the behavior is even more complex. In the case of $CaH_2(PO_3)_4$ [4], no decomposition occurs up to 703 K, when heated in a dry atmosphere. Above this temperature, a two-step process occurs. The first step leads to the ultraphosphate:

$$CaH_2(PO_3)_4 \longrightarrow CaP_4O_{11} + H_2O$$

and the second to the normal polyphosphate after a departure of P_2O_5:

$$CaP_4O_{11} \longrightarrow Ca(PO_3)_2 + P_2O_5$$

For $BiH(PO_3)_4$ which was investigated by Hilmer *et al.* [5], a first decomposition occurs at 743 K:

$$BiH(PO_3)_4 \longrightarrow Bi(PO_3)_3 + HPO_3$$

This first step is reversible, but, on prolonged heating at 773–973 K, bismuth polyphosphate reacts with HPO_3 to form the ultraphosphate BiP_5O_{14}:

$$Bi(PO_3)_3 + 2HPO_3 \longrightarrow BiP_5O_{14} + H_2O$$

For $EuH(PO_3)_4$, Chudinova *et al.* [6] observed a decomposition into $Eu(PO_3)_3$ and EuP_5O_{14} at 843 K in a first step and a further evolution to the polyphosphate at 1243 K:

$$2EuH(PO_3)_4 \xrightarrow{843\,K} Eu(PO_3)_3 + EuP_5O_{14} + H_2O \xrightarrow{1243\,K} 2\,Eu(PO_3)_3 + P_2O_5$$

All long-chain polyphosphates produce glasses when heated up to their melting points and quenched.

A valuable survey of these various studies is given by Bagieu [7] who investigated also the effect of pressure on some of these compounds.

5.3.2. The Present State Chemistry of Long-Chain Polyphosphates

Many long-chain polyphosphates are well characterized today. Thus, it is impossible to present an exhaustive survey of these compounds in this section, especially since an almost encyclopedic review of this field was recently done by one of us [8]. Instead purpose is to illustrate this chapter of condensed phosphate chemistry with some selected examples in order to provide the reader with a general idea of this domain.

5.3.2.1. Monovalent-cation long-chain polyphosphates

Polyphosphates have been characterized for all monovalent cations. Most of them are polymorphic, but, in a number of cases, some polymorphs are not yet accurately characterized.

For $LiPO_3$, proved to be dimorphous with a transition temperature at 523 K by Thilo and Grunze [9] and Benkhoucha and Wunderlich [10], only the high-temperature form was investigated by Grenier and Durif [11] and Guitel and Tordjman [12].

Two crystalline forms of $NaPO_3$ are clearly characterized. They were the subject of many crystallographic studies by Corbridge [13], Jost [14–15], and McAdam et al. [16]. Recently, Immirzi and Porzio [17] claimed the existence of a third form, but this investigation which is based on powder data needs to be confirmed by more reliable data.

$AgPO_3$ was recognized by Jost [18] as an isotype of one form of $NaPO_3$.

KPO_3 was also the subject of many investigations. More than thirty years ago, Andress and Fischer [19] and also Corbridge [13] performed the first crystallographic investigations and Jost [20] determined the atomic arrangement of the room-temperature form. This polyphosphate undergoes several phase transitions that we shall examine at the end of this section.

Many attempts were made to produce NH_4PO_3. In many cases, the resulting products proved to be impure. In a recent and careful investigation, Shen et al. [21] obtained five crystalline forms with chain length greater than 50. One form seems to be closely related to KPO_3. The first structural investigation of one of these forms was recently performed by Brüne and Jansen [22].

$RbPO_3$, $CsPO_3$, and $TlPO_3$ are isotypic. The atomic arrangement was determined with the rubidium salt by Corbridge [13, 23] and later re-examined by

Table 5.3.2

Mixed monovalent-cation long-chain polyphosphates characterized during the elaboration of various MPO_3–$M'PO_3$ phase-equilibrium diagrams or by solid state reactions.

System	Compound	mp (K)	Ref.
$LiPO_3$–KPO_3	α-$LiK(PO_3)_2$ β-$LiK(PO_3)_2$	827	[27, 29–31]
$LiPO_3$–$RbPO_3$	$LiRb(PO_3)_2$ $Li_2Rb(PO_3)_3$	893	[28, 32]
$LiPO_3$–$CsPO_3$	$LiCs(PO_3)_2$ $Li_2Cs(PO_3)_3$	963	[27–28, 32–33]
$LiPO_3$–$TlPO_3$	$LiTl(PO_3)_2$ $Li_2Tl(PO_3)_3$	723 (d) 645 (d)	[28, 32]
$AgPO_3$–KPO_3	$AgK(PO_3)_2$	703 (d)	[34–35]
$AgPO_3$–$RbPO_3$	$AgRb(PO_3)_2$	708 (d)	[34, 36]

d: incongruent melting

Cruickshank [24]. $RbPO_3$ and $CsPO_3$ undergo several phase transitions which will be discussed at the end of this section. $TlPO_3$ is obtained by heating the corresponding cyclotetraphosphate, $Tl_4P_4O_{12}$, above 690 K which is the stable form at low temperature [25–26]. This reaction is not reversible.

Many MPO_3–$M'PO_3$ systems were investigated. Some of them, like $LiPO_3$–$NaPO_3$ [27] or $LiPO_3$–$AgPO_3$ [28], proved to be of simple solid-solution type or of the eutectic type, whereas most of the others revealed the existence of poly-phosphates as intermediate compounds. We summarize these studies in Table 5.3.2. It is worth noting that, in most of the presently investigated $NaPO_3$–M^IPO_3 phase-equilibrium diagrams, the intermediate compounds are cyclophospha-tes, mainly cyclotriphosphates. They will be reported in the corresponding section.

Some other mixed monovalent-cation polyphosphates were prepared by flux methods. $AgCs(PO_3)_2$ and $AgTl(PO_3)_2$ were recognized by Averbuch-Pouchot [34] as isotypes of the corresponding silver–potassium salt and the preparation as well as crystal structure of $Li_2NH_4(PO_3)_3$ reported by Averbuch-Pouchot *et al.* [37].

5.3.2.2. Divalent-cation long-chain polyphosphates

Long-chain polyphosphates are known for all divalent cations. Like the monovalent-cation long-chain polyphosphates, most of them are polymorphic.

Jaulmes [38–39] identified two crystalline forms of $Be(PO_3)_2$. Later, Bagieu-Beucher and Durif [40] and Averbuch-Pouchot *et al.* [41] determined the atomic arrangement of one of them. A third form of $Be(PO_3)_2$ was characterized by Schultz [42] who observed that this form has two modifications itself with a tran-sition temperature at 369 K. Crystal structure of the high-temperature modifica-tion performed by Schultz and Liebau [43–44] is closely related to the keatite form of silica.

$M^{II}(PO_3)_2$ polyphosphates, for M^{II} = Ni, Co, Mg, Cu, Fe, Mn, and Zn, are all isotypic. Bagieu-Beucher *et al.* [45] obtained these salts during high-pressure experiments on the corresponding $M_2^{II}P_4O_{12}$ cyclotetraphosphates. Crystal structure was later performed by Averbuch-Pouchot *et al.* [46] using the zinc salt. Another form of $Zn(PO_3)_2$, characterized by Katnack and Hummel [47], was later reproduced by Schultz [42] by heating a glass of $Zn(PO_3)_2$ at 1073 K for several days. During this investigation, the author observed a strong endo-thermic peak at 1048 K by DTA analysis which is attributed to the irreversible transformation:

$$Zn_2P_4O_{12} \longrightarrow 2Zn(PO_3)_2$$

The CaO–P_2O_5 diagram, established by Hill *et al.* [48–49], shows the exis-tence of two crystalline forms for $Ca(PO_3)_2$. Various investigations by Thilo and

Grunze [50], Morin [51], and Ohashi and Van Wazer [52] report the existence of several crystalline modifications for $Ca(PO_3)_2$, but up to now the only valuable structural data reported by Lehr et al. [53], Schneider et al. [54], Rothammel et al. [55], concern the β-form which is stable at room temperature. β-$Ca(PO_3)_2$ is isotypic with the corresponding strontium and lead salts. The hydrolytic degradation of glassy $Ca(PO_3)_2$ was investigated by Brown et al. [56] and by Huffman and Fleming [57]. These types of investigations were run mainly in order to explore the possible use of various condensed calcium phosphates as fertilizers.

Kreidler and Hummel [58] characterized two crystalline forms for $Sr(PO_3)_2$ and Ropp et al. [59] and Ropp and Aia [60] identified three. Thilo and Grunze [50] and Durif et al. [61] recognized the β-form as isotypic with the corresponding calcium and lead polyphosphates.

Preparation of $Ba(PO_3)_2$ polyphosphate by thermal dehydration condensation of $Ba(H_2PO_4)_2$ was analyzed by Thilo and Grunze [50] and by Kuz'menkov, et al. [62]. As shown by Grenier and Martin [63], this polyphosphate is trimorphic. The atomic arrangements of the orthorhombic β-form was performed by Grenier et al. [64] and that of the γ-form by Coing-Boyat et al. [65].

Brown and Hummel [66] and Thilo and Grunze [67] were the first to report the existence of two crystalline forms for $Cd(PO_3)_2$. The low-temperature form was investigated by Beucher and Tordjman [68], Tordjman et al. [69], and Bagieu-Beucher et al. [70]. Single crystals of the high-temperature form were obtained by Laügt et al. [71] and its crystal structure determined by Bagieu-Beucher et al. [72].

$Pb(PO_3)_2$ was first identified as a long-chain polyphosphate by Andress and Fischer [73] and later its atomic arrangement was determined by Jost [74]. The corresponding calcium and strontium polyphosphates are isotypic.

$Hg(PO_3)_2$ was recognized as a long-chain polyphosphate by Thilo and Grunze [67] who suggested a possible isomorphy with the low-temperature form of $Cd(PO_3)_2$. Beucher [75] and Beucher and Tordjman [68] confirmed this assumption.

Palkina et al. [76] determined the crystal structure of $Pd(PO_3)_2$ and soon after Watanabe et al. [77] show this polyphosphate to be dimorphous. The HT-form is monoclinic and the transformation temperature seems to be very dependent on the heating rate (1181 K for 10 K/min, 933–1008 K for 20 K/min).

Averbuch-Pouchot [4] showed the existence of an acidic calcium long-chain polyphosphate, $CaH_2(PO_3)_4$, very recently and determined its atomic arrangement. This rare example of acidic polyphosphate will be discussed in more details in Chapter 6.

Several systems involving two different divalent-cation phosphates were investigated. Some of them, like $Cu_2P_4O_{12}$–$Pb(PO_3)_2$, $Cu_2P_4O_{12}$–$Sr(PO_3)_2$

elaborated by Laügt [78] and $Sr(PO_3)_2$–$Ba(PO_3)_2$ investigated by Tokman and Bukhalova [79], are of the *eutectic type*. Some others like $Zn(PO_3)_2$–$Mg(PO_3)_2$ investigated by Sarver and Hummel [80], $Cd(PO_3)_2$–$Ca(PO_3)_2$ by Tokman and Bukhalova [81], and $Ca(PO_3)_2$–$Sr(PO_3)_2$ by Bukhalova and Tokman [82], show the existence of a continuous series of *solid solutions*. Nevertheless, a number of them show the existence of *intermediate salts*. $Ba_2Cu(PO_3)_6$, was characterized by Laügt [78] in the $Cu_2P_4O_{12}$–$Ba(PO_3)_2$ system and later investigated by Laügt and Guitel [83]. $Ba_2Zn(PO_3)_6$, isotypic with $Ba_2Cu(PO_3)_6$, was characterized by Bagieu-Beucher and El-Horr [84] during the revision of the $Zn(PO_3)_2$–$Ba(PO_3)_2$ phase-equilibrium diagram first elaborated by Mardirosova *et al.* [85]. $BaCd(PO_3)_4$ and $BaCa(PO_3)_4$ were identified by Bukhalova *et al.* [86] in the $Ba(PO_3)_2$–$Cd(PO_3)_2$ and $Ba(PO_3)_2$–$Ca(PO_3)_2$ phase-equilibrium diagrams. Their existence was later confirmed by Averbuch-Pouchot [87] who prepared two isotypic compounds, $BaMn(PO_3)_4$ and $BaHg(PO_3)_4$ in addition. The atomic arrangement in this class of phosphates was determined by Averbuch-Pouchot *et al.* [88] using the cadmium salt. We summarize some of these investigations and their results in Table 5.3.3.

A description of the atomic arrangement in the zinc polyphosphate will illustrate this section.

Table 5.3.3
Summary of the investigations of $M(PO_3)_2$–$M'(PO_3)_2$ phase-equilibrium diagrams.

System	Ref.	Compound	mp (K)	Ref.
$Ba(PO_3)_2$–$Cu_2P_4O_{12}$	[78]	$Ba_2Cu(PO_3)_6$	1100	[83]
$Ba(PO_3)_2$–$Cd(PO_3)_2$	[86]	$BaCd(PO_3)_4$	1089	[87–88]
$Ba(PO_3)_2$–$Ca(PO_3)_2$	[86]	$BaCa(PO_3)_4$	1153	[87–88]

The atomic arrangement in zinc polyphosphate – $Zn(PO_3)_2$ is monoclinic [46], $C2/c$, with $Z = 4$, and has the following unit-cell dimensions:

$$a = 9.734(2), \quad b = 8.889(2), \quad c = 4.963(1) \text{ Å}, \quad \beta = 108.49(5)°$$

The two basic components of this arrangement are two infinite chains both developping parallel to the **c** direction. The first one is a phosphoric chain whith a repetition period of two PO_4 tetrahedra, whereas the second is built by an assembly of edge-sharing ZnO_6 octahedra. These two components are in Figures 5.3.1 and 5.3.2 separately and respectively presented.

In contrast with what is commonly observed in most of the long-chain polyphosphates, the phosphoric chain is not significantly corrugated. In spite of its apparent regularity, the chain of ZnO_6 octahedra is built by the repetition of a rather distorted octahedron since the O–Zn–O angles range in rather wide limits:

$$80.3 < O–Zn–O < 103.0° \text{ and } 157.4 < O–Zn–O < 167.5°$$

In Figure 5.3.3 a projection along the **c** direction shows the stacking of these

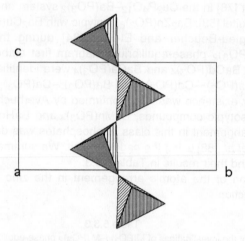

Figure 5.3.1. General aspect of an isolated infinite $(PO_3)_n$ chain in $Zn(PO_3)_2$.

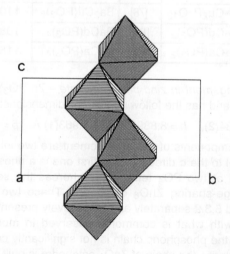

Figure 5.3.2. General aspect of an isolated infinite chain of ZnO_6 octahedra in $Zn(PO_3)_2$.

two components. In spite of the presence of the apparently large rectangular empty channels parallel to the **c** axis, this arrangement is one of the most compact ever to be observed in condensed phosphate chemistry since the volume occupied by one oxygen atom is only 16.97 Å^3.

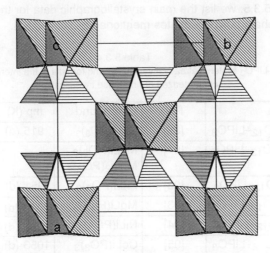

Figure 5.3.3. Projection of the atomic arrangement in Zn(PO$_3$)$_2$, along the **c** direction, showing the packing of the two types of chains.

5.3.2.3. Divalent-monovalent cation long-chain polyphosphates

More than one hundred divalent-monovalent cation polyphosphates are well characterized presently so that this domain is probably the best known among condensed phosphate chemistry today. This abundance of results can be easily explained by the fact that, during the 1970's, numerous M^{II}(PO$_3$)$_2$–M^{I}PO$_3$ phase-equilibrium diagrams were elaborated by various research groups both in France and in the former Soviet Union.

The stoichiometries of the intermediate salts characterized in these systems are various. The most common formulas are M^{II}M^I(PO$_3$)$_3$ and M^{II}M$_2^I$(PO$_3$)$_4$, but M$_2^{II}$M^I(PO$_3$)$_5$ and M$_2^{II}$M$_3^I$(PO$_3$)$_7$ polyphosphates were also observed.

Since it is impossible to give a comprehensive examination of this class of compounds in such a book we restrict our survey to a discussion of the results obtained in the M^{II}(PO$_3$)$_2$–LiPO$_3$ sytems in which the four types of stoichiometries were observed.

Divalent cation-lithium polyphosphates – We report the M^{II}(PO$_3$)$_2$–LiPO$_3$ sytems elaborated presently and the formulæ of the intermediate compounds characterized during these investigations in Table 5.3.4, whereas Figures 5.3.4 and 5.3.5 present two of these phase-equilibrium diagrams.

In addition to the compounds listed in Table 5.3.4, one must report the existence of $FeLi(PO_3)_3$ prepared by hydrothermal synthesis [107] and investigated by Genkina *et al.* [108].

In Table 5.3.5, we list the main crystallographic data for the divalent cation-lithium long-chain polyphosphates mentioned in Table 5.3.4.

Table 5.3.4

The $M^{II}(PO_3)_2$–$LiPO_3$ systems already investigated and formula of the intermediate compounds characterized during these investigations.

System	Ref.	Compound	mp (K)	Ref.
$Cu_2P_4O_{12}$–$LiPO_3$	[89]	$CuLi(PO_3)_3$	915 (d)	[90]
$Mn_2P_4O_{12}$–$LiPO_3$	[91]	$MnLi(PO_3)_3$ $MnLi_2(PO_3)_4$		[91–92]
$Zn_2P_4O_{12}$–$LiPO_3$	[93]	$ZnLi(PO_3)_3$	951 (d)	[93]
$Mg_2P_4O_{12}$–$LiPO_3$	[93]	$MgLi(PO_3)_3$	1040 (d)	[93]
$Ni_2P_4O_{12}$–$LiPO_3$	[94]	$NiLi(PO_3)_3$	1087 (d)	[94]
$Co_2P_4O_{12}$–$LiPO_3$	[95]	$CoLi(PO_3)_3$	1053 (d)	[95]
$Ca(PO_3)_2$–$LiPO_3$	[96]	$CaLi(PO_3)_3$	1040 (d)	[96]
$Sr(PO_3)_2$–$LiPO_3$	[97]	$SrLi(PO_3)_3$	1025 (d)	[97]
$Pb(PO_3)_2$–$LiPO_3$	[98]	$PbLi(PO_3)_3$ (a) $Pb_2Li(PO_3)_5$	893	[99–100]
$Cd(PO_3)_2$–$LiPO_3$	[101]	$CdLi_2(PO_3)_4$	983	[102]
$Hg(PO_3)_2$–$LiPO_3$	[103]	$HgLi_2(PO_3)_4$	828	[102]
$Ba(PO_3)_2$–$LiPO_3$	[104–105]	$Ba_2Li_3(PO_3)_7$	?	[106]

a: $PbLi(PO_3)_3$ does not appear in the phase-equilibrium diagram.
d: incongruent melting

One can mention that one of this lithium derivatives, $CuLi(PO_3)_3$, has a low temperature form, a cyclohexaphosphate $Cu_2Li_2P_6O_{18}$. The transformation:

$$Cu_2Li_2P_6O_{18} \longrightarrow 2CuLi(PO_3)_3$$

is not reversible and ranges over a wide interval of temperature because of the important rearrangement of the structure of the anion.

We shall illustrate this section with the description of the structure of $PbLi(PO_3)_3$ whose main crystal data are reported in Table 5.3.5.

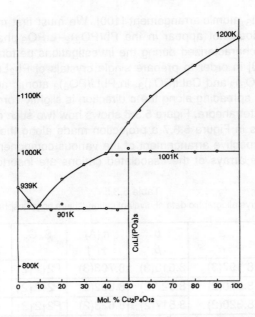

Figure 5.3.4. The $Cu_2P_4O_{12}$–$LiPO_3$ phase-equilibrium diagram.

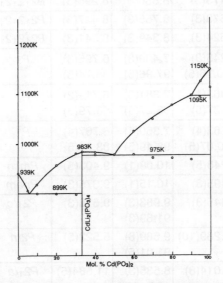

Figure 5.3.5. The $Cd(PO_3)_2$–$LiPO_3$ phase-equilibrium diagram. The transformation $\alpha \rightarrow \beta$ of $Cd(PO_3)_2$ is observed at 1095 K on the right-hand side of the diagram.

– The PbLi(PO$_3$)$_3$ atomic arrangement [100]. We must first mention that this polyphosphate does not appear in the Pb(PO$_3$)$_2$–LiPO$_3$ phase-equilibrium diagram. It was characterized during the investigations performed by Grenier and Mahama [98] in order to prepare single crystals of Pb$_2$Li(PO$_3$)$_5$. It is an isotype of SrLi(PO$_3$)$_3$ and CaLi(PO$_3$)$_3$. In PbLi(PO$_3$)$_3$ atomic arrangement, the phosphoric chain spreading along the **c** direction is slightly corrugated and has a period of three tetrahedra. Figure 5.3.6 shows how two such chains cross the unit cell, whereas in Figure 5.3.7 a projection made along the chain direction illustrates the respective arrangement of the various components in this structure and how the arrays of the associated cations are inserted between the

Table 5.3.5
Main crystallographic data of divalent cation-lithium polyphosphates.

Formula	a α	b β	c (Å) γ (°)	S. G.	Z	Ref.
CuLi(PO$_3$)$_3$	8.197(3)	8.613(3)	8.703(3)	$P2_12_12_1$	4	[89–90]
MgLi(PO$_3$)$_3$	8.350(2)	8.527(2)	8.602(2)	$P2_12_12_1$	4	[93]
ZnLi(PO$_3$)$_3$	8.320(2)	8.517(2)	8.623(2)	$P2_12_12_1$	4	[93]
FeLi(PO$_3$)$_3$	8.362(2)	8.573(1)	8.690(2)	$P2_12_12_1$	4	[107–108]
NiLi(PO$_3$)$_3$	8.473(3)	8.550(3)	8.295(3)	$P2_12_12_1$	4	[94]
MnLi(PO$_3$)$_3$	8.673(3)	8.755(3)	8.447(3)	$P2_12_12_1$	4	[92]
CoLi(PO$_3$)$_3$	8.624(3)	8.349(3)	8.541(3)	$P2_12_12_1$	4	[95]
PbLi(PO$_3$)$_3$	7.245(3) 100.76(5)	7.409(3) 97.96(5)	6.795(3) 83.74(5)	$P\bar{1}$	2	[98]
CaLi(PO$_3$)$_3$	6.963(1) 99.37(5)	7.381(3) 98.27(5)	6.719(2) 83.79(5)	$P\bar{1}$	2	[96]
SrLi(PO$_3$)$_3$	7.163(4) 100.07(5)	7.360(4) 98.49(5)	6.767(4) 83.59(5)	$P\bar{1}$	2	[97]
MnLi$_2$(PO$_3$)$_4$	9.248(5)	10.09(1)	9.403(5)	Pmcn	4	[92]
CdLi$_2$(PO$_3$)$_4$	9.495(3)	10.15(1)	9.375(3)	Pnam	4	[101–102]
HgLi$_2$(PO$_3$)$_4$	9.445(3)	9.983(3) 91.89(3)	9.518(3)	$P2_1/c$	4	[102–103]
Pb$_2$Li(PO$_3$)$_5$	12.289(10)	9.689(8) 91.01(5)	5.523(5)	P2/n	2	[100]
Ba$_2$Li$_3$(PO$_3$)$_7$	18.014(8)	8.535(3) 104.48(2)	11.584(5)	$P2_1/a$	4	[104–107]

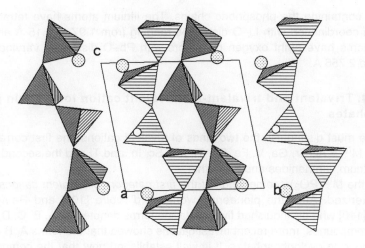

Figure 5.3.6. Projection, along the **a** direction, of the atomic arrangement in PbLi(PO₃)₃ showing how the phosphoric chain, shown by polyhedral representation, develops along the **c** direction. Small open circles represent the lithium atoms and dotted ones the lead atoms.

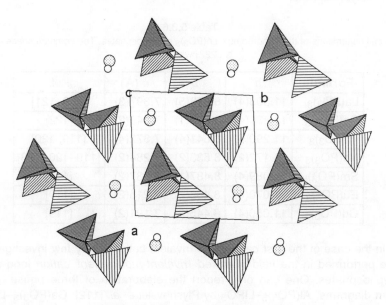

Figure 5.3.7. Projection, along the **c** axis, of the structure of PbLi(PO₃)₃ showing the layer arrangement of this structure. Open circles represent the Li atoms and dotted ones the Pb atoms.

planes containing the phosphoric chains. The lithium atoms have tetrahedral oxygen coordination with Li–O distances ranging from 1.97 to 2.15 Å and the lead atoms have eight oxygen neighbors with Pb–O distances varying from 2.492 to 2.756 Å.

5.3.2.4. Trivalent and trivalent-monovalent cation long-chain poly-phosphates

One must distinguish the two kinds of trivalent cations: the first corresponding to M^{III} = Al, Cr, Ga, V, Fe, Rh, Ti, Mo, Sc, In, and Tl and the second including yttrium, lanthanides and bismuth.

All the $M^{III}(PO_3)_3$ compounds of the *first category of trivalent cations* were characterized. After the pioneering works of d'Yvoire [109] and Rémy and Boullé [110] who distinguished five crystalline forms denoted by A, B, C, D, E for these compounds, more recent investigations showed that the forms A, B and E correspond to cyclophosphates. It is well established now that the compounds belonging to the C-form are long-chain polyphosphates, whereas for the D-form no structural study is available to date. Moreover, a sixth form was characterized by Bagieu-Beucher [111] for the manganese salt. This last form is a long-chain polyphosphate.

Table 5.3.6

Unit-cell dimensions of the orthorhombic $Ln(PO_3)_3$ polyphosphates. The common space group is $C222_1$ with $Z = 4$.

Formula	a	b	c (Å)	Ref.
$La(PO_3)_3$	11.303(4)	8.648(5)	7.397(3)	[119–121]
$Ce(PO_3)_3$	11.236(4)	8.602(1)	7.349(2)	[122]
$Pr(PO_3)_3$	11.290(5)	8.641(4)	7.372(4)	[119, 120]
$Nd(PO_3)_3$	11.172(2)	8.533(2)	7.284(2)	[119–120, 123]
$Sm(PO_3)_3$	11.089(4)	8.487(1)	7.237(2)	[119]
$Eu(PO_3)_3$	11.070(4)	8.463(2)	7.215(2)	[119]
$Gd(PO_3)_3$	11.050(4)	8.445(2)	7.191(2)	[119]

In the case of the *first category* of trivalent cations, very few investigations were performed in the field of *mixed trivalent-monovalent cation* long-chain polyphosphates. One can only report the elaboration of three phase-equilibrium diagrams, $Al(PO_3)_3$–$LiPO_3$ by Plyshevskii *et al.* [112], $Ga(PO_3)_3$–$LiPO_3$ by Bukhalova *et al.* [113], and $Ga(PO_3)_3$–$NaPO_3$ by Bukhalova *et al.* [113]. Only three compounds are clearly characterized, $GaNa(PO_3)_4$ [114], $GaLi(PO_3)_4$

[114–115], and CrLi(PO$_3$)$_4$ [116].

The long-chain polyphosphates including *the trivalent cations of the second category* were rather neglected till the discovery of the lanthanide ultraphosphates and of their interesting properties as efficient laser materials or phosphors. Then, a great amount of systematic investigations were performed in this field. It was established rapidly that trivalent-cation long-chain polyphosphates can exist with two different stoichiometries: $M^{III}H(PO_3)_4$ and $M^{III}(PO_3)_3$.

Most of the $M^{III}(PO_3)_3$ *compounds* can be classified in two main structural types. For M^{III} = Y and Ln from Sm to Lu, they are monoclinic and adopt the

Table 5.3.7

Crystallographic data of the Y and Ln polyphosphates belonging to the monoclinic C-form. The common space group is *Ic* with Z = 12.

Formula	a	b	c (Å)	β (°)	Ref.
Y(PO$_3$)$_3$	11.263(2)	19.93(1)	10.033(2)	97.62(4)	[120]
Sm(PO$_3$)$_3$	11.245(5)	19.86(1)	9.957(5)	97.57(4)	[120]
Eu(PO$_3$)$_3$	11.286(8)	19.74(2)	10.015(7)	97.23(4)	[120]
Gd(PO$_3$)$_3$	11.394(6)	20.31(1)	10.181(6)	96.81(4)	[119–120]
Dy(PO$_3$)$_3$	11.358(7)	20.02(2)	10.169(7)	96.54(4)	[120]
Ho(PO$_3$)$_3$	11.288(3)	19.98(1)	10.147(3)	97.08(4)	[119–120]
Er(PO$_3$)$_3$	11.306(4)	20.10(1)	10.076(3)	97.03(4)	[120, 124]
Tm(PO$_3$)$_3$	11.278(3)	20.01(1)	10.077(3)	97.02(4)	[120]
Yb(PO$_3$)$_3$	11.219(2)	19.983(3)	9.999(3)	97.30(2)	[119–120, 123]
Lu(PO$_3$)$_3$	11.249(4)	19.86(1)	10.018(4)	97.42(4)	[120]

Table 5.3.8

Main crystallographic data of various Bi and Ln polyphosphates.

Formula	a α	b β	c (Å) γ (°)	S. G.	Z	Ref.
Yb(PO$_3$)$_3$	20.974(4)	20.974(4)	12.134(3)	$R3$	24	[125]
Yb(PO$_3$)$_3$	8.361(3) 86.02(3)	7.508(1) 103.35(6)	6.237(5) 90.92(1)	$P\bar{1}$ or $P1$	2	[126]
Er(PO$_3$)$_3$	10.943(3)	6.971(2) 91.82(2)	9.670(2)	Pm	4	[127]
Bi(PO$_3$)$_3$	13.732(2)	6.933(1) 93.35(1)	7.152(1)	$P2_1/a$	4	[128]

arrangement of the C-form already mentioned for the first category of trivalent cations. Their main crystallographic data are reported in Table 5.3.6. For M^{III} = Ln from La to Gd, the $M^{III}(PO_3)_3$ compounds adopt an orthorhombic type of structure. Their main crystallographic data are reported in Table 5.3.7. The examination of these two tables shows that $Sm(PO_3)_3$, $Eu(PO_3)_3$, and $Gd(PO_3)_3$ crystallize with the two forms. A few members of this series adopt other structure types. They are listed in Table 5.3.8.

Acidic polyphosphates are rather rare. Some examples of such compounds were characterized during the numerous investigations of the $M_2O_3-P_2O_5-H_2O$ systems. They all correspond to the formula $M^{III}H(PO_3)_4$ and were observed for M^{III} = Bi, Sm, Eu, Gd, Tb, Dy, Ho, and Er by Palkina *et al.* [117]. They belong to two different structural types. $M^{III}H(PO_3)_4$ are triclinic for M^{III} = Bi, Sm, Eu, and Gd, monoclinic for M^{III} = Gd, Tb, Dy, and Er. The gadolinium derivative can adopt the two types of structure. The two corresponding structures were determined by Palkina and Jost [118] for the bismuth derivative and by Palkina *et al.* [117] for the erbium salt. These atomic arrangements will be discussed in the chapter devoted to the acidic anionic frameworks.

A large number of *trivalent-monovalent cation polyphosphates* including the trivalent cations of the second category were characterized during the last 25 years. Most of them were discovered during the elaboration of $M^{III}(PO_3)_3-M^IPO_3$ phase-equilibrium diagrams or the investigation of numerous $M_2O_3-M_2O-P_2O_5$ systems by various flux methods. We report, in Table 5.3.9, a selection of results obtained during the elaboration of some of these phase-equilibrium diagrams

Table 5.3.9

A selection of results obtained during the investigations of some $M^{III}(PO_3)_3-M^IPO_3$ phase-equilibrium diagrams.

System	Ref.	Compound	mp (K)	Ref.
$La(PO_3)_3-LiPO_3$	[130]	$LaLi(PO_3)_4$	1233 (d)	[130]
$Nd(PO_3)_3-LiPO_3$	[131]	$NdLi(PO_3)_4$	1243 (d)	[131–133]
$Ce(PO_3)_3-NaPO_3$	[134]	$CeNa(PO_3)_4$	1138 (d)	[134]
$Bi(PO_3)_3-AgPO_3$	[135]	$BiAg(PO_3)_4$	863	[135]
$Ce(PO_3)_3-KPO_3$	[136]	$CeK(PO_3)_4$ $CeK_2(PO_3)_5$	1153 (d) 1014 (d)	[136–137]
$Bi(PO_3)_3-KPO_3$	[138]	$BiK(PO_3)_4$	899 (d)	[138]
$Ce(PO_3)_3-CsPO_3$	[139]	$CeCs(PO_3)_4$	1163 (d)	[139]
$Bi(PO_3)_2-CsPO_3$	[140]	$BiCs(PO_3)_4$	853	[140]

d: decomposition

With the exception of a few of them having a stoichiometry corresponding to the formula $LnM_2^I(PO_3)_5$, the majority of the (Y, Bi, Ln)–M^I polyphosphates correspond to the general formula $M^{III}M^I(PO_3)_4$. The literature dealing with these compounds was rather confusing for a long time, but it is well established today that all compounds of $LnM^I(PO_3)_4$ formula can be classified into seven different structural types which are usually denoted by roman numbers. This nomenclature, first proposed by Palkina *et al.* [129], is today generally accepted. Among these seven forms, one of them (form V) corresponds to a series of $M^{III}M^IP_4O_{12}$ cyclotetraphosphates. Table 5.3.10 presents the repartition of these forms in this family of polyphosphates. This table does not include all the assumptions found in scientific literature. For instance, attribution of a compound to a given form based on the comparison of unindexed powder patterns has been rejected.

Table 5.3.10
Distribution of the seven structure types among the $M^{III}M^I(PO_3)_4$ compounds.

M^{III}	Bi	Y	La	Ce	Pr	Nd	Sm	Eu	Gd	Tb	Dy	Ho	Er	Tm	Yb	Lu
Li			I	I	I	I	I	I	I	I	I	I	I	I	I	I
Na	II		II	II	II	II	II	II	II	II	II	II				
Ag							II									
K	III		III	III	III	III	III	III	III	III	III	III	III IV VII	V	V	V
NH₄	IV	VII														
Rb			VI	IV	VI IV	VI IV	VII IV	VI IV	VI IV	VII IV	VI IV	VI IV	VI IV	IV	IV	IV
Cs			IV VI	VI	IV VI	IV VI	IV VI	IV VI	IV VI	IV VI	IV VI	IV VI	IV VI			
Tl						IV										

5.3.2.5. Various topics involving long-chain polyphosphates

Many claims for the identification of hydrated polyphosphates have been reported in the literature. No structural data exist to support these assumptions to date.

Long-chain anions also appear in some mixed anion condensed phosphates. According to us, these compounds must be considered as adducts and will be reported in the appropriate section.

Up to now, organic derivatives have never been observed in this family of phosphates.

Many investigations involving long-chain polyphosphates mainly M^IPO_3 compounds were performed: chemical interactions with other phosphates or non-phosphate compounds, measurements of their thermal stability, surface tensions and various other properties. We report below a selection of these studies.

Surface tension, density, viscosity and conductivity of molten monovalent-cation polyphosphates as $LiPO_3$, $NaPO_3$, KPO_3, and $CsPO_3$ were studied by Sokolova [141] and by Sokolova *et al.* [142].

Many systems involving monovalent-cation long-chain polyphosphates were investigated.

Reaction of KPO_3 with solutions of various metal salts was investigated by Ohashi and Yamagishi [143].

The reaction between germanium dioxide and molten lithium, sodium, or potassium polyphosphates was investigated by Slobodyanik *et al.* [144]. In all these cases, they obtained the rhombohedral $Ge_2M^I(PO_4)_3$ monophosphates.

The reaction of molten $NaPO_3$ and KPO_3 with NaCl at 1073 K was investigated by Markina and Voskresenskaya [145].

The kinetics of the reaction between molten $LiPO_3$, $NaPO_3$, KPO_3 and the corresponding sulphates was investigated by Kochergin *et al.* [146] between 973 and 1023 K.

In the NaF–$NaPO_3$ and KF–KPO_3 phase-equilibrium diagrams elaborated by Bukhalova and Mardirosova [147], four intermediate salts, $NaF.NaPO_3$, $2NaF.NaPO_3$, $KF.KPO_3$, and $2KF.KPO_3$ are observed. The same authors [148] investigated the $LiCl$–$LiPO_3$, $NaCl$–$NaPO_3$, KCl–KPO_3 and $CsCl$–$CsPO_3$ systems. In the lithium system, $2LiCl.LiPO_3$ is formed in the solid state at 777 K and has a polymorphic transformation at 638 K. The $NaCl$–$NaPO_3$ system is of the eutectic type with a compound, $2NaCl.NaPO_3$, forming in the solidus at 778 K. For the KF–KPO_3 system, the authors also found an eutectic diagram, but show the existence of an intermediate compound, $2KCl.KPO_3$ which forms at 673 K in the solidus. For the cesium system, the results are close to those obtained in the case of potassium with the probable formation of $2CsCl.CsPO_3$ at 673 K.

Equilibrium chemical transformations in $NaCl$–$NaPO_3$ melts were investigated by Kovarskaya and Rodionov [149]. When the NaCl content in the mixture is increased, the sodium polyphosphate gradually depolymerizes to tri-, di-, and monophosphate and the composition of the equilibrium melt is dependent only on the starting composition.

The reaction of metal oxides with molten $NaCl$–$NaPO_3$ mixtures were studied by Kovarskaya *et al.* [150]. Detailed results are given in the case of $NaCl$–$NaPO_3$–Fe_2O_3 mixtures.

5.3.2.6. Chemical preparation of long-chain polyphosphates

The insolubility of all long-chain polyphosphates in water limits the possibilities of synthesis considerably, in addition, their glassy melting often prohibits the use of some of the classical processes of crystal growth. Some more or less general processes of preparation can nevertheless be described.

Thermal dehydration of dihydrogenmonophosphates – Most of the monovalent or divalent-cation polyphosphates can be prepared as polycrystalline or glassy specimens by thermal dehydration of the corresponding dihydrogenmonophosphates:

$$M^IH_2PO_4 \longrightarrow M^IPO_3 + H_2O \ [I]$$

$$M^{II}(H_2PO_4)_2 \longrightarrow M^{II}(PO_3)_2 + 2H_2O \ [II]$$

But, as we have mentioned when describing the general properties of dihydrogenmonophosphates, these two types of reactions are often very complex and various intermediate steps are possible. Nevertheless, one can say that if reaction [I] is performed close to the melting point of the polyphosphate or some tens of degrees below a long-chain anion is obtained. Reaction [II] is of a more restricted use, because very dependent on the nature of the divalent cation. For M^{II} = Mg, Co, Mn, Cu, Ni, Zn, it leads to cyclotetraphosphates, $M^{II}_2P_4O_{12}$, stable up to their melting points and to long-chain polyphosphates for M^{II} = Be, Ca, Sr, Ba, Cd, and Hg.

Thermal methods – Higher-valency cation polyphosphates or mixed-cation polyphosphates are often prepared as polycrystalline samples by firing appropriate stoichiometric mixtures. For instance, $CoK_2(PO_3)_4$ [151] can be easily prepared at 923–973 K through the following reaction:

$$CoCO_3 + K_2CO_3 + 4(NH_4)_2HPO_4 \longrightarrow CoK_2(PO_3)_4 + 2CO_2 + 8NH_3 + 6H_2O$$

The reaction temperature for such reactions can be optimized by successive attempts or better determined from the phase-equilibrium diagram, when it exists.

Flux methods – As in many other classes of compounds, this process is widely used for the production of single crystals of long-chain polyphosphates. As in the case of condensed phosphates, the flux is constituted of a large excess of H_3PO_4 or $(NH_4)_2HPO_4$ most of time. On progressive heating, they transform into more or less viscous substances corresponding approximately to the NH_4PO_3 and HPO_3 formulæ. At the end of the crystallization process, the excess flux is removed by hot water. All we have explained concerning the difficulties of the optimization of a flux process when this method was first mentioned at the beginning of this chapter remains valuable for polyphosphates.

We report a flux method as it was used for the production of the trigonal modification of $Yb(PO_3)_3$ [152] in more details. The starting mixture prepared from Yb_2O_3, $NaH_2PO_4 \cdot 2H_2O$, $(NH_4)_2HPO_4$ and H_3PO_4 was established to obtain the following atomic ratio:

$$P: Na: NH_4: Yb = 75: 30: 30: 1$$

This mixture is then heated slowly from room temperature to 653 K and kept at this temperature for two or three weeks. At the end of this process, the excess phosphoric flux is removed by water.

Hydrothermal methods – A small number of long-chain polyphosphates were prepared by hydrothermal processes. For instance, $FeLi(PO_3)_3$ was obtained by Genkina *et al.* [107] during an investigation of the $FeO–LiF–P_2O_5–H_2O$ system between 523 and 723 K at 1500 atmospheres.

Phase transitions in long-chain polyphosphates – Many long-chain polyphosphates undergo polymorphic transformations. This quasi-general property can probably be attributed to the great flexibility of the phosphoric chain.

All the alkali long-chain polyphosphates exhibit several crystalline forms. Shown initially by Thilo and Grunze [49] and later confirmed by Benkhoucha and Wunderlich [10], *LiPO3* undergoes a transition at about 523 K. Three forms of *NaPO3* were investigated, but no clear data on the transitions are available to date. The first valuable investigations on *KPO3* were performed by Jost [20], Jost and Schulze [153–154], and Schmahl [155]. According to these authors, four crystalline forms exist for this salt corresponding to three reversible transitions.

$$KPO_3 \text{ (T)} \xleftrightarrow{548\,K} KPO_3 \text{ (Z)} \xleftrightarrow{733\,K} KPO_3 \text{ (H)} \xleftrightarrow{923\,K} KPO_3 \text{ (HT)}$$

The terminology used for the different modifications of KPO_3 is that given by the authors originally. Bekturov *et al.* [156] also investigated the thermal behavior of KPO_3 and confirmed a phase transition for this compound at 733 K. After the first experiments performed on *RbPO3* and *CsPO3* by Chudinova *et al.* [157], it appears clearly that the thermal behavior of these salts is rather similar to that of the potassium salt. A very recent investigation of the rubidium salt by Holst *et al.* [158] confirms these assumptions. The monoclinic, $P2_1/n$, room-temperature T-form transforms into an orthorhomic, Pbnm, H-form at 661 K. Then, a continuous transition (Tc = 914 K) leads to another orthorhombic, *Bbmm*, HT-form. In addition, on cooling the H-form down to 591 K, an intermediate monoclinic, $P2_1/n$, Z-form occurs which transforms into the room-temperature T-form at 550 K.

$$RbPO_3 \text{ (T)} \xleftrightarrow{661\,K} RbPO_3 \text{ (H)} \xleftrightarrow{914\,K} RbPO_3 \text{ (HT)}$$

Crystal structures of the four forms were performed by the authors who discussed clearly the nature of these transitions. The ammonium derivative, NH_4PO_3, also has several crystalline modifications [21], but it is only very recently that one of them was investigated [22].

Most of divalent-cation polyphosphates are also polymorphic.

At least four crystalline forms of $Be(PO_3)_2$ are known to date. During a general study of beryllium phosphates, Jaulmes [38–39] has shown the existence of two different crystalline forms of $Be(PO_3)_2$; the first form denoted by (I) is prepared at temperatures lower than 673 K, the other by (II) observed at higher temperatures. Bagieu-Beucher and Durif [40] reported a chemical preparation for $Be(PO_3)_2$ (II) single crystals and from a twinned crystal determined the unit cell and space group of this compound. Averbuch-Pouchot *et al.* [41] described a process to obtain untwinned crystals and performed the determination of the atomic arrangement. A third form of $Be(PO_3)_2$ was characterized by Schultz [42] and denoted by (III). During this study, it was observed that this form has two modifications itself with a transition temperature at 369 K. Crystal structure of the high-temperature modification was performed at 392 K by Schultz and Liebau [43–44]; its atomic arrangement is closely related to that of the keatite form of silica. For the last form, the substitution scheme is evidently:

$$3Si^{IV} \longrightarrow 2P^V + Be^{II}$$

For $Ca(PO_3)_2$ [48–52], four modifications were suspected, but only the β-form is well characterized [55].

When investigating the phase relationships in the $SrO-P_2O_5$ system, Kreidler and Hummel [58] characterized two crystalline forms for $Sr(PO_3)_2$. They observed the transition temperature at 1078 K. During studies of the thermal evolution of $Sr(H_2PO_4)_2$ by Ropp *et al.* [59] and by Ropp and Aia [60], three forms of $Sr(PO_3)_2$ were identified. According to these authors, the following schemes are observed:

$$\overset{463-483 \text{ K}}{Sr(H_2PO_4)_2 \longrightarrow SrH_2P_2O_7 + H_2O}$$

$$\overset{593-603 \text{ K}}{SrH_2P_2O_7 \longrightarrow} \overset{673 \text{ K}}{\gamma\text{-}Sr(PO_3)_2 \longrightarrow} \overset{1123 \text{ K}}{\beta\text{-}Sr(PO_3)_2 \longrightarrow \alpha\text{-}Sr(PO_3)_2}$$

The low temperature form is possibly a cyclotriphosphate [59–60]. The β-form, which is isotypic with calcium and lead polyphosphates is the only one well characterized [55].

Grenier and Martin [63] reported the existence of three forms for $Ba(PO_3)_2$: the β-form, already described by Grenier *et al.* [64], a high temperature form called "α" and a third form "γ" which was accidentally produced during their investigations. According to these authors, the transitions occur as follows:

$$\beta\text{-Ba(PO}_3)_2 \xrightarrow{\text{1058 K}} \alpha\text{-Ba(PO}_3)_2$$

$$\gamma\text{-Ba(PO}_3)_2 \xrightarrow{\text{978 K}} \beta\text{-Ba(PO}_3)_2$$

Atomic arrangement of the γ-form was determined by Coing-Boyat *et al.* [65].

Brown and Hummel [66] were the first to report the existence of two crystalline forms for *Cd(PO₃)₂*. The transformation of α-Cd(PO$_3$)$_2$ to β-Cd(PO$_3$)$_2$ occurs at 1008 K and is reversible. Single crystals of the high-temperature β-form, were obtained by Laügt *et al.* [71] by annealing a glass of Cd(PO$_3$)$_2$ for some days at 1113 K. This form is stable for several months at room-temperature, but transforms rapidly to the low-temperature form by heating at 673 K. Its crystal structure was later determined by Bagieu-Beucher *et al.* [70]. The low-temperature form, isotypic with the corresponding mercury salt, was investigated by Beucher *et al.* [72], Tordjman *et al.* [69] and Bagieu-Beucher *et al.* [70].

As can be seen in the section devoted to M^{III}(PO$_3$)$_3$ and $M^{III}M^{I}$(PO$_3$)$_4$ polyphosphates, many of them are also polymorphic, but very little is known about their transitions to date.

The structural aspects of the long-chain polyphosphate anion – The long-chain anions found in this category of condensed phosphates can adopt many various geometries. The chain that we observed when describing Zn(PO$_3$)$_2$ is an almost straight linkage of tetrahedra (Figure 5.3.2), that found in PbLi(PO$_3$)$_3$ is slightly corrugated (Figure 5.3.6), but, in some cases, the chains can adopt very accentuated zigzag configurations of various amplitudes. We present, in Figure 5.3.8, the phosphoric chain found in ZnAg(PO$_3$)$_3$ [159]. Often chains are spiralled around a 2_1 helical axis or, less often, around a 3_1 helical axis like those

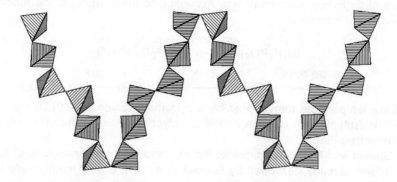

Figure 5.3.8. The infinite phosphoric chain observed in ZnAg(PO$_3$)$_3$.

observed in $NiNH_4(PO_3)_3$ or $Yb(PO_3)_3$. The great flexibility of this type of anion can probably explain the wide number of the observed configurations.

These purely visual considerations are not sufficient, even in a first attempt, for an elementary geometrical classification of long-chain anions. As we have mentioned above, these chains are very long. In a crystalline polyphosphate, characterized as in all other materials by the periodic repetition of a motif, it is evident that the chains themselves must be constituted by the repetition of a fundamental unit built by a group of PO_4 tetrahedra. This repetition unit is often called the *chain period* and one can distinguish or classify chains according to the number of tetrahedra in this basic unit. It must be mentioned that a chain with a period of n tetrahedra may be built up by less than n crystallographically independent tetrahedra when some of its components are located on symmetry elements, usually a twofold axis or a mirror plane, or its arrangement around a 2_1 or 3_1 helical axis. The part of the chain corresponding to the linkage of the crystallographically independent tetrahedra is usually denoted as the *subperiod* or *pseudoperiod*. Attempts were made to classify long-chain polyphosphates according to this concept, mainly in order to establish possible correlations between the chain period and the nature of the associated cations. To date, no clear correlation was found, but, nevertheless, this concept remains useful in illustrating some geometrical aspects of the chain conformation. The largest chain period (16 tetrahedra) that is observed up to now is that found in $YNH_4(PO_3)_4$.

As we have already shown, the infinite phosphoric chains appear as more or less corrugated. Attempts were done in silicate chemistry where a similar type of anion exists to evaluate the degree of stretching of a given chain by comparing the number of tetrahedra in the repetition unit to the length of this unit. Liebau [160] applied this concept to silicates and was able to find correlations between this *stretching factor* and the nature of the associated cations. In the field of phosphates and probably because of the more restricted number of examples, our attempts were not conclusive. Nevertheless we report the definition of this factor as given by Liebau [160]. The Fs stretching factor as defined by Liebau is expressed by the following formula:

$$Fs = I/L_t \cdot n$$

where n is the period of the chain, I its length in angströms, and L_t a constant taken as the value of I/n for the most highly stretched chain. For silicates, L_t was found to have a value of 2.70 Å, whereas in the case of phosphates we have determined a value of 2.48 Å in the atomic arrangement of $Zn(PO_3)_2$. It can easily be seen from this formulation that Fs will decrease from 1.00 for the most highly stretched chain to lower values for more and more corrugated chains. In the case of silicates, Liebau reported a minimum value of 0.234 after the exami-

Table 5.3.11
Main geometrical features in some chain anions.

Formula	P–P (Å)	P–O–P (°)	P–P–P (°)	Period	Fs	Ref.
$NaPO_3$ (A)	2.870 2.967	124.8 136.1	106.1 106.6	4	0.625	[14–16]
$AgK(PO_3)_2$	2.919 2.957	129.6 134.5	92.6 111.9	4	0.755	[34]
$Be(PO_3)_2$ (II)	2.957 2.887	129.5 139.9	98.7 108.1	4	0.702	[41]
$Be(PO_3)_2$ (III)	2.933 2.951	139.8 137.4	105.2 103.5	8	0.438	[43–44]
$Zn(PO_3)_2$	2.946	135.5	114.7	2	1.000	[46]
$Ca(PO_3)_2$	2.947 2.978 2.939 2.980	140.6 135.9 135.9 141.5	111.6 93.1 88.2 116.7	4	0.705	[55]
$Cd(PO_3)_2$ (LT)	2.934 2.875	133.3 128.0	102.0 105.8	4	0.709	[68–70]
$Cd(PO_3)_2$ (HT)	2.982 3.002	150.2 154.2	113.1 94.9	4	0.742	[71–72]
β-$Ba(PO_3)_2$	2.898	124.6 133.9	102.2	2	0.909	[64]
γ-$Ba(PO_3)_2$	2.818 3.022	120.7 143.4	127.3 93.0	4	0.695	[65]
$Pd(PO_3)_2$	2.872	129.6	94.9	2	0.853	[76]
$ZnAg(PO_3)_3$	2.981 2.912 2.892 2.889	139.7 131.3 129.8 131.0	104.8 110.6 105.0	12	0.468	[159]
$Mn(PO_3)_3$	3.008 2.872	146.2 128.7	130.2 94.1	3	0.855	[111]
$YNH_4(PO_3)_4$	2.895 2.897 2.839 2.869	130.8 132.3 126.6 130.0	137.8 117.0 133.6 97.1 108.2	16	0.436	[161]
$Ba_2Cu(PO_3)_6$	2.866 2.883 2.943 2.920 2.896 3.017	127.6 133.8 131.3 131.0 146.2 129.0	140.6 113.1 113.6 94.3 109.6 99.5	12	0.245	[83]

nation of a great number of them. In phosphate chemistry, the minimum value of 0.245 was observed for $Ba_2Cu(PO_3)_6$ (Table 5.3.11).

In Table 5.3.11, we present the P–P distances, the P–O–P and P–P–P angles as well as the periods and the stretching coefficients for a limited number of phosphoric chains. A more exhaustive study involving the examination of the geometrical data for all the long-chain polyphosphates to be accurately investigated is in preparation. Preliminary results obtained during the course of this investigation show that, the observed values never depart significantly from those given for the selected examples in Table 5.3.11 in spite of the great diversity of the long-chain anions. Moreover, these preliminary results confirm that there are no apparent meaningful correlations between the geometry of a chain and the nature of the associated cations.

5.3.3. An Ersatz of Asbestos: a Promising Application of Polyphosphates

Among the many applications of long-chain polyphosphates, it is worth reporting with some details the attempt made by Griffith *et al.* [163–164] some years ago at the Monsanto Company to produce phosphate fibers that are able to be substituted to asbestos.

This last material, which is well-known for the antiquity of its remarkable thermo-resistant and insulating properties, is used industrially with a production increasing rapidly (approximately 2.10^6 tons a year in the recent years) since 1868. No other known material, either natural or synthetic could exhibit properties able to compete with the remarkable qualities of asbestos till a recent date. Unfortunately, asbestos has been proved to be dangerous, provoking various very serious diseases. These health problems are not connected to a somewhat chemical activity, but due to a prolonged mechanical action on living organisms which ingested asbestos for this substance cannot be degraded by these organisms.

Natural asbestos materials are found in two different kinds of silicate, the amphibole family and the chrysotile family. In the amphibole family, the anionic framework is a double infinite chain of SiO_4 tetrahedra, whereas, in the chrysotile family, it is an infinite two-dimensional layer. Both have similar properties and a well-known fibrous aspect. Most of the extracted asbestos is of the chrysotile type, although the fibers are not so long and silky as those of the amphibole type.

A harmless asbestos ersatz should have properties that are evidently very similar to those of natural asbestos and be rapidly biodegradable by living organisms. Condensed phosphates are good examples because they satisfy these two conditions. The chemistry of long-chain polyphosphates shows the existence of infinite anions clearly, not exactly similar to those found in natural asbestos, but, nevertheless, leading to a more or less accentuated fibrous morphology for many of them. Most of the alkali metal Kurrol's salts are fibrous but unfortunately degraded rapidly by water. Nevertheless, among the many synthetic long-chain polyphosphates a good number of them are very stable in normal conditions and when ingested or inhaled these compounds are degraded by the phosphatase enzymes existing in all living organisms very rapidly. These enzymes accelerate by a factor of at least 10^6 the normal rate of degradation of the phosphate chains, transforming them into derivatives used currently as animal nutrients.

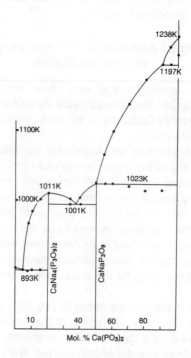

Figure 5.3.9. The $Ca(PO_3)_2$–$NaPO_3$ phase-equilibrium diagram as revised by Grenier et al. [162].

The examination of the CaO–Na_2O–P_2O_5 system shows the existence of several very stable long-chain polyphosphates: among them, $Ca(PO_3)_2$ and $CaNa(PO_3)_3$, the latter observed in the portion $Ca(PO_3)_2$–$NaPO_3$ of this system. The $Ca(PO_3)_2$–$NaPO_3$ phase-equilibrium diagram is represented in Figure 5.3.9. The other compound observed in this diagram, $CaNa_4(P_3O_9)_2$, is a cyclotriphosphate. Both $Ca(PO_3)_2$ and $CaNa(PO_3)_3$ were selected by Griffith and co-workers [163–164] as potential materials for the production of heat-resistant fibers.

$CaNa(PO_3)_3$ is isotypic with the lead-lithium polyphosphate that we have described on p. 208 and $Ca(PO_3)_2$ is isotypic with the corresponding lead and strontium salts. To describe its structure, we use the data recently reported by Rothhammel *et al.* [55]. $Ca(PO_3)_2$ is monoclinic, $P2_1/a$, with $Z = 8$ and the following unit-cell dimensions:

$$a = 16.96(1), \quad b = 7.7144(2), \quad c = 6.9963(2) \text{ Å}, \quad \beta = 90.394(5)°$$

The phosphoric chain has a period of four tetrahedra. As shown by Figure 5.3.10, four of them cross the unit cell parallel to the **c** direction. They are mode-

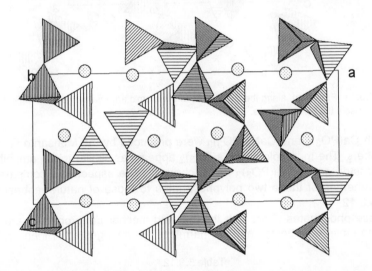

Figure 5.3.10. Projection, along the **b** dierection, of the atomic arrangement in $Ca(PO_3)_2$. Dotted circles represent the calcium atoms.

rately corrugated and arranged in such a way to create oblong channels parallel to the **b** direction in which are located the calcium atoms. The two crystallographically independent calcium atoms have different oxygen coordinations. One of them has seven oxygen neighbors ($2.338 < Ca–O < 2.637$ Å), the second has eight ($2.381 < Ca–O < 2.696$ Å). Figure 5.3.11 shows this arrangement which is in projection along the chain direction, exhibiting clearly how the zigzag arrays of calcium atoms are inserted between the phosphoric chains.

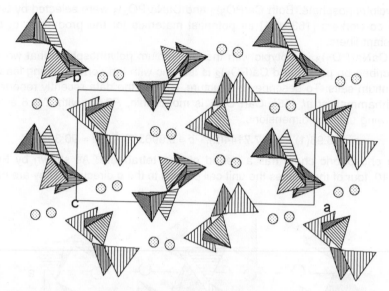

Figure 5.3.11. Projection, along the **c** direction, of the atomic arrangement in Ca(PO$_3$)$_2$. Dotted circles represent the calcium atoms.

Both Ca(PO$_3$)$_2$ and CaNa(PO$_3$)$_3$ were produced by the Monsanto Company as fibers. The morphology of Ca(PO$_3$)$_2$ appears as similar to the amphibole-like asbestos and CaNa(PO$_3$)$_3$ has a chrysotile-like aspect. We compare the main properties of these two polyphosphates to those of natural asbestos in Table 5.3.12.

These phosphates fibers with thermal and mechanical properties are very similar to those of asbestos and are harmless. When ingested or inhaled, they

Table 5.3.12
Comparison of the main properties of asbestos with those of Ca(PO$_3$)$_2$ and CaNa(PO$_3$)$_3$.

Characteristic	Asbestos	Ca(PO$_3$)$_2$	CaNa(PO$_3$)$_3$
Diameter (μ)	0.1	10	0.1
Tensile strengths (psi)	2.0 to 10x10^5	2.0 to 3x10^5	1.5 to 3.3x10^5
Modulus (psi)	2.1 to 2.5x10^7	6. to 16x10^6	13. to 21x10^6
Thermal resistance	673–1273 K	1243 K	1043 K
Corrosion resistance	Base	Acid	Acid
Toxicity	Carcinogen Lung irritant	Nutrient	Nutrient

become a nutrient for the phosphatase enzymes which exists in all living systems accelerated by at least a factor of 10^6 the degradation of the chains.

Unfortunately, in spite of these evident properties and of their harmless behavior, the production of these two materials has been stopped on the faith of arguments which are more emotional than scientifically based. The irrational and unjustified fear of phosphorus compounds seems prevailing on the danger of asbestos diseases.

References

1. J. Majling and F. Hanic, *Phase Chemistry of Condensed Phosphates,* in Topics in Phosphorus Chemistry, Vol. 10, ed. E. J. Griffith and M. Grayson, (Interscience Publishers, New York, 1980), p. 341–502.

2. G. A. Bukhalova, I. V. Mardirosova, and M. M. Ali, *Russ. J. Inorg. Chem.,* **33**, (1988), 1438–1441.

3. H. D. Park and E. R. Kreidler, *J. Am. Ceram. Soc.,* **67**, (1984), 23–26.

4. M. T. Averbuch-Pouchot, *Z. anorg. allg. Chem.,* **621**, (1995), 506–509.

5. N. Hilmer, N. N. Chudinova, and K. H. Jost, *Izv. Akad. Nauk SSSR, Neorg. Mater.,* **14**, (1978), 1507–1515.

6. N. N. Chudinova, L. P. Shklover, L. I. Shkol'nikova, A. E. Balanevskaya, and G. M. Balagina, *Izv. Akad. Nauk SSSR, Neorg. Mater.,***14**, (1978), 1324–1328.

7. M. Bagieu, *Thesis,* (Univ. of Grenoble, France, 1980).

8. A. Durif, *Crystal Chemistry of Condensed Phosphates,* (Plenum Press, New York, 1995).

9. E. Thilo and H. Grunze, *Z. anorg. allg. Chem.,* **281**, (1955), 262–283.

10. R. Benkhoucha and B. Wunderlich, *Acta Crystallogr.,* **B35**, (1979), 265–267.

11. J. C. Grenier and A. Durif, *Z. Kristallogr.,* **137**, (1973), 10–16.

12. J. C. Guitel and I. Tordjman, *Acta Crystallogr.,* **B32**, (1976), 2960–2966.

13. D. E. C. Corbridge, *Acta Crystallogr.,* **8**, (1955), 520.

14. K. H. Jost, *Acta Crystallogr.,* **14**, (1961), 844–847.

15. K. H. Jost, *Acta Crystallogr.,* **16**, (1963), 640–642.

16. A. McAdam, K. H. Jost, and B. Beagley, *Acta Crystallogr.,* **B24**, (1968),

1621–1622.

17.   A. Immirzi and W. Porzio, Acta Crystallogr., **B38**, (1982), 2788–2792.

18.   K. H. Jost, Acta Crystallogr., **14**, (1961), 779–784.

19.   K. R. Andress and K. Fischer, Z. anorg. allg. Chem., **273**, (1953), 193–199.

20.   K. H. Jost, Acta Crystallogr., **16**, (1963), 623–626.

21.   C. Y. Chen, N. E. Stahlheber, and D. R. Dyroff, J. Am. Chem. Soc., **91**, (1969), 62–67.

22.   B. Brüne and M. Jansen, Z. anorg. allg. Chem., **620**, (1994), 931–935.

23.   D. E. C. Corbridge, Acta Crystallogr., **9**, (1956), 308–314.

24.   D. W. J. Cruickshank, Acta Crystallogr., **17**, (1964), 681–682.

25.   K. Dostal, V. Kocman, and V. Ehrenbergrova, Z. anorg. allg. Chem., **367**, (1969), 80–91.

26.   N. El-Horr, J. Solid State Chem., **90**, (1991), 386–387.

27.   I. V. Mardirosova and G. A. Bukhalova, Russ. J. Inorg. Chem., **11**, (1966), 1275–1277.

28.   C. Cavero-Ghersi, Thesis, (Univ. of Grenoble, France, 1975).

29.   N. El-Horr, C. Cavero-Ghersi, and M. Bagieu-Beucher, C. R. Acad. Sci., Sér. 2, **297**, (1983), 479–482.

30.   N. El-Horr, M. Bagieu, and I. Tordjman, Acta Crystallogr., **C39**, (1983), 1597–1599.

31.   N. El-Horr and M. Bagieu, Acta Crystallogr., **C41**, (1985), 1157–1159.

32.   N. El-Horr and M. Bagieu, C. R. Acad. Sci., Sér. 2, **312**, (1991), 373–375.

33.   N. El-Horr and M. Bagieu, Acta Crystallogr., **C43**, (1987), 603–605.

34.   M. T. Averbuch-Pouchot, J. Solid State Chem., **102**, (1993), 93–99.

35.   M. A. Savenkova, I. V. Mardirosova, and E. V. Poletaev, Russ. J. Inorg. Chem., **20**, (1975), 1374–1376.

36.   M. A. Savenkova, L. V. Kubasova, I. V. Mardirosova, and E. V. Poletaev, Izv. Akad. Nauk SSSR, Neorg. Mater., **11**, (1975), 2200–2202.

37.   M. T. Averbuch-Pouchot, A. Durif, and J. C. Guitel, Acta Crystallogr., **B32**, (1976), 2440–2443.

38.   S. Jaulmes, Thesis, (Univ. of Paris, France, 1964).

39   S. Jaulmes, Rev. Chim. Minér., **1**, (1964), 617–671.

40. M. Bagieu-Beucher and A. Durif, *Bull. Soc. fr. Minéral. Cristallogr.*, **93**, (1970), 129–130.

41. M. T. Averbuch-Pouchot, A. Durif, and I. Tordjman, *Acta Crystallogr.*, **B33**, (1977), 3462–3464.

42. E. Schultz, *Thesis*, (Univ. of Kiel, Germany, 1974).

43. E. Schultz and F. Liebau, *Naturwiss.*, (1973), 429–430.

44. E. Schultz and F. Liebau, *Z. Kristallogr.*, **154**, (1981), 115–126.

45. M. Bagieu-Beucher, M. Gondrand, and M. Perroux, *J. Solid State Chem.*, **19**, (1976), 353–357.

46. M. T. Averbuch-Pouchot, A. Durif, and M. Bagieu-Beucher, *Acta Crystallogr.*, **C39**, (1983), 25–26.

47. F. L. Katnack and F. A. Hummel, *J. Electrochem.*, **105**, (1958), 125–133.

48. W. L. Hill, G. T. Faust, and D. S. Reynolds, *Am. J. Sci.*, **242**, (1944), 457–477.

49. W. L. Hill, G. T. Faust, and D. S. Reynolds, *Am. J. Sci.*, **242**, (1944), 542–562.

50. E. Thilo and I. Grunze, *Z. anorg. Chem.*, **290**, (1957), 223–237.

51. C. Morin, *Bull. Soc. Chim. Fr.*, (1961), 1726–1734.

52. S. Ohashi and J. R. Van Wazer, *J. Am. Chem. Soc.*, **81**, (1959), 830–832.

53   J. R. Lehr, E. H. Brown, A. W. Frazier, J. P. Smith, and R. D. Thrasher, *Crystallographic Properties of Fertilizer Compounds*, (Chemical Engineering Bulletin, Tennessee Valley Authority, 6, 1967).

54. M. Schneider, K. H. Jost, and P. Leibnitz, *Z. anorg. allg. Chem.*, **527**, (1985), 99–104.

55. W. Rothammel, F. H. Burzlaff, and R. Specht, *Acta Crystallogr.*, **C45**, (1989), 551–553.

56. E. H. Brown, J. R. Lehr, J. P. Smith, W. E. Brown, and A. W. Frazier, *J. Phys. Chem.*, **61**, (1957), 1669–1670.

57. E. O. Huffman and J. D. Fleming, *J. Phys. Chem.*, **64**, (1960), 240–244.

58. E. R. Kreidler and F. A. Hummel, *Inorg. Chem.*, **6**, (1967), 884–891.

59. R. C. Ropp, M. A. Aia, C. W. W. Hoffman, T. J. Veleker, and R. W. Mooney, *Anal. Chem.*, **31**, (1959), 1164–1167.

60. R. C. Ropp and M. A. Aia, *Anal. Chem.*, **34**, (1962), 1288–1291.

61. A. Durif, M. Bagieu-Beucher, C. Martin, and J. C. Grenier, *Bull. Soc. fr. Minéral. Cristallogr.*, **95**, (1972), 146–148.

62.   M. I. Kuz'menkov, V. V. Pechkovskii, and S. V. Plyshevskii, *Russ. J. Inorg. Chem.*, **17**, (1972), 985–987.

63.   J. C. Grenier and C. Martin, *Bull. Soc. fr. Minéral. Cristallogr.*, **98**, (1975), 107–110.

64.   J. C. Grenier, C. Martin, A. Durif, D. Tranqui, and J. C. Guitel, *Bull. Soc. fr. Minéral. Cristallogr.*, **90**, (1967), 24–31.

65.   J. Coing-Boyat, M. T. Averbuch-Pouchot, and J. C. Guitel, *Acta Crystallogr.*, **B 34**, (1978), 2689–2692.

66.   J. J. Brown and F. A. Hummel, *J. Electrochem. Soc.*, **111**, (1964), 660–665.

67.   E. Thilo and I. Grunze, *Z. anorg. allg. Chem.*, **290**, (1957), 209–222.

68.   M. Beucher and I. Tordjman, *Bull. Soc. fr. Minéral. Cristallogr.*, **91**, (1968), 207–209.

69.   I. Tordjman, M. Beucher, J. C. Guitel, and G. Bassi, *Bull. Soc. fr. Minéral. Cristallogr.*, **91**, (1968), 344–349.

70.   M. Bagieu-Beucher, J. C. Guitel, I. Tordjman, and A. Durif, *Bull. Soc. fr. Minéral. Cristallogr.*, **97**, (1974), 481–484.

71.   M. Laügt, M. Bagieu-Beucher, and J. C. Grenier, *C. R. Acad. Sci.*, Sér. C, **275**, (1972), 1283–1285.

72.   M. Bagieu-Beucher, M. Brunel-Laügt, and J. C. Guitel, *Acta Crystallogr.*, **B35**, (1979), 292–295.

73.   K. R. Andress and K. Fischer, *Z. anorg. allg. Chem.*, **273**, (1953), 193–199.

74.   K. H. Jost, *Acta Crystallogr.*, **17**, (1964), 1539–1544.

75.   M. Beucher, *Dissertation*, (Univ. of Grenoble, France, 1968).

76.   K. K. Palkina, S. I. Maksimova, A. V. Lavrov, and N. A. Chalisova, *Sov. Phys. Dokl.*, **23**, (1978), 691–692.

77.   A. Watanabe, S. Takenouchi, J. P. Vignancourt, P. Conflant, M. Drache, and J. C. Boivin, *J. Solid State Chem.*, **107**, (1993), 93–100.

78   M. Laügt, *C. R. Acad. Sci.*, Sér. C, **275**, (1972), 1197–1200.

79   I. A. Tokman and G. A. Bukhalova, *Russ. J. Inorg. Chem.*, **22**, (1977), 578–580.

80.   J. F. Sarver and F. A. Hummel, *J. Electrochem. Soc.*, **106**, (1959), 500–504.

81.   I. A. Tokman and G. A. Bukhalova, *Russ. J. Inorg. Chem.*, **22**, (1977), 578–580.

82. G. A. Bukhalova and I. A. Tokman, *Russ. J. Inorg. Chem.*, **22**, (1977), 1051–1052.

83. M. Laügt and J. C. Guitel, *Acta Crystallogr.*, **B 31**, (1975), 1148–1153.

84. M. Bagieu-Beucher and N. El-Horr, unpublished results.

85. I. V. Mardirosova, V. A. Matrosova, M. A. Savenkova, and G. A. Bukhalova, *Izv. Akad. Nauk SSSR, Neorg. Mater.*, **15**, (1979), 2079–2081.

86. G. A. Bukhalova, I. A. Tokman, and V. M. Shpakova, *Russ. J. Inorg. Chem.*, **15**, (1970), 865–866.

87. M. T. Averbuch-Pouchot, *J. Appl. Crystallogr.*, **8**, (1975), 389–390.

88. M. T. Averbuch-Pouchot, A. Durif, and J. C. Guitel, *Acta Crystallogr.*, **B31**, (1975), 2453–2456.

89. M. Laügt, *C. R. Acad. Sci.*, Sér. C, **269**, (1969), 1122–1124.

90. M. Laügt, I. Tordjman, J. C. Guitel, and M. Roudaut, *Acta Crystallogr.*, **B28**, (1972), 2352–2358.

91. G. A. Bukhalova, I. G. Rabkina, I. V. Mardirosova, and V. N. Mirnyl, *Ukrain. Khimi. Zhur.*, **41**, (1975), 1144–1147.

92. M. T. Averbuch-Pouchot and A. Durif, *J. Appl. Crystallogr.*, **5**, (1972), 307–308.

93. M. T. Averbuch-Pouchot and E. Rakotomahanina-Ralaisoa, *Bull. Soc. fr. Minéral. Cristallogr.*, **93**, (1970), 394–396.

94. P. de Pontcharra and A. Durif, *C. R. Acad. Sci.*, Sér. C, **278**, (1974), 175–178.

95. E. Rakotomahanina-Rolaisoa, *Thesis*, (Univ. of Grenoble, France, 1972).

96. Y. Henry and A. Durif, *C. R. Acad. Sci.*, Sér. C, **270**, (1970), 423–425.

97. C. Martin and A. Durif, *Bull. Soc. fr. Minéral. Cristallogr.*, **92**, (1969), 489–490.

98. J. C. Grenier and I. Mahama, *C. R. Acad. Sci.*, Sér. C, **274**, (1972), 1063–1065.

99. J. C. Guitel and M. Brunel-Laügt, *Acta Crystallogr.*, **B33**, (1977), 2713–2716.

100. N. El-Horr and M. Bagieu-Beucher, *Acta Crystallogr.*, **C42**, (1986), 647–651.

101. M. T. Averbuch-Pouchot and A. Durif, *Mat. Res. Bull.*, **4**, (1969), 859–868.

102. M. T. Averbuch-Pouchot, I. Tordjman, and J. C. Guitel, *Acta Crystallogr.*, **B32**, (1976), 2953–2956.

103. C. Raholison and M. T. Averbuch-Pouchot, *C. R. Acad. Sci.*, Sér. C, **274**, (1972), 1066–1068.

104. C. Martin, *Thesis*, (Univ. of Grenoble, France, 1972).

105. N. El-Horr, *Thesis*, (Univ. of Grenoble, France, 1988).

106. N. El-Horr, M. Bagieu, J. C. Guitel, and I. Tordjman, *Z. Kristallogr.*, **169**, (1984), 73–82.

107. E. A. Genkina, B. A. Maksimov, Yu. K. Kabalov, and O. K. Mel'nikov, *Dokl. Akad. Nauk SSSR*, **270**, (1983), 1113–1116.

108. E. A. Genkina, N. S. Triodina, O. K. Mel'nikov, and B. A. Maksimov, *Izv. Akad. Nauk SSSR, Neorg. Mater.*, **24**, (1988), 1158–1162.

109. F. d'Yvoire, *Bull. Soc. Chim.*, (1962), 1237–1243.

110. P. Rémy and A. Boullé, *Bull. Soc. Chim.*, (1972), 2215–2221.

111. M. Bagieu-Beucher, *Acta Crystallogr.*, **B34**, (1978), 1443–1446.

112. S. V. Plyshevskii, M. I. Kuz'menkov, and V. V. Pechkovskii, *Russ. J. Inorg. Chem.*, **22**, (1977), 475–476.

113. G. A. Bukhalova, I. V. Mardirosova, N. P. Vassel, and M. A. Savenkova, *Izv. Akad. Nauk SSSR, Neorg. Mater.*, **27**, (1991), 828–831.

114. N. N. Chudinova, M. A. Avaliani, and I. V. Tananaev, *Izv. Akad. Nauk SSSR, Ser. Khim.*, **5**, (1979), 373–375.

115. K. K. Palkina, S. I. Maksimova, and N. T. Chibiskova, *Izv. Akad. Nauk SSSR, Neorg. Mater.*, **17**, (1981), 95–98.

116. N. N. Chudinova, M. A. Avaliani, L. S. Guzeeva, and I. V. Tananaev, *Izv. Akad. Nauk SSSR, Neorg. Mater.*, **14**, (1978), 2054–2060.

117. K. K. Palkina, N. N. Chudinova, G. M. Balagina, S. I. Maksimova, and N. T. Chibiskova, *Izv. Akad. Nauk SSSR, Neorg. Mater.*, **18**, (1982), 1561–1566.

118. K. K. Palkina and K. H. Jost, *Acta Crystallogr.*, **B31**, (1975), 2285–2290.

119. K. K. Palkina, S. I. Maksimova, and N. T. Chibiskova, *Russ. J. Inorg. Chem.*, **38**, (1993), 750–779.

120. P. P. Mel'nikov, L. N. Komissarova, and T. A. Butuzova, *Izv. Akad. Nauk SSSR, Neorg. Mater.*, **17**, (1981), 2110–2112.

121. J. Matuszewski, J. Kropiwnicka, and T. Znamierowska, *J. Solid State Chem.*, **75**, (1988), 285–290.

122. E. N. Deichman, I. V. Tananaev, Zh. A. Ezhova, and K. K. Palkina, *Izv. Akad. Nauk SSSR, Neorg. Mater.*, **6**, (1970), 1645–1649.

123. H. Y-P. Hong, *Acta Crystallogr.*, **B30**, (1974), 1857–1861.

124. O. S. Tarasenkova, G. I. Dorokhova, N. N. Chudinova, B. N. Litvin, and N. V. Vinogradova, *Izv. Akad. Nauk SSSR, Neorg. Mater.*, **21**, (1985), 452–458.

125. N. Yu. Anisimova, V. K. Trunov, N. B. Karmanovskaya, and N. N. Chudinova, *Neorg. Mater.*, **28**, (1992), 441–444.

126. M. Rzaigui and N. Kbir-Ariguib, *Bull. Soc. Chim. Belg.*, **94**, (1985), 619–620.

127. G. I. Dorokhova and O. G. Karpov, *Sov. Phys. Crystallogr.*, **29**, (1984), 400–402.

128. K. K. Palkina and K. H. Jost, *Acta Crystallogr.*, **B31**, (1975), 2281–2285.

129. K. K. Palkina, N. N. Chudinova, B. N. Litvin, and N. V. Vinogradova, *Izv. Akad. Nauk SSSR, Neorg. Mater.*, **17**, (1981), 1501–1503.

130. M. F. Moktar, N. Kbir-Ariguib, and M. Trabelsi, *J. Solid State Chem.*, **38**, (1981), 133–137.

131. J. Nakano, S. Miyazawa, and T. Yamada, *Mater. Res. Bull.*, **14**, (1979), 21–26.

132. H. Koizumi, *Acta Crystallogr.*, **B 32**, (1976), 266–268.

133. H. Y-P. Hong, *Mater. Res. Bull.*, **10**, (1975), 635–640.

134. M. Rzaigui, M. Trabelsi, and N. Kbir-Ariguib, *C. R. Acad. Sci.*, Sér. 2, **292**, (1981), 505–508.

135. G. A. Bukhalova, I. V. Mardirosova, and M. M. Ali, *Dokl. Akad. Nauk SSSR, Neorg. Mater.*, **20**, (1984), 120–122.

136. M. Rzaigui, M. Dabbabi, and N. Kbir-Ariguib, *J. Chim. Phys.*, **78**, (1981), 563–566.

137. S. A. Linde, Yu. E. Gorbunova, and A. V. Lavrov, *Russ. J. Inorg. Chem.*, **28**, (1983), 804–807.

138. G. A. Bukhalova, R. S. Faustova, and M. A. Savenkova, *Zh. Prikl. Khim.*, **50**, (1977), 171–173.

139. G. A. Bukhalova, I. V. Mardirosova, and M. M. Ali, *Izv. Akad. Nauk SSSR, Neorg. Mater.*, **20**, (1984), 1405–1408.

140. G. A. Bukhalova, R. S. Faustova, and M. A. Savenkova, *Russ. J. Inorg. Chem.*, **2**, (1977), 778–779.

141. I. D. Sokolova, *Russ. J. Inorg. Chem.*, **11**, (1966), 502–503.

142. I. D. Sokolova, E. L. Krivovyazov, and N. K. Voskresenskaya, *Russ. J. Inorg. Chem.*, **8**, (1963), 1375–1378.

143. S. Ohashi and K. Yamagishi, *J. Jpn. Chem. Soc.*, **33**, (1960), 1431–

1435.

144. N. S. Slobodyanik, P. G. Nagornyi, and T. I. Zhunkovskaya, *Russ. J. Inorg. Chem.*, **26**, (1981), 838–839.

145. I. B. Markina and N. K. Voskresenskaya, *Russ. J. Inorg. Chem.*, **12**, (1967), 407–411.

146. V. P. Kochergin, Z. A. Shevrina, L. V. Paderova, and A. N. Kruglov, *Russ. J. Inorg. Chem.*, **22**, (1977), 22–23.

147. G. A. Bukhalova and I. V. Mardirosova, *Russ. J. Inorg. Chem.*, **11**, (1966), 85–87.

148. G. A. Bukhalova and I. V. Mardirosova, *Russ. J. Inorg. Chem.*, **12**, (1967) 1158–1161.

149. E. N. Kovarskaya and Yu. I. Rodionov, *Izv. Akad. Nauk SSSR, Neorg. Mater.*, **24**, (1988), 642–645.

150. E. N. Kovarskaya, V. S. Mityakhina, Yu. I. Rodionov, and M. Yo. Silin, *Izv. Akad. Nauk SSSR, Neorg. Mater.*, **24**, (1988), 655–660.

151. B. Thonnerieux, D. Tranqui, A. Durif, and M. T. Averbuch-Pouchot, *C. R. Acad. Sci., Sér. C*, **266**, (1968), 208–210.

152. N. Yu. Anisimova, V. K. Trunov, N. B. Karmanovskaya, and N. N. Chudinova, *Izv. Akad. Nauk SSSR, Neorg. Mater.*, **28**, (1992), 441–444.

153. K. H. Jost and H. J. Schülze, *Acta Crystallogr.*, **B25**, (1969), 1110–1118.

154. K. H. Jost and H. J. Schülze, *Acta Crystallogr.*, **B27**, (1971), 1345–1353.

155. W. W. Schmahl, *Z. Kristallogr.*, **178**, (1987), 197–198.

156. A. B. Bekturov, D. Z. Serazetdinov, E. V. Poletaev, and S. M. Divnenko, *Russ. J. Inorg. Chem.*, **13**, (1968), 20–23.

157. N. N. Chudinova, L. A. Borodina, U. Schülke, and K. H. Jost, *Izv. Akad. Nauk SSSR, Neorg. Mater.*, **25**, (1989), 459–465.

158. C. Holst, W. W. Schmahl, and H. Fuess, *Z. Kristallogr.*, **209**, (1994), 322–327.

159. M. T. Averbuch-Pouchot and A. Durif, *J. Solid State Chem.*, **49**, (1983), 341–352.

160. F. Liebau, *Structural Chemistry of Silicates*, (Springer-Verlag, Berlin, 1985).

161. M. Bagieu-Beucher and J. C. Guitel, *Z. anorg. allg. Chem.*, **559**, (1988), 123–130.

162. J. C. Grenier, C. Martin, and A. Durif, *Bull. Soc. fr. Minéral. Cristallogr.*,

93, (1970), 52–55.

163. E. J. Griffith, *Phosphorus Chemistry*, (1981), 361–365.

164. E. J. Griffith, J. A. Hinkebein, R. L. Hansen, M. M. Crutchfield, W. C. McDaniel, and T. Ngo, *AIChE meeting*, (New Orleans, USA, 1981).

5.4. CYCLOPHOSPHATES

5.4.1. Introduction

As we have already said in Chapter 4 on phosphate nomenclature, condensed phosphates belonging to this family are characterized by cyclic $[P_nO_{3n}]^{n-}$ anions, built by n corner-sharing PO_4 tetrahedra. Ring anions of this type are known to exist for n = 3, 4, 5, 6, 8, 9, 10, and 12.

Systematic studies of cyclotri- and cyclotetraphosphates were carried out for the last thirty years. On the contrary, the studies of cyclohexaphophates did not start seriously before 1985, that of cycloocta- and cyclodecaphosphates before 1990, and that of cyclododecaphosphates before 1994. The discovery of the existence of cyclononaphosphates only occured some weeks ago.

In this field and in almost all the other classes of condensed phosphates, with some rare exceptions, the basic properties of most of the characterized compounds have not been investigated still.

For each family of cyclophosphates, the chemical preparations are very specific and tributary of a flux method for non-soluble salts and of the elaboration of an appropriate starting material for the other ones. In all cases, the optimizations of the processes were very long. Thus, we shall report the specific chemical preparations for each family of cyclophosphates separately.

As in the family of long-chain polyphosphates, the acidic anions are very rare. One can only report $Na_2HP_3O_9$ characterized with a good accuracy and may be $Na_2H_2P_4O_{12}$ still not yet clearly investigated.

Two reviews dealing with some topics of crystal chemistry of cyclophosphates were recently published [1–2]. Therefore, we will give in this section more details on some recent novelties in the development of cyclodeca- and cyclododecaphosphate chemistry and the recent discovery of cyclononaphosphates.

5.4.2. General Properties

One could expect specific properties for each family of cyclophosphates. In fact and with rare exceptions, the general properties of cyclophosphates do not vary considerably with the ring size. Their chemical behavior seems mainly dependent on the nature of the associated cations as in many other chemical families.

Thus, the *water solubility* of alkali cyclophosphates seems to be independent on the ring size. For instance, the recently characterized potassium cyclodecaphosphate and potassium cyclododecaphosphate, $K_{10}P_{10}O_{30}.4H_2O$ and $K_{12}P_{12}O_{36}.19/2H_2O$, show very high water solubility compared with the potassium cyclotri- or cyclotetraphosphates. A study of the solubilities of the various monovalent-cation cyclooctaphosphates [3] shows that the solubility of the sodium derivative (0.029 mole/liter) is considerably lower than those of the other alkali and ammonium derivatives which vary from 0.43 to 1.3 mole/liter. For cyclophosphates with higher-valency associated cations, despite the lack of quantitative measurements we can say that their solubility is considerably lower.

The *thermal behavior* of cyclophosphates also seems to be very dependent on the nature of the associated cations. As far as we know, all monovalent-cation cyclophosphates transform into long chain polyphosphates when heated. The evolution is different for salts of divalent cations. Some of them, like calcium, cadmium, strontium, barium, and lead cyclophosphates, transform also into long-chain polyphosphates, but the magnesium, cobalt, nickel, copper, manganese salts are stable as cyclotetraphosphates up to their melting points.

For mixed-cation cyclophosphates, the thermal evolution does not obey any rule. Some of them, as $BaNaP_3O_9$ or $SrK_2P_4O_{12}$, for instance, are stable as cyclophosphates up to their melting points, whereas others including all the incongruent melting salts, decompose. For instance, calcium-potassium cyclotetraphosphate, $CaK_2P_4O_{12}$, decomposes into a cyclotriphosphate and a long-chain polyphosphate when heated up to 973 K:

$$CaK_2P_4O_{12} \longrightarrow CaKP_3O_9 + KPO_3$$

As all condensed phosphoric anions, the P_nO_{3n} rings decondense by *hydrolysis*. The water-soluble cyclophosphates are generally stable in an approximately neutral aqueous solution at room temperature. The rate of hydrolysis is rather low under these conditions and it seems that this stability increases with the ring size. Nevertheless, as in all condensed phosphates, the cyclophosphates hydrolyze upon departure from these conditions. This decyclization process can take various aspects depending on the nature of the associated cation. For instance, potassium cyclotriphosphate is rapidly transformed into KH_2PO_4 when heated slowly in a wet atmosphere or simply kept in hot water,

$$K_3P_3O_9 + 3H_2O \longrightarrow 3KH_2PO_4$$

but, for lanthanum cyclotriphosphate, this hydrolysis is accompanied by the formation of phosphoric acid:

$$LnP_3O_9.3H_2O \longrightarrow LnPO_4 + 2H_3PO_4$$

It was thought for a time, that the opening of the phosphoric rings by hydrolysis in a basic medium could be a proper way to prepare oligophosphates according to the following schematic process:

$$M_nP_nO_{3n} + 2MOH \longrightarrow M_{(n+2)}P_nO_{(3n+1)} + H_2O$$

Probably because it was difficult to master, this type of reaction is most of time deceptive. Only a few examples of oligophosphates prepared by this way can be reported.

Sotnikova-Yuzhik et al. [4] prepared $Li_5P_3O_{10}.5H_2O$, a crystalline lithium triphosphate, by hydrolytic decyclization of $Li_3P_3O_9$:

$$\overset{313-353\,K}{Li_3P_3O_9 + 2LiOH \longrightarrow Li_5P_3O_{10} + H_2O}$$

For a long time, the only suggested method for the preparation of water-soluble tetraphosphates was the alkaline hydrolysis of cyclotetraphosphates based on the following scheme:

$$H_4P_4O_{12} + H_2O \longrightarrow H_6P_4O_{13}$$

Various investigations were performed by Thilo and Ratz [5], Westman and Scott [6], Quimby [7] and Watters et al. [8] using this process. The conditions of preparation by this method are tedious, time consuming and deceptive most of time. For instance, Thilo and Ratz [5] described an attempt to prepare $Na_6P_4O_{13}$ by hydrolytic decyclization of sodium cyclotetraphosphate,

$$Na_4P_4O_{12} + 2NaOH(H_2O) \longrightarrow Na_6P_4O_{13}.xH_2O$$

but could not characterize the final oily product which decomposes rapidly:

$$Na_6P_4O_{13}.xH_2O \longrightarrow 2Na_3HP_2O_7.H_2O$$

It was observed that the addition of zinc chloride to an aqueous solution of sodium cyclotriphosphate provokes a rapid opening of the ring anion and that crystals of a zinc-sodium triphosphate, $Zn_2NaP_3O_{10}.9H_2O$, appear in the solution after some hours.

$[Co(NH_3)_6]_2P_4O_{13}.5H_2O$ reported by Schulz and Jansen [9] is the only example of preparation of a crystalline tetraphosphate by controlled hydrolysis of a P_4O_{12} ring anion till now.

5.4.3. Cyclotriphosphates

5.4.3.1. Introduction

During the last century, several chemists have suspected the existence of cyclic phosphoric anions and have produced several crystalline compounds possessing this type of anions. For instance, sodium cyclotriphosphate seems to have been prepared as early as 1833 by Graham [10]. In the middle of the last century, Fleitmann and Henneberg [11], using the Graham's principles, prepared several salts and double salts, today recognized as cyclotriphosphates. They also suggested the trimeric nature of the anion. During the second part of the last century, chemists, like Lindbom [12], Tammann [13–14], Von Knorre [15], described a good number of additional cyclotriphosphates. It must be recognized that most of them were reproduced and clearly characterized as cyclotriphosphates recently. During the first half of our century, this part of chemistry did not develop significantly because no proper starting material was available for syntheses in sufficient amounts and of sufficient purity. The chemical preparations reported for the elaboration of sodium cyclotriphosphate, the parent compound of almost all the compounds prepared during the last century, were uneasy, tedious, and in many cases not reproducible. Nevertheless, as early as 1938, Boullé [16] succeeded in preparing silver cyclotriphosphate monohydrate, $Ag_3P_3O_9.H_2O$, and showed how this salt can be used through a metathesis reaction to synthesize any kind of water-soluble cyclotriphosphates. This method has been extensively used since then and is known today as the Boullé's process. This process will be described in details in the part of the review devoted to the general methods of preparation of cyclotriphosphates. It was not before the early fifties that a careful study of the thermal reorganization of NaH_2PO_4 by Thilo and Grunze [17] opened the way to a reproducible production of pure $Na_3P_3O_9$, according to the scheme:

$$3NaH_2PO_4 \longrightarrow Na_3P_3O_9 + 3H_2O$$

The first structural evidence for the cyclic nature of the anion, already clearly suggested by Lindbom [12], was reported by Eanes and Ondik [18] in 1962, when performing the crystal structure of $LiK_2P_3O_9.H_2O$.

During the last 30 years, more than one hundred cyclotriphosphates were characterized.

5.4.3.2. Present state of cyclotriphosphate chemistry

Monovalent-cation cyclotriphosphates – Cyclotriphosphates are known for all monovalent cations to date. The sodium derivatives were relatively well investigated, but very little is known concerning the basic properties of the others outside their atomic arrangements, generally determined with accuracy. They crystallize with various hydration states. In Table 5.4.1, we report the list of these compounds and their methods of preparation.

The potassium and ammonium salts are isotypic, as well as the rubidium and cesium monohydrates.

The four anhydrous $Na_2M^IP_3O_9$ are also isotypic and their atomic arrangements are identical with the margarosanite silicates. We shall come back further on this isomorphy when discussing the rare analogies encountered between cyclotriphosphates and silicates.

Table 5.4.1
Methods of preparation of monovalent cyclotriphosphates.

Formula	Preparation method	Ref.
$Li_3P_3O_9.3H_2O$	Boullé's process	[19–21, 22]
$Na_2HP_3O_9$	Flux method	[23–24]
$Na_3P_3O_9$	Thermal process	[25–26]
$Na_3P_3O_9.H_2O$	Crystallization in H_2O	[25–26]
$Na_3P_3O_9.6H_2O$	Crystallization in H_2O	[25, 27–28]
$Ag_3P_3O_9.H_2O$	Aqueous chemistry	[11–12, 15, 29]
$(NH_4)_3P_3O_9$	Boullé's process	[30–31]
$K_3P_3O_9$	Boullé's process	[30–33]
$Rb_3P_3O_9.H_2O$	Boullé's process	[30]
$Cs_3P_3O_9.H_2O$	Boullé's process	[30, 34]
$Tl_3P_3O_9$	Boullé's process	[30, 35]
$Na_2KP_3O_9$	Phase-equilibrium diagram	[36–41]
$Na_2NH_4P_3O_9$	Boullé's process	[42]
$Na_2RbP_3O_9$	Phase-equilibrium diagram	[43]
$Na_2TlP_3O_9$	Boullé's process	[42]
$LiK_2P_3O_9.H_2O$	Boullé's process	[18]
$LiNa_2P_3O_9.4H_2O$	Exchange-resin technique	[44]
$Na_2CsP_3O_9.2H_2O$	Exchange-resin technique	[35]

The elaboration of phase-equilibrium diagrams played an important rôle in the development of the crystal chemistry of condensed phosphates. Thus, it seems interesting to illustrate this section by giving (Figure 5.4.1) the $RbPO_3$–$NaPO_3$ phase-equilibrium diagram elaborated by Cavero-Ghersi and Durif [43].

To illustrate this class of compounds, it is worth reporting the atomic arrangement of sodium cyclotriphosphate. This salt is the first known cyclophosphate. It was characterized as early as 1833 by Graham [10] and was known, for a long time, as the soluble form of "sodium metaphosphate" but it is not before 1965 that its atomic arrangement was determined by Ondik [26]. This salt, whose chemical preparation is described with details at the end of this

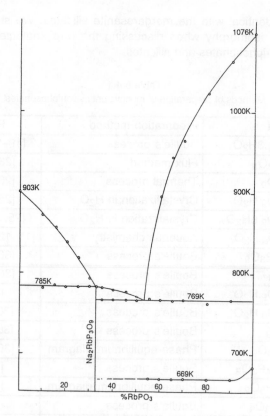

Figure 5.4.1. The $RbPO_3$–$NaPO_3$ phase-equilibrium diagram as elaborated by Cavero-Ghersi and Durif [43].

section, remains the basic material for the elaboration of most of cyclotriphosphates today.

$Na_3P_3O_9$ is orthorhombic, *Pnma*, with $Z = 4$ and the following unit-cell dimensions:

$$a = 13.214, \quad b = 7.708, \quad c = 7.928 \text{ Å}$$

Figure 5.4.2 shows its arrangement in projection along the **c** axis. The phosphoric ring anion has a mirror symmetry. One phosphorus atom, one bonding and two external oxygen atoms are located in the mirror plane. In addition, the author [26] noticed that this ring has a strong pseudo-3m symmetry, probably induced by the pseudo-hexagonal character of the arrangement reflected by the metrics of the unit cell. These ring anions form corrugated festoons spreading along the **b** direction and lined by the sodium atoms. There are

Figure 5.4.2. Projection, along the **c** direction, of the atomic arrangement in $Na_3P_3O_9$. Phosphoric ring anions are given by polyhedral representation and Na atoms by open circles.

two crystallographically independent sodium atoms, one located on the mirror plane, the second in general crystallographic position. Both sodium atoms are fivefold coordinated and the NaO_5 polyhedra build a three-dimensional network through edge and corner sharing. The Na–O distances, within the NaO_5 polyhedra, range between 2.337 and 2.466 Å.

Divalent-cation cyclotriphosphates – Up to now, all the divalent-cation cyclotriphosphates that have been clearly characterized are hydrated. In Table 5.4.2, we list them together with their main crystallographic data. In spite of repetitive claims for the preparation of anhydrous divalent-cation cyclotriphosphates, no structural evidence is given for their existence. Some hydrated divalent-cation cyclotriphosphates were reported and clearly characterized during the last century. For instance, the case of the manganese salt described successively by Tammann [14] and by Von Knorre [15] and of the lead salt whose existence was reported for the first time in 1848 by Fleitman and Henneberg [11].

All the derivatives listed in Table 5.4.2 were prepared using the Boullé's process. Among them, $Cd_3(P_3O_9)_2.14H_2O$ was the object of several physico-chemical investigations due to its interesting zeolitic properties. Therefore, we have selected to illustrate it in this section.

This compound was first characterized and investigated by Averbuch-Pouchot and Durif [45] and Averbuch-Pouchot *et al.* [46–47]. During the structural investigations, it was clearly established that eight of the fourteen water molecules are of a zeolitic nature and the six remaining ones are involved in the cadmium coordination polyhedron.

Table 5.4.2

Main crystallographic data for the divalent-cation cyclotriphosphates.

Formula	a α	b β	c (Å) γ(°)	S. G.	Z	Ref.
$Cd_3(P_3O_9)_2.14H_2O$	12.228(3)	12.228(3)	5.451(3)	$P\bar{3}$	1	[45–47]
$Cd_3(P_3O_9)_2.10H_2O$	9.424(8)	17.87(1) 107.72(1)	7.762(7)	$P2_1/n$	2	[48]
$Ca_3(P_3O_9)_2.10H_2O$	9.332(7)	18.13(1) 106.69(5)	7.841(5)	$P2_1/n$	2	[49]
$Mn_3(P_3O_9)_2.10H_2O$	9.219(4)	17.733(8) 107.37(2)	7.644(3)	$P2_1/n$	2	[49]
$Sr_3(P_3O_9)_2.7H_2O$	16.05(1)	12.33(1)	10.87(1)	$Pnma$	4	[50–51]
$Ba_3(P_3O_9)_2.6H_2O$	7.547(4) 108.58(8)	11.975(6) 100.35(8)	13.068(8) 95.54(8)	$P\bar{1}$	2	[52–53]
$Ba_3(P_3O_9)_2.4H_2O$	16.09(1)	8.368(5) 95.38(5)	7.717(3)	$C2/m$	2	[54]
$Pb_3(P_3O_9)_2.3H_2O$	11.957(5)	11.957(5)	12.270(5)	$P4_12_12$	4	[55–56]
$Ba_2Zn(P_3O_9)_2.10H_2O$	26.52(3)	7.625(5) 100.93(5)	12.92(1)	$C2/c$	4	[57]

Figure 5.4.3 gives the projection, along the **c** direction, of a part of its arrangement mainly characterized by the existence of large channels containing the zeolitic water molecules around the ternary axes. The phosphoric ring anions located around the internal threefold axes are interconnected by $CdO_4(H_2O)_2$ octahedra to build, around the $\bar{3}$ axes, large channels parallel to the **c** direction. Within these channels are located the zeolitic water molecules. Only six of them could be clearly localized, the last two are probably located on the threefold axes. In Figure 5.4.4 a projection along the [110] direction shows how the cohesion of the arrays of $CdO_4(H_2O)_2$ octahedra and phosphoric rings occurs along the **c** direction. Each cadmium atom shares its four oxygen atoms with four different ring anions thus forming the cohesion both in the (a, a) plane and along the **c** direction. Within the $CdO_4(H_2O)_2$ octahedron the Cd–O distances range from 2.20 to 2.28 Å and the Cd–H_2O distance is 2.29 Å.

The zeolitic properties of these eight water molecules have been carefully

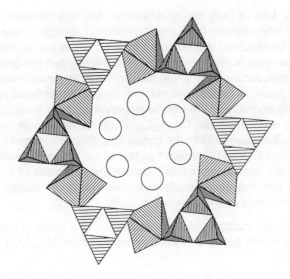

Figure 5.4.3. The atomic organization of $Cd_3(P_3O_9)_2.14H_2O$ around the $\bar{3}$ axis, in projection along the **c** direction. PO_4 tetrahedra and $CdO_4(H_2O)_2$ octahedra are given by polyhedral representation and zeolitic water molecules represented by open circles.

Figure 5.4.4. Projection, along [110], of a part of the atomic arrangement in $Cd_3(P_3O_9)_2.14H_2O$, showing how the cohesion along the **c** direction is performed. The unshared corners of the $CdO_4(H_2O)_2$ octahedra correspond to the non-zeolitic water molecules.

investigated by Michot [58], Simonot-Grange [59] and Simonot-Grange and Michot [60–61] who showed that, under dynamical vacuum, these eight water molecules can be removed without any alteration of the atomic framework and that this phenomenon is reversible. They also investigated the decondensation scheme of the tetradecahydrate in wet atmospheres, at various temperatures.

For some reasons which are not yet clearly understood, the Boullé's process, leading normally to the tetradecahydrate at room temperature, some-times produces a decahydrate, $Cd_3(P_3O_9)_2.10H_2O$, isotypic of the correspon-ding calcium or manganese salts. Its crystal structure determined by Averbuch-Pouchot, Durif and Guitel [48] shows that two of the ten water molecules are of zeolitic nature. No relationship exists between the structures of the two hydra-tes, explaining why the decahydrate was never observed during dehydration experiments performed on the tetradecahydrate.

Divalent-monovalent cation cyclotriphosphates – Two types of stoichiometries were observed in this family of cyclotriphosphates: $M^{II}M^{I}_4(P_3O_9)_2.nH_2O$ and $M^{II}M^{I}P_3O_9.nH_2O$.

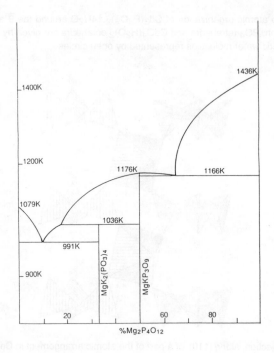

Figure 5.4.5. The $Mg_2P_4O_{12}$–KPO_3 system as elaborated by Averbuch-Pouchot *et al.* [74].

Most of the *anhydrous salts* were characterized during the elaborations of $M^I PO_3$–$M^{II}(PO_3)_2$ phase-equilibrium diagrams and prepared as single crystals by various flux methods, whereas the great majority of the hydrates was syn-

Table 5.4.3
$M^{II}(PO_3)_2$–$M^I PO_3$ systems and cyclotriphosphates found in them.

System	Ref.	Compound	mp (K)	Ref.
$Ca(PO_3)_2$–$NaPO_3$	[62–63]	$CaNa_4(P_3O_9)_2$	1006	[62, 64]
$Cd(PO_3)_2$–$NaPO_3$	[65]	$CdNa_4(P_3O_9)_2$	968	[65]
$Pb(PO_3)_2$–$NaPO_3$	[66]	$PbNa_4(P_3O_9)_2$	913	[67]
		$PbNaP_3O_9$ (a)		
$Ba(PO_3)_2$–$NaPO_3$	[68]	$BaNaP_3O_9$	939	[69]
		$BaNa_4(P_3O_9)_2$	956 (d)	[67]
$Cd(PO_3)_2$–$AgPO_3$	[70]	$CdAgP_3O_9$	923 (d)	[71]
$Sr(PO_3)_2$–$AgPO_3$	[72]	$SrAgP_3O_9$ (c)	926 (d)	
$Mg_2P_4O_{12}$–KPO_3	[73–74]	$MgKP_3O_9$	1176	[75–76]
$Ca(PO_3)_2$–KPO_3	[73]	$CaKP_3O_9$ (b)	1123	[75]
$Zn(PO_3)_2$–KPO_3	[74, 77]	$ZnKP_3O_9$	1111	[75]
$Co_2P_4O_{12}$–KPO_3	[78]	$CoKP_3O_9$ (b)	1069	[75]
$Cd(PO_3)_2$–KPO_3	[79]	$CdKP_3O_9$	1021	[75]
$Mg_2P_4O_{12}$–$RbPO_3$	[80]	$MgRbP_3O_9$	1139 (d)	[80]
$Cd(PO_3)_2$–$RbPO_3$	[79]	$CdRbP_3O_9$	1024	[71]
$Ca(PO_3)_2$–$RbPO_3$	[81]	$CaRbP_3O_9$	1180	[81]
$Cd(PO_3)_2$–$CsPO_3$	[82]	$CdCsP_3O_9$	996	[83]
$Sr(PO_3)_2$–$CsPO_3$	[84]	$SrCsP_3O_9$ (c)	1078 (d)	[84]
$Mg_2P_4O_{12}$–$TlPO_3$	[85]	$MgTlP_3O_9$	1031 (d)	[85]
		$MgTl_4(P_3O_9)_2$	763 (d)	[78, 85]
$Co_2P_4O_{12}$–$TlPO_3$	[85]	$CoTl_4(P_3O_9)_2$	754 (d)	[85]
$Zn(PO_3)_2$–$TlPO_3$	[86]	$ZnTl_4(P_3O_9)_2$	713 (d)	[86]
$Cd(PO_3)_2$–$TlPO_3$	[70, 87]	$CdTlP_3O_9$	962 (d)	[71]
		$CdTl_4(P_3O_9)_2$	749 (d)	[87]
$Ca(PO_3)_2$–$TlPO_3$	[85]	$CaTlP_3O_9$	1089	[85]
		$CaTl_4(P_3O_9)_2$	793 (d)	[85]

a: does not appear in the phase-equilibrium diagram
b: dimorphic salt
c: no structural proof for the nature of the anion
d: decomposition temperature for non-congruent meltings

thesized by the Boullé's process [16] or by the classical methods of aqueous chemistry.

As mentioned above, the great majority of the anhydrous salts were characterized while studying the phase-equilibrium diagrams. As an example, we report in Figure 5.4.5, the $Mg_2P_4O_{12}-KPO_3$ phase-equilibrium diagram as

Table 5.4.4
Chemical preparation of some various cyclotriphosphates.

Formula	Preparation	Ref.
$BaAgP_3O_9$	Dehydration of the tetrahydrate	[88]
$MnKP_3O_9$	Thermal process	[75]
$HgKP_3O_9$	Flux method	[89]
$CaCsP_3O_9$	Flux method	[90]
$M^{II}NH_4P_3O_9$ (*)	Flux or thermal processes	[76–77, 89, 91]
$HgNa_2(NH_4)_2(P_3O_9)_2$	Flux method	[92]

* M^{II} = Zn, Co, Ca, Cd, Mg, and Mn

Table 5.4.5
$M^IM^{II}P_3O_9.H_2O$ cyclotriphosphates presently characterized.

Formula	Preparation	Ref.
$CaNaP_3O_9.3H_2O$	Classical aqueous chemistry	[93–94]
$SrNaP_3O_9.3H_2O$	Classical aqueous chemistry	[68, 95]
$BaNaP_3O_9.4H_2O$	Classical aqueous chemistry	[68]
$BaNaP_3O_9.3H_2O$	Dehydration of $BaNaP_3O_9.4H_2O$	[96]
$BaAgP_3O_9.4H_2O$	Boullé's process	[97–98]
$SrKP_3O_9.3H_2O$	Boullé's process	[99]
$BaKP_3O_9.H_2O$	Boullé's process	[99–100]
$CaNH_4P_3O_9.3H_2O$	Boullé's process	[101]
$SrNH_4P_3O_9.4H_2O$	Boullé's process	[99]
$SrNH_4P_3O_9.3H_2O$	Boullé's process	[102]
$BaNH_4P_3O_9.H_2O$	Boullé's process	[103]
$SrRbP_3O_9.4H_2O$	Boullé's process	[99]
$SrTlP_3O_9.4H_2O$	Boullé's process	[99]
$BaTlP_3O_9.H_2O$	Boullé's process	[103]
$BaCsP_3O_9.H_2O$	Boullé's process	[104]

revised by Averbuch-Pouchot *et al.* [74]. Table 5.4.3 reports the list of the systems studied and of the cyclotriphosphates found as intermediate compounds in these systems. The table also includes the temperature and the nature of the melt.

In addition to these results obtained through studying phase-equilibrium diagrams, some other anhydrous cyclotriphosphates were characterized by various other methods. We list them in Table 5.4.4 with their methods of preparation indicated. This part of the section will be illustrated further when we discuss the structural analogies between cyclotriphosphates and silicates in the case of the benitoite-like compounds $M^{II}M^{I}P_3O_9$.

All the *hydrated compounds* that we list in Tables 5.4.5 and 5.4.6 were prepared either by the Boullé's process or by the classical methods of aqueous chemistry.

Table 5.4.6

$M_4^{I}M^{II}P_3O_9.H_2O$ cyclotriphosphates presently characterized.

Formula	Ref.	Formula	Ref.
$NiNa_4(P_3O_9)_2.6H_2O$	[105–106]	$CdK_4(P_3O_9)_2.2H_2O$	[111]
$CuNa_4(P_3O_9)_2.4H_2O$	[107]	$Cu(NH_4)_4(P_3O_9)_2.4H_2O$	[110]
$NiAg_4(P_3O_9)_2.6H_2O$	[105–106]	$Ni(NH_4)_4(P_3O_9)_2.4H_2O$	[112]
$ZnK_4(P_3O_9)_2.6H_2O$ (*)	[108]	$Co(NH_4)_4(P_3O_9)_2.4H_2O$	[113]
$NiK_4(P_3O_9)_2.7H_2O$	[109]	$CoRb_4(P_3O_9)_2.6H_2O$	[114]
$CoK_4(P_3O_9)_2.7H_2O$	[109]	$NiCs_4(P_3O_9)_2.6H_2O$	[105]
$CuK_4(P_3O_9)_2.4H_2O$	[110]		

* erroneously described as a tetrahydrate

We will describe the atomic arrangement of *NiK₄(P₃O₉)₂.7H₂O* as an example of a hydrated cyclotriphosphate.

$NiK_4(P_3O_9)_2.7H_2O$ is orthorhombic, *Fm2m*, with $Z = 4$ and the following unit-cell dimensions:

$$a = 23.03(1), \quad b = 11.882(4), \quad c = 8.732(4) \text{ Å}$$

In this structure, as in $Na_3P_3O_9$, the phosphoric ring anion has mirror symmetry with one phosphorus, two external oxygen and one bonding oxygen atom located in the mirror plane. The nickel atom, at the origin of the unit cell, is octahedrally coordinated by six water molecules. As shown by Figure 5.4.6, these $Ni(H_2O)_6$ groups do not share an edge or a corner. Inside these groups, the Ni–H_2O distances spread between 2.023 and 2.082 Å. The two crystallographically independent potassium atoms have a sixfold coordination, with K–O distances

Figure 5.4.6. Projection, along the **b** axis, of the atomic arrangement of NiK$_4$(P$_3$O$_9$)$_2$.7H$_2$O. The phosphoric anions and the Ni(H$_2$O)$_6$ octahedra are shown as polyhedra. Open circles are the potassium atoms.

ranging from 2.694 to 2.826 Å. Through edge- and corner-sharing, these KO$_6$ polyhedra build layers parallel to the (a, c) plane. Such a layer is represented in Figure 5.4.7.

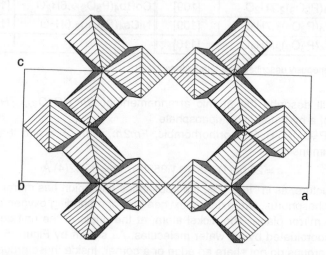

Figure 5.4.7. Projection, along the **b** axis, of a layer of edge- and corner-sharing KO$_6$ octahedra in NiK$_4$(P$_3$O$_9$)$_2$.7H$_2$O.

Trivalent and trivalent-monovalent cation cyclotriphosphates – This class of compounds has not been extensively investigated yet and only a small number of representatives are well characterized to date.

A series of $LnP_3O_9.3H_2O$ isotypic compounds (Ln = La, Ce, Nd, Pr) were prepared and characterized by Serra and Giesbrecht [115] and Bagieu-Beucher and Durif [119] and their crystal structure determined by Bagieu-Beucher et al. [120]. The thermal behavior of these compounds was carefully investigated by Gobled [121]. They are not stable in air and decompose irreversibly, according to the following scheme:

$$LnP_3O_9.3H_2O \longrightarrow LnPO_4 + P_2O_5 + 3H_2O$$

Two pentahydrates (Nd, Sm) were also characterized by Bagieu-Beucher [122] and Birke and Kempe [116–118] claimed the existence of other hydrates, $PrP_3O_9.4H_2O$, $LaP_3O_9.4H_2O$, and $ErP_3O_9.4.4H_2O$.

$M^{III}Na_3(P_3O_9)_2.9H_2O$ (M^{III} = Sm, Eu, Gd, Dy, Ho, Er, Y, and Bi) were prepared by Bagieu-Beucher [124] and their atomic arrangement determined by Bagieu-Beucher and Durif [123]. These salts are not stable for a long time under normal conditions.

Table 5.4.7
Crystal data for organic and alkali-organic cation cyclotriphosphates.

Formula	a α	b β	c (Å) γ (°)	S. G.	Z	Ref.
$(eda)_3(P_3O_9)_2$	15.558(8) 104.14(5)	10.450(6) 102.73(5)	7.639(4) 86.71(5)	$P\bar{1}$	2	[125]
$(ipa)_3P_3O_9$	25.22(2)	12.22(2) 123.90(2)	15.45(2)	$C2$	8	[126]
$(eda)KP_3O_9$	20.850(8)	9.044(4)	11.653(5)	$Ccca$	8	[127]
$(ma)_3P_3O_9$	12.144(7)	5.361(5) 97.32(8)	7.203(7)	$P2_1/n$	4	[128]
$(gly)_3P_3O_9$	12.223(8)	14.52(1) 100.47(5)	10.229(7)	$P2_1/c$	4	[129]
$(gua)_3P_3O_9.2H_2O$	12.140(8)	15.183(8) 97.49(5)	10.706(5)	$P2_1/n$	4	[130]

eda = $[NH_3–(CH_2)_2–NH_3]^{2+}$ or ethylenediammonium
ipa = $[(CH_3)_2–CH–NH_3]^+$ or isopropylammonium
ma = $(CH_3–NH_3)^+$ or methylammonium
gly = $(NH_3–CH_2–COOH)^+$ or glycinium
gua = $[C(NH_2)_3]^+$ or guanidinium

Organic and akali-organic cation cyclotriphosphates – It is only very recently, that the Boullé's metathesis reaction was extended by the authors to the syntheses of organic cation or mixed metal-organic cation cyclotriphosphates. This new chapter of condensed phosphate chemistry is still under development. We list, in Table 5.4.7, the first results obtained to date. All these salts are stable under normal conditions and their atomic arrangements have been accurately determined for all of them.

Remark – P_3O_9 ring anions also appear in many adducts between monovalent-cation cyclotriphosphates and telluric acid. These compounds will be reviewed in the section devoted to adducts (p. 344–346).

5.4.3.3. Analogies between cyclotriphosphates and silicates

When reviewing the family of monophosphates, we encountered several examples of structural analogies between phosphates and silicates and even between phosphates and various forms of silica. On the contrary, in the area of condensed phosphates, in spite of numerous analogies in the geometry of condensed silicate and phosphate anions, very few cases of isomorphism between silicates and phosphates were reported.

Thus, we think it is worth noticing that, in the family of cyclotriphosphates, two groups of compounds appeared as isostructural with two previously well established silicate-structure types.

In *the first group*, we find a series of four compounds, $Na_2KP_3O_9$,

Table 5.4.8

Unit-cell dimensions of a margarosanite-like silicate and of the four cyclotriphosphates isotypic of this silicate.

Formula	a α	b β	c (Å) γ (°)	Ref.
$Ca_2PbSi_3O_9$	6.768 110.36	9.575 102.98	6.718 83.02	[125]
$Na_2KP_3O_9$	6.886(2) 110.07(2)	9.494(3) 104.69(2)	6.797(2) 86.68(2)	[40–41]
$Na_2RbP_3O_9$	7.010(2) 108.98(5)	9.542(3) 104.26(5)	6.783(3) 87.37(5)	[43]
$Na_2TlP_3O_9$	6.977 108.70	9.511 104.40	6.787 86.82	[42]
$Na_2NH_4P_3O_9$	6.918 106.87	9.412 106.87	7.006 88.09	[42]

$Na_2RbP_3O_9$, $Na_2TlP_3O_9$, and $Na_2NH_4P_3O_9$, all isotypic of the margarosanite-like silicates: $Ca_2BaSi_3O_9$, $Ca_2PbSi_3O_9$... Their bimolecular unit-cell dimensions are given in Table 5.4.8. All these cyclotriphosphates are triclinic, $P\bar{1}$. The structure of $Na_2KP_3O_9$ was determined as an example of the atomic arrangement in margarasonite-like phosphates. This atomic arrangement is represented in projection along the **a** direction in Figure 5.4.8. As can be expected in a triclinic cyclotriphosphate, the ring anion has no internal symmetry. One of the two independent sodium atoms has a sixfold coordination, with Na–O distances ranging from 2.383 to 2.555 Å. The second one has only five nearest neighbors, with Na–O distances varying from 2.370 to 2.574 Å. The potassium atom has ten nearest neighbors, with K–O distances within the range: 2.690–3.356 Å. The NaO_6 and KO_{10} polyhedra build layers in $y = \pm 0.83$ planes alternating with layers of NaO_5 polyhedra located in planes $y = (2n+1)b/2$. The centers of the phosphoric rings are situated half-way between these layers.

In *the second group*, one finds that fifteen compounds are isotypic with the

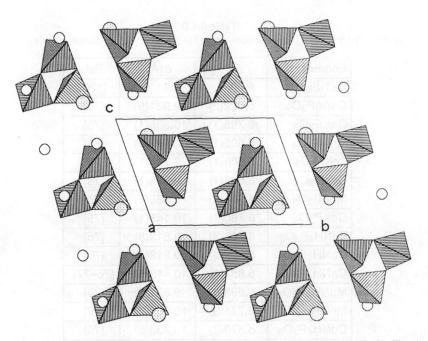

Figure 5.4.8. Projection, along the **a** direction, of the atomic arrangement in $Na_2KP_3O_9$. The ring anions are given by polyhedral representation, the larger circles are the potassium and the smaller the sodium atoms.

well-known silicate benitoite, $BaTiSi_3O_9$. The substitution scheme is:

$$Si_3O_9 \longrightarrow P_3O_9$$

$$Ti + Ba \longrightarrow M^{II} + M^{I}$$

All of them are hexagonal, $P\bar{6}c2$, and their unit cells gathered in Table 5.4.9.

We describe this type of atomic arrangement using the structural data obtained in the case of the calcium-ammonium salt, $CaNH_4P_3O_9$ [76–77].

As depicted by Figure 5.4.9, a projection along the **c** axis, the ring anions are built around the $\bar{6}$ axis, with the P atoms and their bonding oxygen atoms at $z = 1/4$ and $3/4$ on the mirror plane and therefore having $3/m$ internal symmetry. The divalent atoms, located on one of the $\bar{6}$ internal axis at $z = 0$ and $1/2$, have a regular octahedral coordination with 32 symmetry, whereas the monovalent cations are sited on the other internal $\bar{6}$ axis, at the same heights. Therefore, the structure can be described as built by layers of phosphoric groups alternating with layers of associated cations. Figure 5.4.10 shows clearly the alternance of these layers. In the present example, the Ca–O distance is 2.329 and the N–O distance 2.921 Å.

Table 5.4.9
Unit-cell dimensions of the benitoite and of the fifteen isotypic cyclotriphosphates.

Formula	a	c (Å)	Ref.
$BaTiSi_3O_9$	6.60	9.71	[126]
$CdAgP_3O_9$	6.622(2)	9.921(5)	[71]
$CaKP_3O_9$	6.795(1)	10.336(1)	[75]
$MgKP_3O_9$	6.605(1)	9.772(1)	[75]
$ZnKP_3O_9$	6.606(1)	9.743(1)	[75]
$CoKP_3O_9$	6.637(1)	9.895(1)	[75]
$MnKP_3O_9$	6.686(1)	9.958(1)	[75]
$CdKP_3O_9$	6.780(1)	10.148(1)	[75]
$ZnNH_4P_3O_9$	6.718(3)	9.819(5)	[76]
$CoNH_4P_3O_9$	6.695(3)	9.819(5)	[76]
$CaNH_4P_3O_9$	6.887(3)	10.448(5)	[76–77]
$MgNH_4P_3O_9$	6.698(3)	9.831(5)	[76]
$MnNH_4P_3O_9$	6.771(3)	10.026(5)	[76]
$CdNH_4P_3O_9$	6.870(3)	10.233(5)	[76]
$CdRbP_3O_9$	6.858(2)	10.211(5)	[71]
$CdTlP_3O_9$	6.845(2)	10.154(5)	[71]

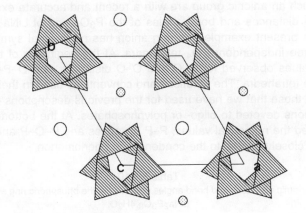

Figure 5.4.9. Projection, along the **c** direction, of the atomic arrangement in CaNH₄P₃O₉. The ring anions are given by polyhedral representation, the open circles are the ammonium groups, the grey ones the calcium atoms.

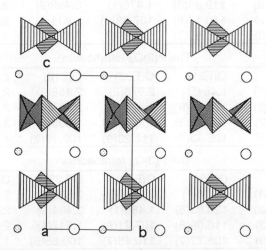

Figure 5.4.10. Projection, along the **a** direction, of the atomic arrangement of CaNH₄P₃O₉. Conventions used for this view are identical to those in Figure 5.4.9.

5.4.3.4. The P₃O₉ anion

Before starting any discussion on the general geometrical features of this type of rings, let us examine what the numerical values for the various components of such an anionic group are with a recent and accurate example. Table 5.4.10 lists distances and bond angles of the P_3O_9 ring of $LiNa_2P_3O_9.4H_2O$ [44]. In the present example, the ring anion has no internal symmetry and is built by three independent PO_4 tetrahedra. At the beginning of the table, we give the values observed for P–O and O–O distances and O–P–O angles in these three tetrahedra. The notations and conventions used in these tables are identical to those that we have used for the previous descriptions of tetrahedra in the sections devoted to oligo- or polyphosphates. At the bottom of the table are gathered the numerical values, P–P distances and P–O–P and P–P–P angles, more closely related to the condensation phenomenon.

Table 5.4.10
Interatomic distances and bond angles observed in the phosphoric ring anion of $LiNa_2P_3O_9.4H_2O$.

The $P(1)O_4$ tetrahedron

P(1)	O(E11)	O(E12)	O(L12)	O(L13)
O(E11)	1.477(1)	2.555(2)	2.533(2)	2.487(2)
O(E12)	119.89(9)	1.475(1)	2.492(2)	2.534(2)
O(L12)	110.17(8)	107.68(8)	1.611(1)	2.476(2)
O(L13)	106.99(7)	110.09(7)	100.27(7)	1.616(1)

The $P(2)O_4$ tetrahedron

P(2)	O(E21)	O(E22)	O(L12)	O(L23)
O(E21)	1.484(1)	2.576(2)	2.456(2)	2.476(2)
O(E22)	120.95(9)	1.477(1)	2.503(2)	2.537(2)
O(L12)	105.25(8)	108.56(9)	1.605(1)	2.491(2)
O(L23)	106.93(8)	111.23(9)	102.19(7)	1.596(1)

The $P(3)O_4$ tetrahedron

P(3)	O(E31)	O(E32)	O(L23)	O(L13)
O(E31)	1.476(1)	2.571(2)	2.536(2)	2.459(2)
O(E32)	120.83(8)	1.480(1)	2.494(2)	2.536(2)
O(L23)	110.09(8)	107.21(8)	1.617(1)	2.481(2)
O(L13)	105.77(7)	110.45(7)	100.65(6)	1.606(1)

P(1)–P(2)	2.8835(6)	P(1)–O(L12)–P(2)	127.48(8)
P(1)–P(3)	2.8959(6)	P(1)–O(L13)–P(3)	127.98(8)
P(2)–P(3)	2.8642(6)	P(2)–O(L23)–P(3)	126.12(8)

P(1)–P(2)–P(3)	60.51(1)
P(2)–P(3)–P(1)	60.08(1)
P(2)–P(1)–P(3)	59.42(1)

Inside a given tetrahedron, the observed values depart significantly from those that one could expect in a regular polyhedron. Thus, in the present example, the *P–O distances* corresponding to the oxygen atoms O(L) of the P–O–P bonds, range between 1.596 and 1.617 Å, whereas the P–O distances corresponding to the unshared oxygen atoms O(E) vary from 1.475 to 1.484 Å.

The *O–P–O angles* are also very dispersed. Those involving two unshared (or external) O(E) oxygen atoms are close to 120°. Those corresponding to two bonding O(L) oxygen atoms are, as usually, observed close to 100°. They range between 100.27 and 102.19° in the present example. For the last category of O–P–O angles, O(E)–P–O(L), the observed values generally vary between 105 and 110°. A range of 105.25–111.23° is observed in the present ring. The O–P–O average value in a given PO_4 tetrahedron never departs significantly from the theoretical one. Here, we observe three very similar values, 109.18, 109.18 and 109.17°. Quasi-identical observations can be done for any kind of PO_4 tetrahedron involved in the constitution of a condensed anion.

The three other numerical values found in Table 5.4.10, P–P distances, P–O–P and P–P–P angles are, on the contrary, slightly more dependent on the geometry of the anion, mainly the last one. The three *P–P distances* (2.864, 2.884 and 2.896 Å) give a good idea of the values generally observed. Due to the strains imposed by the geometrical nature of such a small ring, the *P–P–P angles* cannot depart significantly from 60°. The observed values range between 59.42 and 60.51°. The three *P–O(L)–P angles* of the ring, 126.12, 127.48, and 127.98°, correspond to values that we shall find repeatedly along our survey of ring anions.

In a recent detailed survey of cyclophosphates [2], involving the study of more than one hundred tetrahedra in cyclotriphosphates, the authors report the following ranges of values and averages for some distances and angles in P_3O_9 anions. We report these values in Table 5.4.11.

Table 5.4.11
Range and average for P–P distances, P–P–P and P–O(L)–P angles in cyclotriphosphates.

	Range	Average
P–P (Å)	2.818–2.974	2.890
P–P–P (°)	58.1–62.6	60 (*)
P–O(L)–P (°)	122.6–137.2	127.9

* by definition

P_3O_9 ring anions can adopt various *internal symmetries*. Up to now, rings with 3/m, 3, 2, and m symmetry were observed. One can note three main features:

   i) The distribution among these various symmetries is not regular. Thus, only two rings adopt the 3/m symmetry, two the 2 symmetry, five a 3 symmetry, six a m symmetry, and all the remaining ones (about forty) have no proper symmetry.

   ii) The averages and ranges given in the table are not significantly different for a set of rings with the same symmetry. Thus, the authors report [2] for the P–O–P angles an average value of 127.5° for rings with no internal symmetry and 128.2° for rings with a mirror symmetry.

   iii) The distribution of these values is regular. For instance, we report in Figure 5.4.11, the distribution of the P–O–P angles among the 111 values examined by the authors.

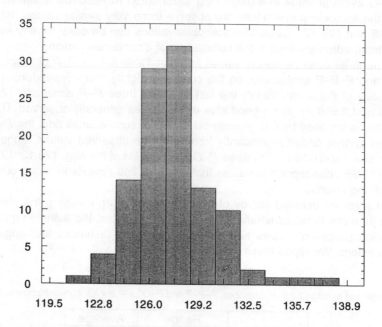

Figure 5.4.11. Distribution of the P–O–P angles in the P_3O_9 rings.

5.4.3.5. Chemical preparation of cyclotriphosphates

   Most of these preparations involve the use of sodium cyclotriphosphate as starting material.

Chemical preparation of sodium cyclotriphosphate – By heating NaH_2PO_4, in a temperature range of 803–823 K and for at least five hours, one obtains

$Na_3P_3O_9$ according to the following scheme:

$$3NaH_2PO_4 \longrightarrow Na_3P_3O_9 + 3H_2O$$

The cyclotriphosphate obtained is sometimes contaminated with a small amount (most of time less than 1%) of some insoluble sodium phosphates, mainly long-chain polyphosphates. A dissolution in water followed by a filtration eliminates these polyphosphates. In the resulting solution, kept at room temperature, crystals of the hexahydrate, $Na_3P_3O_9.6H_2O$, appear after some days. Firing the hexahydrate at 623 K leads to a very pure $Na_3P_3O_9$.

The Boullé's process – This metathesis reaction is widely used for the preparation of water-soluble cyclotriphosphates. The starting material is silver cyclotriphosphate monohydrate, $Ag_3P_3O_9.H_2O$. This salt, sparingly water soluble, is prepared by adding an aqueous solution of silver nitrate (~ M/10) to an aqueous solution of sodium cyclotriphosphate of approximately the same concentration. The reaction is:

$$Na_3P_3O_9 + 3AgNO_3 \longrightarrow Ag_3P_3O_9.H_2O + 3NaNO_3$$

An excess of silver nitrate is recommended to avoid the formation of a sodium containing compound. The precipitation is achieved within one day. The precipitate appears as nice monoclinic crystals which are non-light sensitive and stable at room temperature for years.

A typical run of the Boullé's process is now described: the preparation of potassium cyclotriphosphate, $K_3P_3O_9$. A slurry of $Ag_3P_3O_9.H_2O$ in water is slowly added to an aqueous solution of potassium chloride in the stoichiometric ratio. The reaction is:

$$Ag_3P_3O_9.H_2O + 3KCl \longrightarrow K_3P_3O_9 + 3AgCl + H_2O$$

After stirring for about an hour, the silver chloride is removed by filtration. The resulting solution is evaporated to obtain crystals of $K_3P_3O_9$ or added with ethanol until the precipitation of polycrystalline potassium cyclotriphosphate. These operations must be run at room or lower temperature to avoid the hydrolysis of the ring. This process seems rather simple, but the chemical preparation of a sodium-free silver cyclotriphosphate is not so easy and needs some precautions as we have reported above.

Many water-soluble cyclotriphosphates were prepared by this process.

More classical methods – The classical methods of aqueous chemistry are, in many cases, valuable for the production of cyclotriphosphates. For instance, $CaNaP_3O_9.3H_2O$ is very easily prepared by adding, in the stoichiometric ratio, a M/10 aqueous solution of $CaCl_2$ to an aqueous solution of $Na_3P_3O_9$ of same concentration. $CaNaP_3O_9.3H_2O$ is then precipitated by adding ethyl alcohol to the resulting solution.

Thermal methods – Many cyclotriphosphates were prepared as polycrystalline samples through what is commonly called "thermal methods". A typical example is the preparation of most of the $M^{II}M^{I}P_3O_9$ benitoite-like compounds. They are obtained in a good state of purity by heating at relatively low temperatures (most of time less than 773 K) a stoichiometric mixture of $(NH_4)_2HPO_4$ with the corresponding carbonates. The reaction scheme is:

$$3(NH_4)_2HPO_4 + CoCO_3 + 1/2K_2CO_3 \longrightarrow$$
$$CoKP_3O_9 + 3/2CO_2 + 6NH_3 + 9/2H_2O$$

The starting mixture must be slowly heated up to the optimum temperature with frequent homogeneization grindings.

Flux methods – This type of process which is very common in solid state chemistry, is frequently used, mainly for growing single crystals. In the case of phosphates, the basic principle of this process is to use a starting mixture containing a large excess of H_3PO_4 or $(NH_4)_2HPO_4$. The composition of the starting mixture and the optimum temperature to be used are parameters difficult to determine. The optimization of a reproducible flux process for synthetizing a given material is in most cases a boring and long enterprise. After some hours or some days of heating at the appropriate temperature, a crystalline formation occurs in the melt. Then, the excess of the phosphoric flux is removed with hot water and the crystalline part separated by filtration. When the elaborated compound appears as water soluble, special care has to be taken for the final operation. In some cases, water can be replaced by alcohol or a manual or mechanical extraction used.

Ion-exchange resins – The cyclotriphosphoric acid, $H_3P_3O_9$, can be obtained by using ion-exchange resins. Amberlite IR220 or IRN77 are extensively used. An aqueous solution of $Na_3P_3O_9$ is slowly passed through a column of resins and the $H_3P_3O_9$ produced is immediately neutralized by the required carbonates or hydroxides. For instance, $Na_2LiP_3O_9$, $Mn_3(P_3O_9)_2.10H_2O$, $Ca_3(P_3O_9)_2.10H_2O$, and $K_3P_3O_9$ were prepared by this process.

Non-conventional methods – Rather unconventional methods of preparation are sometimes reported with more or less details in chemical literature. Most of them were elaborated by crystallographers in search of a single-crystal production.

$Ba_3(P_3O_9)_2.4H_2O$ was prepared by Averbuch-Pouchot and Durif [54] when investigating the reactions between $Na_3P_3O_9$ and $BaCl_2$ solutions at various concentrations. The authors reported a reproducible procedure for producing this tetrahydrate as polycrystalline samples or as single crystals. If a very concentrated aqueous solution of $BaCl_2$ (15 g of $BaCl_2.2H_2O$ in 35 cm^3 of water) is slowly added to a concentrated aqueous solution of $Na_3P_3O_9$ (3 g of

$Na_3P_3O_9$ in 25 cm^3 of water), without any mechanical stirring, one observes the formation of a large amount of small tufts of lath-like needles almost immediately. After filtration, to eliminate the mother liquor, an equivalent volume of water is added to the solid phase, which within a few minutes is transformed into small crystals of $Ba_3(P_3O_9)_2.4H_2O$. If the same experiment is performed with less concentrated solutions, the same quantities of starting materials being respectively dissolved in 50 cm^3 of water, one observes the formation of large crystals of the tetrahydrate after some hours. In this case, the preparation is sometimes contaminated by 10 to 20% of large elongated prisms of the hexahydrate. All the experiments are run at room temperature. Crystals of $Ba_3(P_3O_9)_2.4H_2O$ appear as stout monoclinic prisms, stable at room temperature. The authors [54] report a description of the atomic arrangement.

A series of $Ln(Bi)Na_3(P_3O_9)_2.9H_2O$ were prepared by Bagieu-Beucher and Durif [123] and Bagieu-Beucher [124] by adding solid lanthanide nitrate or chloride to a saturated solution of $Na_3P_3O_9$. In such a process, the crystal growth is very rapid. In some experiments, crytals up to 1 mm long were obtained within half an hour. These salts are not stable for a long time. This unstability can probably be explained by the fact that some of the water molecules are of a zeolitic nature.

Organic-cation cyclotriphosphates – The recent development of the crystal chemistry of organic-cation condensed phosphates did not imply the use of new specific processes for their preparation. Most of them were synthesized using processes reported above. The Boullé's process was used when the corresponding organic chlorhydrate is available. For instance, guanidinium cyclotriphosphate dihydrate was prepared by this way:

$$3[C(NH_2)_3]Cl + Ag_3P_3O_9 \longrightarrow [C(NH_2)_3]_3P_3O_9 + 3AgCl$$

The direct action of cyclotriphosphoric acid, produced through ion-exchange resins, on the required amine has also been used sometimes:

$$3R-NH_2 + H_3P_3O_9 \longrightarrow (R-NH_3)_3P_3O_9$$

References

1. M. T. Averbuch-Pouchot and A. Durif, *Eur. J. Solid State Inorg. Chem.*, **28**, (1991), 9–22.

2. M. T. Averbuch-Pouchot and A. Durif, *Crystal Chemistry of Cyclophosphates* in Stereochemistry of Organometallic and Inorganic Compounds, Vol. 5, (Elsevier, Amsterdam, 1994).

3.   U. Schülke and N. N. Chudinova, *Izv. Akad. Nauk SSSR, Neorg. Mater.*, **10**, (1974), 1697–1703.

4.   V. A. Sotnikova-Yuzhik, G. V. Peslyak, and E. A. Prodan, *Russ. J. Inorg. Chem.*, **32**, (1987), 1505–1507.

5.   E. Thilo and R. Ratz, *Z. anorg. Chem.*, **260**, (1949), 255–266.

6.   A. E. R. Westman and A. E. Scott, *Nature*, **168**, (1951), 740.

7.   O. T. Quimby, *J. Phys. Chem.*, **58**, (1954), 615–624.

8.   J. I. Watters, P. E. Sturrock, and R. E. Simonaitis, *Inorg. Chem.*, **2**, (1963), 765–767.

9.   F. Schulz and M. Jansen, *Z. anorg. allg. Chem.*, **543**, (1986), 152–160.

10.   T. Graham, *Phil. Trans. Roy. Soc.*, **123A**, (1833), 253–284.

11.   Th. Fleitmann and W. Henneberg, *Liebig's Annalen der Chemie*, **65**, (1848), 304–334 and 387–390.

12.   C. G. Lindbom, *Ber.*, **8**, (1875), 122–124.

13.   G. Tammann, *Z. Phys. Chem.*, **6**, (1890), 122–140.

14.   G. Tammann, *J. Prak. Chem.*, **45**, (1892), 417–474.

15.   G. V. Knorre, *Z. anorg. Chem.*, **24**, (1900), 369–401.

16.   A. Boullé, *C. R. Acad. Sci.*, **206**, (1938), 517–519.

17.   E. Thilo and H. Grunze, *Z. anorg. allg. Chem.*, **281**, (1955), 263–283.

18.   E. D. Eanes and H. M. Ondik, *Acta Crystallogr.*, **15**, (1962), 1280–1285.

19.   E. D. Eanes, N. B. S. Monograph, **25**, Sect. 2, p 20.

20.   R. Masse, J. C. Grenier, G. Bassi, and I. Tordjman, *Z. Kristallogr.*, **137**, (1973), 17–23.

21.   R. Masse, J. C. Grenier, G. Bassi, and I. Tordjman, *Cryst. Struct. Comm.*, **1**, (1972), 239–241.

22.   J. C. Grenier and A. Durif, *Z. Kristallogr.*, **137**, (1973), 10–16.

23.   E. J. Griffith, *J. Am. Chem. Soc.*, **78**, (1956), 3867–3870.

24.   M. T. Averbuch-Pouchot, J. C. Guitel, and A. Durif, *Acta Crystallogr.*, **C39**, (1983), 809–810.

25.   H. M. Ondik and J. W. Gryder, *J. Inorg. Nucl. Chem.*, (1960), 240–246.

26.   H. M. Ondik, *Acta Crystallogr.*, **18**, (1965), 226–232.

27.   I. Tordjman and J. C. Guitel, *Acta Crystallogr.*, **B32**, (1976), 1871–1874.

28.   E. A. Prodan and S. I. Pytlev, *Izv. Akad. Nauk SSSR, Neorg. Mater.*, **19**, (1983), 639–643.

29. M. Bagieu-Beucher, A. Durif, and J. C. Guitel, *Acta Crystallogr.*, **B31**, (1975), 2264–2267.

30. J. C. Grenier, *Bull. Soc. fr. Minér. Cristallogr.*, **96**, (1973), 171–178.

31. J. C. Grenier and A. Durif, *Rev. Chim. Minér.*, **9**, (1972), 351–355.

32. M. Bagieu-Beucher, I. Tordjman, A. Durif, and J. C. Guitel, *Acta Crystallogr.*, **B32**, (1976), 1427–1430.

33. N. M. Dombrovskii and V. A. Koval, *Izv. Akad. Nauk SSSR, Neorg. Mater.*, **12**, (1976), 738–741.

34. I. Tordjman, R. Masse, and J. C. Guitel, *Acta Crystallogr.*, **B33**, (1977), 585–586.

35. N. Boudjada, *Thesis*, (Univ. of Grenoble, France, 1985).

36. G. Tammann and A. Ruppelt, *Z. anorg. allg. Chem.*, **197**, (1931), 65–89.

37. G. Morey, *J. Am. Chem. Soc.*, **76**, (1954), 4724–4726.

38. E. J. Griffith and J. R. Van Wazer, *J. Am. Chem. Soc.*, **77**, (1955), 4222.

39. G. A. Bukhalova and I. V. Mardirosova, *Russ. J. Inorg. Chem.*, **11**, (1966), 495–497.

40. C. Cavero-Ghersi and A. Durif, *C. R. Acad. Sci.*, Sér. C, **278**, (1974), 459–461.

41. I. Tordjman, A. Durif, and C. Cavero-Ghersi, *Acta Crystallogr.*, **B30**, (1974), 2701–2704.

42. C. Cavero-Ghersi, *Thesis*, (Univ. of Grenoble, France, 1975).

43. C. Cavero-Ghersi and A. Durif, *C. R. Acad. Sci.*, Sér. C, **280**, (1975), 579–581.

44. M. T. Averbuch-Pouchot and A. Durif, *Eur. J. Solid State Inorg. Chem.*, **30**, (1993), 1075–1082.

45. M. T. Averbuch-Pouchot and A. Durif, *Z. Kristallogr.*, **135**, (1972), 318–319.

46. M. T. Averbuch-Pouchot, A. Durif, and I. Tordjman, *Cryst. Struct. Comm.*, **2**, (1973), 89–90.

47. M. T. Averbuch-Pouchot, A. Durif, and J. C. Guitel, *Acta Crystallogr.*, **B32**, (1976), 1533–1535.

48. M. T. Averbuch-Pouchot, A. Durif, and J. C. Guitel, *Acta Crystallogr.*, **B32**, (1976), 1894–1896.

49. N. El-Horr and A. Durif, *C. R. Acad. Sci.*, Sér. 2, **296**, (1983), 1185–1187.

50. A. Durif, M. Bagieu-Beucher, C. Martin, and J. C. Grenier, *Bull. Soc. fr.*

Minér. Cristallogr., **95**, (1972), 146–148.

51. I. Tordjman, A. Durif, and J. C. Guitel, *Acta Crystallogr.*, **B32**, (1976), 205–208.

52. J. C. Grenier and C. Martin, *Bull. Soc. fr. Minér. Cristallogr.*, **98**, (1975), 107–110.

53. R. Masse, J. C. Guitel, and A. Durif, *Acta Crystallogr.*, **B32**, (1976), 1892–1894.

54. M. T. Averbuch-Pouchot and A. Durif, *Z. Kristallogr.*, **174**, (1986), 219–224.

55. A. Durif and M. Brunel-Laügt, *J. Appl. Crystallogr.*, **9**, (1976), 154.

56. M. Brunel-Laügt, I. Tordjman, and A. Durif, *Acta Crystallogr.*, **B32**, (1976), 3246–3249.

57. A. Durif, M. T. Averbuch-Pouchot, and J. C. Guitel, *Acta Crystallogr.*, **B31**, (1975), 2680–2682.

58. D. Michot, *Thesis*, (Univ. of Dijon, France, 1975).

59. M. H. Simonot-Grange, *J. Solid State Chem.*, **46**, (1976), 76–86.

60. M. H. Simonot-Grange and D. Michot, *Phosphorus*, **6**, (1976), 103–105.

61. M. H. Simonot-Grange and D. Michot, *Phosphorus and Sulfur*, **4**, (1978), 35–38.

62. J. C. Grenier, C. Martin, and A. Durif, *Bull. Soc. fr. Minér. Cristallogr.*, **93**, (1970), 52–55.

63. G. W. Morey, *J. Am. Chem. Soc.*, **74**, (1952), 5783–5784.

64. E. J. Griffith, *Inorg. Chem.*, **2**, (1962), 962.

65. M. T. Averbuch-Pouchot and A. Durif, *Mat. Res. Bull.*, **4**, (1969), 859–868.

66. I. Mahama, M. T. Averbuch-Pouchot, and J. C. Grenier, *C. R. Acad. Sci.*, Sér. C, **280**, (1975), 1105–1107.

67. M. T. Averbuch-Pouchot and A. Durif, *Z. Kristallogr.*, **164**, (1983), 307–313.

68. C. Martin and A. Durif, *Bull. Soc. fr. Minér. Cristallogr.*, **95**, (1972), 149–153.

69. C. Martin and A. Mitschler, *Acta Crystallogr.*, **B28**, (1972), 2348–2352.

70. M. T. Averbuch-Pouchot, *C. R. Acad. Sci.*, Sér. C, **268**, (1969), 1253–1255.

71. M. T. Pouchot, I. Tordjman, and A. Durif, *Bull. Soc. fr. Minér. Cristallogr.*, **89**, (1966), 405–406.

72. M. A. Savenkova, I. V. Mardirosova, and V. A. Matrosova, *Izv. Akad. Nauk SSSR, Neorg. Mater.*, **12**, (1976), 1324–1325.

73. R. Andrieu and R. Diament, *C. R. Acad. Sci.*, **259**, (1964), 4708–4711.

74. M. T. Averbuch-Pouchot, C. Martin, E. Rakotomahanina-Rolaisoa, and A. Durif, *Bull. Soc. fr. Minér. Cristallogr.*, **93**, (1970), 282–286.

75. R. Andrieu, R. Diament, A. Durif, M. T. Pouchot, and D. Tranqui, *C. R. Acad. Sci.*, Sér. B, **262**, (1966), 718–721.

76. R. Masse, J. C. Grenier, M. T. Averbuch-Pouchot, D. Tranqui, and A. Durif, *Bull. Soc. fr. Minér. Cristallogr.*, **90**, (1967), 158–161.

77. J. L. Prisset, *Dissertation*, (Univ. of Grenoble, France, 1982).

78. E. Rakotomahanina-Rolaisoa, *Thesis*, (Univ. of Grenoble, France, 1972).

79. A. Mermet, M. T. Averbuch-Pouchot, and A. Durif, *Bull. Soc. fr. Minéral. Cristallogr.*, **92**, (1969), 87–90.

80. E. Rakotomahanina-Ralaisoa, *Bull. Soc. fr. Minér. Cristallogr.*, **95**, (1972), 143–145.

81. Y. Henry and A. Durif, *Bull. Soc. fr. Minér. Cristallogr.*, **92**, (1969), 484–486.

82. M. T. Averbuch-Pouchot, *C. R. Acad. Sci.*, Sér. C, **269**, (1969), 26–29.

83. M. T. Averbuch-Pouchot and A. Durif, *Acta Crystallogr.*, **B33**, (1977), 3114–3116.

84. I. A. Tokman and G. A. Bukhalova, *Izv. Akad. Nauk SSSR, Neorg. Mater.*, **11**, (1975), 1654–1656.

85. E. Rakotomahanina-Rolaisoa, Y. Henry, A. Durif, and C. Raholison, *Bull. Soc. fr. Minér. Cristallogr.*, **93**, (1970), 43–51.

86. M. T. Averbuch-Pouchot, *Bull. Soc. fr. Minér. Cristallogr.*, **95**, (1972), 558–564.

87. M. T. Averbuch-Pouchot, *Thesis*, (Univ. of Grenoble, France, 1974).

88. A. Durif and M. T. Averbuch-Pouchot, *J. Appl. Crystallogr.*, **9**, (1976), 247–248.

89. M. T. Averbuch-Pouchot and A. Durif, *Acta Crystallogr.*, **C42**, (1986), 930–931.

90. R. Masse, A. Durif, and J. C. Guitel, *Z. Kristallogr.*, **141**, (1975), 113–125.

91. J. C. Grenier and R. Masse, *Bull. Soc. fr. Minér. Cristallogr.*, **91**, (1968), 428–439.

92. M. T. Averbuch-Pouchot and A. Durif, *Acta Crystallogr.*, **C42**, (1986), 932–933.

93. A. Durif, *C. R. Acad. Sci.*, Sér. C, **275**, (1972), 1379–1382.

94. M. H. Simonot-Grange and P. Jamet, *Phosphorus and Sulfur*, **3**, (1977), 197–202.

95. R. Zilber, I. Tordjman, A. Durif, and J. C. Guitel, *Z. Kristallogr.*, **140**, (1974), 350–359.

96. M. T. Averbuch-Pouchot and A. Durif, *Acta Crystallogr.*, **C43**, (1987), 390–392.

97. A. Durif and M. T. Averbuch-Pouchot, *J. Appl. Crystallogr.*, **9**, (1976), 247–248.

98. D. Seethanen, A. Durif, and J. C. Guitel, *Acta Crystallogr.*, **B32**, (1977), 2716–2719.

99. C. Martin, *Thesis*, (Univ. of Grenoble, France, 1972).

100. D. Seethanen and A. Durif, *Acta Crystallogr.*, **B34**, (1978), 1091–1093.

101. R. Masse, A. Durif, and J. C. Guitel, *Z. Kristallogr.* **141**, (1975), 113–125.

102. A. Takenaka, I. Motooka, and H. Nariai, *Bull. Chem. Soc. Jpn.*, **62**, (1989), 2819–2823.

103. A. Durif, C. Martin, and G. Bassi, *Bull. Soc. fr. Minér. Cristallogr.*, **98**, (1975), 19–24.

104. R. Masse and M. T. Averbuch-Pouchot, *Mat. Res. Bull.*, **12**, (1977), 13–16.

105. A. Jouini and M. Dabbabi, *C. R. Acad. Sci.*, Sér. 2, **301**, (1985), 1347–1349.

106. A. Jouini and M. Dabbabi, *Acta Crystallogr.*, **C42**, (1986), 268–270.

107. A. Durif and M. T. Averbuch-Pouchot, *Z. anorg. allg. Chem.*, **514**, (1984), 85–91.

108. D. Seethanen, A. Durif, and M. T. Averbuch-Pouchot, *Acta Crystallogr.*, **B34**, (1978), 14–17.

109. D. Seethanen, I. Tordjman, and M. T. Averbuch-Pouchot, *Acta Crystallogr.*, **B34**, (1978), 2387–2390.

110. A. Durif and M. T. Averbuch-Pouchot, *Acta Crystallogr.*, **C43**, (1987), 819–821.

111. M. T. Averbuch-Pouchot, *Acta Crystallogr.*, **B34**, (1978), 20–22.

112. A. Jouini and M. Dabbabi, *Bull. Soc. Chim. Tunisie*, **2**, (1987), 29–34.

113. M. S. Belkhiria, M. Ben-Amara, and M. Dabbabi, *Acta Crystallogr.*, **C43**, (1987), 609–610.

114. M. S. Belkhiria, M. Dabbabi, and M. Ben-Amara, *Acta Crystallogr.*, **C43**, (1987), 2270–2272.

115. O. A. Serra and E. Giesbrecht, *J. Inorg. Nucl. Chem.*, **30**, (1968), 793–799.

116. M. Bagieu-Beucher and A. Durif, *Bull. Soc. fr. Minér. Cristallogr.*, **94**, (1971), 440–441.

117. M. Bagieu-Beucher, I. Tordjman, and A. Durif, *Rev. Chim. Minér.*, **8**, (1971), 753–760.

118. D. Gobled, *Thesis*, (Univ. of Dijon, France, 1973).

119. M. Bagieu-Beucher, *Thesis*, (Univ. of Grenoble, France, 1980).

120. P. Birke and G. Kempe, *Z. Chem.*, **13**, (1973), 151–152.

121. P. Birke and G. Kempe, *Z. Chem.*, **13**, (1973), 65–66.

122. P. Birke and G. Kempe, *Z. Chem.*, **13**, (1973), 110–111.

123. M. Bagieu-Beucher, *C. R. Acad. Sci.*, Sér. 2, **308**, (1989), 377–379.

124. M. Bagieu-Beucher and A. Durif, *Z. Kristallogr.*, **178**, (1987), 239–247.

125. *Crystal Data Determinative Tables*, J. C. P. D. S., Vol. 4, (1978).

126. W. H. Zachariasen, *Z. Kristallogr.*, **74**, (1930), 139–146.

125. M. T. Averbuch-Pouchot, A. Durif, and J. C. Guitel, *Acta Crystallogr.*, **C45**, (1989), 1320–1322.

126. M. T. Averbuch-Pouchot, A. Durif, and J. C. Guitel, *Acta Crystallogr.*, **C44**, (1988), 1907–1909.

127. M. T. Averbuch-Pouchot and A. Durif, *Acta Crystallogr.*, **C44**, (1988), 1909–1911.

128. M. T. Averbuch-Pouchot, A. Durif, and J. C. Guitel, *Acta Crystallogr.*, **C44**, (1988), 97–98.

129. M. T. Averbuch-Pouchot, A. Durif, and J. C. Guitel, *Acta Crystallogr.*, **C44**, (1988), 99–102.

130. M. T. Averbuch-Pouchot and A. Durif, *Eur. J. Solid State Inorg. Chem.* **30**, (1993), 471–482.

5.4.4. Cyclotetraphosphates

5.4.4.1. Introduction

In the absence of any possibility of proper structural characterization for a long time, the development of cyclotetraphosphate chemistry was marked by many controversies related mainly to the degree of condensation of the anion of the existent phosphates at its beginning.

As far as we know, Maddrell [1] was the first to prepare cyclotetraphosphates. He synthesized a series of compounds of general formula $M^{II}O.P_2O_5$, with M^{II} = Mn, Cu, Co, Ni, Mg... He called them "metaphosphates" and all were recognized as cyclotetraphosphates much later. Several other chemists of the last century resumed and developed this investigation. Using the copper cyclotetraphosphate, $Cu_2P_4O_{12}$, they performed the synthesis of the corresponding sodium salt by using the following reaction:

$$Cu_2P_4O_{12} + 2Na_2S \longrightarrow Na_4P_4O_{12} + 2CuS$$

It seems that it is not before Glätzel [2] that the tetrameric and cyclic nature of the anion was suspected. Then, the term "tetrametaphosphates" coined by this scientist was commonly used by chemists. In spite of the uneasiness and that it is time consuming, the process described above for the preparation of the sodium cyclotetraphosphate, was the only possible one for the production of this starting material for a long time. One must wait till the middle of this century for the elaboration of a more convenient process based on the low-temperature (273 K) hydrolytic cleavage of P_4O_{10}, leading to cyclotetraphosphoric acid with very high yields, according to the following scheme:

$$P_4O_{10} + 2H_2O \longrightarrow H_4P_4O_{12}$$

This reaction was mentionned and used several times during the first part of our century and was carefully analysed and optimized finally by Bell et al. [3] and by Thilo and Wicker [4].

The first structural proof of the cyclic nature of the P_4O_{12} anion was given in 1937 by Pauling and Sherman [5] who solved the crystal structure of the aluminium salt: $Al_4(P_4O_{12})_3$. To date, more than ninety cyclotetraphosphates have been clearly characterized.

5.4.4.2. Present state of the cyclotetraphosphate chemistry

Monovalent-cation and mixed-monovalent cation cyclotetraphosphates – With the exception of the silver salt, which is not well characterized, all monovalent-cation cyclotetraphosphates are investigated and, as in many other classes of compounds, the sodium derivatives received more attention than the other monovalent-cation salts. In Table 5.4.12, we report a list of the presently known monovalent-cation cyclotetraphosphates and of their methods of preparation.

A problem remains for the structure of the only lithium salt which has been studied. During its structural investigation [9], five sites of water molecules were located, but the high thermal factors of two of them probably correspond to partly occupied positions and the formulation as a tetrahydrate seems the most probable.

For a long time, the existence of two crystalline forms for the tetrahydrate of the sodium derivative is well established. There were some controversies on the transformation leading from the monoclinic to the triclinic form of $Na_4P_4O_{12}.4H_2O$. Earlier studies claimed a transition temperature at 327 K [12],

Table 5.4.12
Monovalent-cation cyclotetraphosphates and their methods of preparation.

Formula		Preparation	Ref.
$Li_4P_4O_{12}.4H_2O$	(*)	$H_4P_4O_{12} + Li_2CO_3$	[6–9]
$Li_4P_4O_{12}$ (2 forms)		Condensation of LiH_2PO_4	[10]
$Li_4P_4O_{12}.2H_2O$(2 forms)		id.	[10]
$Li_4P_4O_{12}.4H_2O$ (2 forms)		id.	[10]
$Li_4P_4O_{12}.6H_2O$		id.	[10]
$Li_4P_4O_{12}.8H_2O$		id.	[10]
$Na_4P_4O_{12}$	(*)	Dehydration of $Na_4P_4O_{12}.4H_2O$	[11]
$Na_4P_4O_{12}.H_2O$	(*)	id.	[11]
$Na_4P_4O_{12}.4H_2O$ (2 forms)		$H_4P_4O_{12} + Na_2CO_3$	[12–17]
	(*)	Low-temperature crystallization	[18]
$Na_4P_4O_{12}.10H_2O$			
$K_4P_4O_{12}.2H_2O$	(*)	$H_4P_4O_{12} + K_2CO_3$	[17–18]
$K_4P_4O_{12}.4H_2O$	(*)	id.	[18–19]
$(NH_4)_3HP_4O_{12}.H_2O$		$P_4O_{10} + $ conc. NH_4OH (278 K)	[20]
$(NH_4)_4P_4O_{12}$	(*)	$H_4P_4O_{12} + $ dil. NH_4OH (278 K)	[21–25]
$Rb_4P_4O_{12}.4H_2O$ (*)		$H_4P_4O_{12} + Rb_2CO_3$	[26]
$Cs_4P_4O_{12}.4H_2O$	(*)	$H_4P_4O_{12} + Cs_2CO_3$	[26]
$Tl_4P_4O_{12}$	(*)	Condensation of TlH_2PO_4	[27–31]

* Crystal structure determined

but later Ondik [16] could not observe this transformation and concluded, based on a comparison of the two arrangements, that a transition in the solid state seems unlikely. More recently, Griffith [32] resumed these experiments and observed that air-dried crystals of both forms do not transform but that for wet samples containing 13% of water the transition of the monoclinic form into the triclinic one occurs at about 327 K, accompanied by a small thermal effect. The author concludes that the previous observations were in fact correct since when run with the decahydrate as a starting material, this phosphate probably retains a sufficient amount of residual water up to 327 K to initiate the transition. This transformation, as observed by Griffith, is not reversible.

The effect of water vapor on the transformation of anhydrous $Na_4P_4O_{12}$ was investigated by Nariai *et al.* [33]. This salt was prepared by heating the tetrahydrate for two hours at 373 K and one hour at 423 K. The experiments were run in dry and wet atmospheres. In both cases a conversion into a mixture of two forms of $Na_3P_3O_9$ occurs at 723 K.

The preparation of $Na_2H_2P_4O_{12}$ was described by Griffith [34] by the reaction of H_3PO_4 on NaH_2PO_4 at 673 K:

$$2NaH_2PO_4 + 2H_3PO_4 \longrightarrow Na_2H_2P_4O_{12} + 4H_2O$$

According to the author, the melting or more probably, the decomposition point of $Na_2H_2P_4O_{12}$ is close to 673 K. A more detailed description of the preparation was reported later by the same author [35]. This salt is almost insoluble in water and has a fibrous crystal habit. These two characteristics are rather surprising for an alkali cyclotetraphosphate. To explain these properties, Gryder *et al.* [36] performed a crystallographic investigation of this material. They found the so-called $Na_2H_2P_4O_{12}$ to be an intimate intergrowth of two distinct crystalline forms, both monoclinic, and from various considerations concluded that the anionic framework must be built by chains of P_4O_{12} rings sharing an oxygen atom, leading to the formula of an ultraphosphate: $Na_2P_4O_{11}$. On the other hand, Jarchow [37] proposed a structural model based on the $Na_2H_2P_4O_{12}$ formula proposed by Griffith. Owing to the poor quality of the experimental data, this work is not conclusive and needs revision.

A few *mixed-monovalent cation cyclotetraphosphates* were described. These investigations are relatively recent and lead to accurate determinations of the atomic arrangement in all cases. We list them below:

– two crystalline forms of $Na_2K_2P_4O_{12}.2H_2O$ were prepared by Averbuch-Pouchot and Durif [17]. During the same study, the authors described also $Na_2(NH_4)_2P_4O_{12}.2H_2O$ and $Na_2Rb_2P_4O_{12}.2H_2O$, both isotypic of the triclinic form of the potassium–sodium salt.

– $Na_3CsP_4O_{12}.4H_2O$ and $Na_3CsP_4O_{12}.3H_2O$ were characterized and investigated by the same authors [38].

Anhydrous divalent-cation cyclotetraphosphates – This section is mainly dominated by the existence of eight isotypic $M_2^{II}P_4O_{12}$ compounds, with M^{II} = Zn, Ni, Co, Fe, Mg, Cu, Mn, Cd. Some of them were characterized last century and the cyclic tetrameric nature of their anion already suspected at that time. Later, pioneer work in this field was done by Thilo and Grunze [39], during their study of the thermal condensation of the $M^{II}(H_2PO_4)_2$ monophosphates:

$$2M^{II}(H_2PO_4)_2 \longrightarrow M_2^{II}P_4O_{12} + 4H_2O$$

From X-ray experiments, they showed that the resulting compounds are isomorphous and, from paper chromatography, probably cyclotetraphosphates. After this study, these compounds were extensively investigated and their crystal structures repeatedly determined after the first determination by Laügt *et al.* [40]. We report in Table 5.4.13 the main studies of these compounds.

Table 5.4.13
Main investigations on $M_2^{II}P_4O_{12}$.

Nature of the investigation	Ref.
Unit-cell parameter determinations	[41–43]
Structural determinations for	
$\quad Cu_2P_4O_{12}$	[40]
$\quad Mg_2P_4O_{12}$	[44]
$\quad Co_2P_4O_{12}$	[45]
$\quad Ni_2P_4O_{12}$	[46]
$\quad Fe_2P_4O_{12}$	[47–48]
$\quad NiCoP_4O_{12}$ and $NiZnP_4O_{12}$	[49]
Magnetic properties	[50]
Pigment properties	[51–52]
Surface tension for $Zn_2P_4O_{12}$	[53]
Condensation of $M^{II}(H_2PO_4)_2.2H_2O$ for $M^{II} = Mn, Mg$	[54–56]
Hydrothermal synthesis for $Fe_2P_4O_{12}$	[57]

With the exception of the cadmium salt, these compounds are prepared by thermal or flux methods. In most of cases, the crystals produced by flux methods are twinned.

$Cd_2P_4O_{12}$ is a special case in this series. It cannot be prepared by flux method similar to that used for the other $M_2^{II}P_4O_{12}$ salts. Its existence was reported for the first time by Thilo and Grunze [39] during a study of the thermal condensation of $Cd(H_2PO_4)_2.2H_2O$:

$$Cd(H_2PO_4)_2.2H_2O \xrightarrow{373\,K} Cd(H_2PO_4)_2 \xrightarrow{453\,K} CdH_2P_2O_7 \xrightarrow{473\,K} Cd_2P_4O_{12}$$

According to these authors, the cyclotetraphosphate is transformed into a long-chain polyphosphate at about 873 K. Some years later, the same study was resumed by Ropp and Aia [60] by differential thermal analysis. For the first two steps, they found transformation temperatures relatively close to those reported by Thilo and Grunze [39]: i.e. 403 K to 433 K instead of 373 K and 483 K instead of 453 K. For the last step, corresponding to the condensation of the diphosphate into the cyclotetraphosphate, they give a temperature of 683 K.

According to the same authors, the so-obtained cyclotetraphosphate transforms at 823 K into a tetraphosphate $Cd_3P_4O_{13}$. In spite of these data, the preparation of pure $Cd_2P_4O_{12}$ was not possible until the process described by Laügt et al. [43]. These authors used as starting material a not yet characterized hydrate of cadmium cyclotetraphosphate prepared at room temperature by action of $H_4P_4O_{12}$ on cadmium carbonate. By heating this hydrate slowly up to 573 K and keeping it at this temperature for 30 min, they obtained pure specimens of $Cd_2P_4O_{12}$, not contaminated by the cadmium long-chain polyphosphate. In contradiction with the experiments of the previous authors, they observed that prolonged heating at 573 K transforms $Cd_2P_4O_{12}$ into the long-chain cadmium polyphosphate irreversibly.

Except for the cadmium derivative, a high-pressure form was observed for the $M_2^{II}P_4O_{12}$ salts by Bagieu-Beucher et al. [58]. For a time, this high-pressure form was also thought to be a cyclotetraphosphate, but was later recognized as a long-chain polyphosphate by Averbuch-Pouchot et al. [59]. In the case of the zinc salt, the two forms can be obtained under normal conditions of pressure.

Table 5.4.14

Unit-cell dimensions of the $M_2^{II}P_4O_{12}$ compounds isotypic with $Cu_2P_4O_{12}$. Space group is C2/c with $Z = 4$.

Formula	a α	b β	c (Å) γ (°)	Ref.
$Cu_2P_4O_{12}$	12.552(8)	8.083(3) 118.66(1)	9.573(3)	[40, 42]
$Mg_2P_4O_{12}$	11.77(1)	8.287(4) 118.87(2)	9.949(10)	[42]
$Ni_2P_4O_{12}$	11.65(1)	8.241(4) 118.46(2)	9.857(6)	[42]
$Co_2P_4O_{12}$	11.815(5)	8.310(8) 118.46(2)	9.339(10)	[42]
$Zn_2P_4O_{12}$	11.78(1)	8.302(6) 118.81(2)	9.927(8)	[42]
$Mn_2P_4O_{12}$	12.08(1)	8.471(6) 119.29(2)	10.171(8)	[42]
$Fe_2P_4O_{12}$	11.952(2)	8.359(2) 118.76(5)	9.932(2)	[47, 58]
$Cd_2P_4O_{12}$	12.319(5)	8.631(3) 119.33(5)	10.382(6)	[43]

We illustrate the structural part of this section by describing the atomic arrangement in $Cu_2P_4O_{12}$ which is common to the eight cyclotetraphosphates whose unit-cell dimensions are reported in Table 5.4.14. For the description of the present atomic arrangement and its drawings, we used the data reported by Laügt *et al.* [40].

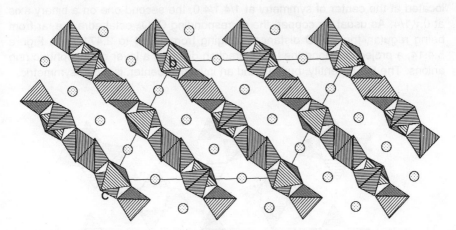

Figure 5.4.12. Projection, along the **b** axis, of the structure of $Cu_2P_4O_{12}$. The layers of phosphoric anions are given by polyhedral representation and the spotted circles are the copper atoms.

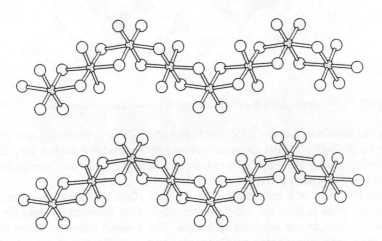

Figure 5.4.13. Projection, along the [10$\bar{1}$] direction, of a layer of copper octahedra.

As shown by Figure 5.4.12, this structure is, as very often in condensed phosphate arrangements, a typical layered one. Planes containing the ring anions alternate with planes of CuO_6 octahedra perpendicular to the $[10\bar{1}]$ direction. Within a plane of copper octahedra, these entities, through edge-sharing, build infinite chains. Figure 5.4.13 represents the projection of such a layer along the $[10\bar{1}]$ direction. One of the two independent copper atoms is located at the center of symmetry at 1/4,1/4,0, the second one on a binary axis at 0,y,1/4. As usual for copper, the corresponding CuO_6 octahedra are far from being regular, the Cu–O distances ranging from 1.914 to 2.471 Å. In Figure 5.4.14, a projection along the $[10\bar{1}]$ direction displays a layer of phosphoric ring anions. The P_4O_{12} entity, built around an inversion center, is centrosymmetric.

Figure 5.4.14. Projection, along the $[10\bar{1}]$ direction, of a layer of phosphoric anions.

Hydrated divalent-cation cyclotetraphosphates – The preparation and crystal structure of $Zn_2P_4O_{12}.8H_2O$ were reported by Averbuch-Pouchot [61]. Later, the corresponding Mg, Fe, Co, and Cu salts were prepared and recognized as isotypes by Foumakoye [62].

The structure of a monoclinic form of $Ca_2P_4O_{12}.4H_2O$ was performed by Schneider *et al.* [63]. A second form was afterwards reported by Schülke [64], but no crystal data are reported up to now. The thermal evolution of the monoclinic form was investigated by Schneider and Jost [65–66]. A monohydrate,

$Ca_2P_4O_{12}.H_2O$, appears at 393 K.

Durif and Averbuch-Pouchot [67] obtained $Sr_2P_4O_{12}.6H_2O$ by the action of $H_4P_4O_{12}$ on a solution of strontium nitrate and described its structure.

The various hydrates of the lead salt, $Pb_2P_4O_{12}.4H_2O$, $Pb_2P_4O_{12}.3H_2O$, and $Pb_2P_4O_{12}.2H_2O$, were carefully investigated because one of them, the tetrahydrate, is a fundamental starting material for the preparation of lead cyclooctaphosphate (see the corresponding section). Crystal structure of the tetrahydrate was performed by Worzala [68]. The same author [69] obtained single-crystals of the dihydrate by heating the tetrahydrate at 373 K and described its structure. The trihydrate was later prepared and described by Klinkert and Jansen [70].

Anhydrous divalent-monovalent cation cyclotetraphosphates – Most of these compounds correspond to the formula, $M^{II}M_2^{I}P_4O_{12}$. A number of them were characterized during the elaboration of various $M^{II}(PO_3)_2$–M^IPO_3 phase equilibrium diagrams and, with the exception of $CaNa_2P_4O_{12}$ and $SrNa_2P_4O_{12}$, are all isotypic with $SrK_2P_4O_{12}$. We report, in Table 5.4.15, some details on the investigations and properties of these compounds.

Table 5.4.15
The eight cyclotetraphosphates isotypic with $SrK_2P_4O_{12}$.

Compound	Diagram or preparation	mp (K)		Ref.
$CaK_2P_4O_{12}$	Thermal process	973	(t)	[71]
$SrK_2P_4O_{12}$	$Sr(PO_3)_2$–KPO_3	977		[72–75]
$PbK_2P_4O_{12}$	Thermal process	810	(t)	[71]
$Sr(NH_4)_2P_4O_{12}$	Thermal process			[74, 76]
$Pb(NH_4)_2P_4O_{12}$	Thermal process			[71]
$SrTl_2P_4O_{12}$	$Sr(PO_3)_2$–$TlPO_3$	816		[73–74]
$PbTl_2P_4O_{12}$	Thermal process			[71]
$SrRb_2P_4O_{12}$	$Sr(PO_3)_2$–$RbPO_3$	1033 (d)		[73–74]
$PbRb_2P_4O_{12}$	Thermal process			[71]

d: decomposition
t: transformation

$CaK_2P_4O_{12}$ was not observed in the $Ca(PO_3)_2$–KPO_3 phase-equilibrium diagram [77–78]. It decomposes at 973 K:

$$CaK_2P_4O_{12} \longrightarrow CaKP_3O_9 + KPO_3$$

$PbK_2P_4O_{12}$ which is also not observed in the corresponding phase diagram [79] is, in fact, the low-temperature form of $PbK_2(PO_3)_4$ [80].

Only two sodium salts are observed with this stoichiometry: $CaNa_2P_4O_{12}$ and $SrNa_2P_4O_{12}$. The strontium salt was not observed in the $Sr(PO_3)_2$–$NaPO_3$ system [73], but was later obtained by Averbuch-Pouchot and Durif [81] during attempts to prepare $SrNa_4(P_3O_9)_2$ single crystals by a flux method. The calcium derivative was very recently obtained by Averbuch-Pouchot [82].

Some compounds of this family have various stoichiometries. This is the case for three isotypic derivatives, $Zn_4Na_4(P_4O_{12})_3$, $Co_4Na_4(P_4O_{12})_3$, and $Ni_4Na_4(P_4O_{12})_3$, characterized by Averbuch-Pouchot and Durif [83] and whose structure is closely related to that of $Al_4(P_4O_{12})_3$. The same authors [84] also found a very similar arrangement when investigating $Sr_3Cs_4H_2(P_4O_{12})_3$ obtained during attempts to produce single crystals of $SrCs_2P_4O_{12}$.

Hydrated divalent-monovalent cation cyclotetraphosphates – The existence of $CaNa_2P_4O_{12}.11/2H_2O$ and $SrNa_2P_4O_{12}.6H_2O$ and their atomic arrangements were reported by Averbuch-Pouchot and Durif [85] and Durif *et al.* [86]. Two isotypic salts, $NiK_2P_4O_{12}.7H_2O$ and $Ni(NH_4)_2P_4O_{12}.7H_2O$, were described by Jouini *et al.* [87] as well as $NiK_2P_4O_{12}.5H_2O$ and $CoK_2P_4O_{12}.5H_2O$, also isotypic, by Jouini *et al.* [88]. $Ca(NH_4)_2P_4O_{12}.2H_2O$ prepared by Cavero-Ghersi [89] was later investigated by Tordjman *et al.* [90] and $Ca_4K_4(P_4O_{12})_3.8H_2O$ prepared and studied by Averbuch-Pouchot and Durif [85].

Trivalent-cation cyclotetraphosphates – All the presently well characterized trivalent-cation cyclotetraphosphates belong to a type of structure that was first investigated by Hendricks and Wyckoff [91] and later described in 1937 by Pauling and Sherman [5] for the aluminium salt, $Al_4(P_4O_{12})_3$. This study was the first structural proof of the existence of cyclic phosphoric anions. All these compounds are cubic, $\bar{I}43d$, with $Z = 4$ and their unit-cell dimensions are reported in Table 5.4.16.

Trivalent-monovalent cation cyclotetraphosphates – Since the characterization of the stable rare-earth ultraphosphates, LnP_5O_{14}, by Beucher [100] and the

Table 5.4.16
Unit-cell dimensions of the six compounds isotypic with $Al_4(P_4O_{12})_3$.

Formula	Unit cell (Å)	Ref.
$Al_4(P_4O_{12})_3$	13.730	[5]
$Fe_4(P_4O_{12})_3$	14.013(7)	[92]
$Cr_4(P_4O_{12})_3$	13.912	[93]
$Ti_4(P_4O_{12})_3$	13.82	[94]
$Sc_4(P_4O_{12})_3$	14.363(5)	[95–98]
$Yb_4(P_4O_{12})_3$	14.66(1)	[99]

discovery of their application as efficient laser materials, a great number of studies of the $Ln_2O_3-P_2O_5$ and $Ln_2O_3-M_2^IO-P_2O_5$ systems were performed. Most of the new materials characterized during these studies are long-chain polyphosphates with the exception of some $LnM^IP_4O_{12}$ compounds that we report in this section.

All these compounds were obtained during the investigations of the corresponding systems by flux methods. Most of them are isotypic of the praseodymium–ammonium salt, first investigated by Masse *et al.* [101]. We report their main crystallographic features in Table 5.4.17.

Table 5.4.17
Main crystallographic features of the eight compounds isotypic with $PrNH_4P_4O_{12}$. The space group is *C2/c* and $Z = 4$.

Formula	a	b β (°)	c (Å)	Ref.
$PrNH_4P_4O_{12}$	7.916(5)	12.647(10) 110.34(8)	10.672(9)	[101]
$NdNH_4P_4O_{12}$	7.881(8)	12.55(1) 110.80(10)	10.65(1)	[101]
$CeNH_4P_4O_{12}$	7.930(3)	12.634(5) 110.05(3)	10.699(5)	[102–104]
$NdKP_4O_{12}$	7.888	12.447 112.70	10.770	[105–106]
$HoKP_4O_{12}$	7.798(1)	12.310(1) 112.63(1)	10.511(1)	[107]
$EuKP_4O_{12}$	7.76	12.03 112.5	10.96	[108]
$NdRbP_4O_{12}$	7.845(2)	12.691(3) 112.34(1)	10.688(3)	[109]
$SmRbP_4O_{12}$	7.868(2)	12.735(3) 111.25(2)	10.589	[110]

A few $LnM^IP_4O_{12}$ compounds, as $NdCsP_4O_{12}$ [111], $CeNH_4P_4O_{12}$ [112], and $NdRbP_4O_{12}$ [113], also adopt a cubic structure type that is closely related to that of $Al_4(P_4O_{12})_3$.

An investigation and a general discussion of the $Ln_2O_3-Rb_2O-P_2O_5-H_2O$ systems was published by Byrappa and Litvin [114] and the $LnCsP_4O_{12}$ compounds discussed by Byrappa *et al.* [115].

Palkina *et al.* [116] prepared several $LnM^IP_4O_{12}.nH_2O$ compounds by the reaction of the monovalent-cation cyclotetraphosphate with the rare-earth nitrate. Among them, the erbium–potassium cyclotetraphosphate hexahydrate, $ErKP_4O_{12}.6H_2O$, was investigated in more details by the authors who report a complete description of the atomic arrangement.

Organic and metal-organic cation cyclotetraphosphates – Recently, a number of organic- and metal-organic cation cyclotetraphosphates were characterized. Two rather similar methods were used for their preparations:

   – action of $H_4P_4O_{12}$ acid, obtained from $Na_4P_4O_{12}.4H_2O$ by ion-exchange resins, on the appropriate amino-compound

   – addition of the proper quantity of P_4O_{10} to an icy aqueous solution of the amine

In both cases, the reaction is of the same type:

$$4R–NH_2 + H_4P_4O_{12} \longrightarrow (R–NH_3)_4P_4O_{12}$$

Most of these compounds are stable for years under normal conditions. In all cases, their atomic arrangements were determined very accurately. These salts are listed in Table 5.4.18.

Table 5.4.18

List of the presently investigated organic-cation cyclotetraphosphates.

Formula	Nature of the organic group	Ref.
$(NH_3CH_2COOH)_4P_4O_{12}$	glycinium	[117]
$[NH_3(CH_2)_2OH]_4P_4O_{12}$	ethanolammonium	[118]
$(C_2N_2H_{10})_2P_4O_{12}.H_2O$	ethylenediammonium	[119]
$(C_3N_2H_{12})_2P_4O_{12}.2H_2O$	1-3 diammonium propane	[120]
$(C_4N_3H_{15})_2P_4O_{12}.4H_2O$	diethylenetriammonium	[121]
$Na_2(C_2N_2H_{10})P_4O_{12}.2H_2O$	ethylenediammonium	[122–123]
$K_2(C_2N_2H_{10})P_4O_{12}.2H_2O$	id.	[122–123]
$NaLi(C_2N_2H_{10})P_4O_{12}.3H_2O$	id.	[124]
$M^{II}(C_2N_2H_{10})_3(P_4O_{12})_2.14H_2O$ M^{II} = Cu, Zn, Mg, Cd, Co, Ni, Mn	id.	[125]
$Ca(C_2N_2H_{10})P_4O_{12}.15/2H_2O$	id.	[126]
$Sr(C_2N_2H_{10})P_4O_{12}.5H_2O$	id.	[127]

5.4.4.3. Adducts

P_4O_{12} ring anions are also observed in some adducts between monovalent cation cyclotetraphosphates and telluric acid and in some mixed-anion phosphates as $Pb_2Cs_3(P_4O_{12})(PO_3)_3$ and $KAl_2H_2(P_3O_{10})(P_4O_{12})$. This type of phosphates will be examined in the section devoted to adducts (p. 353–355).

5.4.4.4. Preparation of cyclotetraphosphates

We will report the most common processes used for the preparation of cyclotetraphosphates. When compared with what we have described for the preparation of cyclotriphosphates, two things must be noted:
– for cyclotriphosphates and other cyclophosphates the most important starting material for syntheses is the sodium salt. But in the case of cyclotetraphosphates, this compound cannot be prepared by thermal condensation of NaH_2PO_4 through the hypothetical reaction scheme:

$$4NaH_2PO_4 \longrightarrow Na_4P_4O_{12} + 4H_2O$$

– the Boullé's metathesis reaction using the silver salt as starting material was almost never used for the production of cyclotetraphosphates.

As for the cyclotriphosphates, the starting material that is mostly used for these preparations is the sodium salt: $Na_4P_4O_{12}.4H_2O$. For a long time, the only cyclotetraphosphates that was possible to prepare in a reproducible way were the members of the $M_2^{II}P_4O_{12}$ series (M^{II} = Cu, Mg, Co...). They are simply prepared by the thermal method leading to well crystallized samples. A mixture of phosphoric acid and a salt of the required divalent cation is prepared with a good excess of phosphoric acid and then heated for approximately one day at a temperature of 627–733 K. After the apparition of a crystalline mass inside this bath, the excess phosphoric flux is removed by hot water and the crystalline, $M_2^{II}P_4O_{12}$, obtained is separated by filtration. One of this compound, $Cu_2P_4O_{12}$, was extensively used to prepare the sodium salt through the procedure that we describe below up to the end of the fifties.

Preparation of $Na_4P_4O_{12}.4H_2O$ by a exchange reaction – Finely divided $Cu_2P_4O_{12}$ is added to an aqueous solution of Na_2S. The mixture is stirred mechanically up to the end of the following reaction:

$$Cu_2P_4O_{12} + 2Na_2S \longrightarrow Na_4P_4O_{12} + 2CuS$$

The temperature of the mixture must be kept close to 273 K during the reaction to avoid the hydrolysis of the P_4O_{12} groups. Then, the insoluble copper sulfide is removed by filtration and the resulting solution is evaporated at room-temperature or added with alcohol or sodium chloride to precipitate $Na_4P_4O_{12}.4H_2O$.

Preparation of $H_4P_4O_{12}$ by hydrolysis of P_4O_{10} – When added to ice water, P_4O_{10} decomposes according to the following equation:

$$P_4O_{10} + 2H_2O \longrightarrow H_4P_4O_{12}$$

producing cyclotetraphosphoric acid. Very good yields (> 75%) can be so obtained.

If a stoichiometric quantity of P_4O_{10} is added to an icy aqueous solution of Na_2CO_3 or $NaHCO_3$, one obtains a solution of sodium cyclotetraphosphate directly. Under these conditions, $Na_4P_4O_{12}.4H_2O$ is obtained with a good yield.

Most of the recently characterized alkali or mixed-alkali cyclotetraphosphates and a good number of organic-cation derivatives were prepared using this procedure.

Thermal processes – Many anhydrous cyclotetraphosphates were prepared by heating, at relatively low temperatures (623–723 K), a stoichiometric mixture of the appropriate starting materials. For instance, $SrK_2P_4O_{12}$ can be obtained according to the following reaction:

$$SrCO_3 + 4(NH_4)_2HPO_4 + K_2CO_3 \longrightarrow SrK_2P_4O_{12} + 2CO_2 + 8NH_3 + 6H_2O$$

Most of the $M^{II}K_2P_4O_{12}$ compounds were simply prepared as polycrystalline samples using this process.

Flux methods – Most of the cyclotetraphosphates characterized during the elaboration of the $M^{II}(PO_3)_2$–M^IPO_3 phase-equilibrium diagrams and, in general, all insoluble cyclotetraphosphates were synthesized as single-crystals by using flux methods similar to those described for the preparations of cyclotriphosphates. As mentioned before, the main feature of this type of process is the usage of a large excess of a phosphoric flux. A typical run of such a process used for the synthesis of $SrNa_2P_4O_{12}$ is reported below. In 8.6 cm^3 of H_3PO_4 (85%) are added 0.5 g of $SrCO_3$ and 3.5 g of Na_2CO_3. The resulting mixture is then heated at 623 K for one day. After removing the excess phosphoric flux by hot water, a crystalline formation containing more than 80% of $SrNa_2P_4O_{12}$ is obtained.

Classical methods of aqueous crystallization – As for cyclotriphosphates, classical methods of aqueous chemistry were frequently used. For instance, $Sr(NH_4)_2P_4O_{12}$ is prepared by mixing N/10 aqueous solutions of $(NH_4)_4P_4O_{12}$ and $SrCl_2$. Similar procedures were used for the preparation of many cyclotetraphosphates.

Ion-exchange resins – Cyclotetraphosphoric acid produced by passing an aqueous solution of $Na_4P_4O_{12}$ through a column of ion-exchange resins is commonly used. For instance, calcium and strontium cyclotetraphosphates were synthesized by adding prepared cyclotetraphosphoric acid to aqueous solutions of the corresponding nitrates. Most of time, the produced acid is added immediately to an appropriate mixture of carbonates or hydroxides to avoid a possible hydrolysis of the anion.

Hydrothermal methods – They are not commonly used in the field of condensed phosphates, but they were employed sometimes. For instance, $Fe_2P_4O_{12}$ was prepared under hydrothermal conditions at 703 K and 1000 atmospheres.

Non-conventional methods – When given by the authors, they are reported in the "Present state of the cyclotetraphosphate chemistry" section.

Organic-cation cyclotetraphosphates – The chemical preparations of organic-cation cyclotetraphosphates are not fundamentally different from those described above. Most of them were prepared by the action of the acid, $H_4P_4O_{12}$, on the corresponding amine or its carbonate. Schematically, the reaction is:

$$4R-NH_2 + H_4P_4O_{12} \longrightarrow (R-NH_3)_4P_4O_{12}$$

Nevertheless, some attempts, sometimes successful, were made to use the ill characterized precipitated silver cyclotetraphosphate in order to generalize the Boullé's process for the preparation of organic-cation cyclotetraphosphates. In one of the preparation described for the synthesis of guanidinium cyclotetraphosphate tetrahydrate, this silver salt was used assuming its chemical formula was that of a tetrahydrate.

5.4.4.5. The P_4O_{12} anion

We do not report here any example of the detailed geometry of a P_4O_{12} ring anion which is similar to the one of P_3O_9. We will only comment on the aspects directly connected to the condensation of the tetrahedra and the various internal symmetries observed in this type of rings.

One can imagine many geometrical conformations for the P_4O_{12} ring anion, but we shall limit the discussion to the presently observed geometries. To date, forty-nine P_4O_{12} groups were investigated with high or acceptable accuracies. Classified according to their geometries, eight types of P_4O_{12} rings are observed up to now with $\bar{1}$, 2, 2/m, m, mm, $\bar{4}$, $\bar{4}$2m or no internal symmetry. The number of representatives in each group is variable; only one ring is observed with a 2/m symmetry, whereas seventeen have a $\bar{1}$ symmetry.

If one excludes some rare exceptions, the *P–P–P angles*, strictly equal to 90° in rings with mm or $\bar{4}$2m symmetries, never depart significantly from this value. Their overall average value is 88.3°. For the calculation of this average, rings with $\bar{1}$, m and 2/m internal symmetries were not taken into account since the two different P–P–P angles are complementary in these cases. Angles of rings with mm and $\bar{4}$2m were also eliminated from the calculation since the P-P-P angles are 90° for symmetry reasons. Therefore, the P–P–P average value reported above is strongly weighted by values observed in rings of low internal symmetry. The largest P–P–P angle (97.8°) was observed in the monoclinic form of $Na_4P_4O_{12}.4H_2O$. As for cyclotriphosphates, the estimated standard deviations for the reported P–P–P values can be evaluated to be less than 0.1 but, for recent structural determinations, this same value is very often less than 0.03.

The *P–P distances,* in cyclotetraphosphates, are quite comparable to those observed in all the other kinds of cyclophosphates with an overall average of 2.938 Å. They vary for cyclotetraphosphates within the following range:

$$2.867 < P–P < 3.015 \text{ Å}$$

The estimated standard deviations for the reported P–P distance values can be evaluated in general to be 0.005, but often close to 0.001 for the most recent data.

The *P–O–P angles* do not also depart significantly from their general average: 131.8° and range from 124.1 to 139.8°. The estimated standard deviation for this value is generally less than 0.1.

As for cyclotriphosphates, no correlation could be established between the ring symmetry and the nature of the associated cations.

References

1.   R. Maddrell, *Liebig Ann.*, **61**, (1847), 53–63.

2.   A. Glätzel, *Inaugural Dissertation*, (Univ. of Würzburg, Germany, 1880).

3.   R. N. Bell, L. F. Audrieth, and O. F. Hill, *Ind. Eng. Chem.*, **44**, (1952), 568–572.

4.   E. Thilo and W. Wicker, *Z. anorg. allg. Chem.*, **277**, (1952), 27–36.

5.   L. Pauling and J. S. Sherman, *Z. Kristallogr.*, **96**, (1937), 481–487.

6.   J. C. Grenier and A. Durif, *Z. Kristallogr.*, **137**, (1973), 10–16.

7.   H. Grunze and E. Thilo, *Z. anorg. allg. Chem.*, **281**, (1955), 284–292.

8.   H. Grunze, *Angew. Chem.*, **67**, (1955), 408.

9.   M. T Averbuch-Pouchot and A. Durif, *Acta Crystallogr.*, **C42**, (1986), 129–131.

10. U. Schülke and R. Kayser, *Z. anorg. allg. Chem.*, **531**, (1985), 167–176.

11. D. M. Wiench and M. Jansen, *Monat. Chem.*, **114**, (1983), 699–709.

12. R. N. Bell, L. F. Audrieth, and O. F. Hill, *Ind. Eng. Chem.*, **44**, (1952), 568–572.

13. K. A. Andress, W. Gehring, and K. Fischer, *Z. anorg. allg. Chem.*, **260**, (1949), 331–336.

14.   D. L. Barney and J. W. Gryder, *J. Am. Chem. Soc.*, **77**, (1955), 3195–3198.

15.   H. M. Ondik, S. Block, and C. H. Mac Gillavry, *Acta Crystallogr.*, **14**, (1961), 555–561.

16.   H. M. Ondik, *Acta Crystallogr.*, **17**, (1964), 1139–1145.

17.   M. T. Averbuch-Pouchot and A. Durif, *J. Solid State Chem.*, **58**, (1985), 119–132.

18.   J. R. Van Wazer, *Phosphorus and its Compounds*, (Interscience, New York, 1966).

19.   M. T. Averbuch-Pouchot and A. Durif, *Acta Crystallogr.*, **C41**, (1985), 1564–1566.

20.   K. R. Waerstad, G. H. McClellan, A. W. Frazier, and R. C. Sheridan, *J. Appl. Crystallogr.*, **2**, (1969), 306–307.

21.   C. Romers, J. A. A. Ketelaar, and C. H. Mac Gillavry, *Nature*, **164**, (1949), 960–961.

22.   K. A. Andress and K. Fischer, *Acta Crystallogr.*, **3**, (1950), 399–400.

23.   C. Romers, J. A. A. Ketelaar, and C. H. Mac Gillavry, *Acta Crystallogr.*, **4**, (1951), 114–120.

24.   D. W. J. Cruickshank, *Acta Crystallogr.*, **17**, (1964), 675–676.

25.   D. A. Koster and A. J. Wagner, *J. Chem. Soc.*, Sect. A, (1970), 435–441.

26.   M. T. Averbuch-Pouchot and A. Durif, *Acta Crystallogr.*, **C42**, (1986), 131–133.

27.   K. Dostal and V. Kocman, *Z. anorg. allg. Chem.*, **367**, (1969), 92–101.

28.   K. Dostal, V. Kocman, and V. Ehrenbergrova, *Z. anorg. allg. Chem.*, **367**, (1969), 80–91.

29.   K. Dostal and V. Kocman, *Z. Chem.*, **5**, (1965), 344.

30.   J. K. Fawcett, V. Kocman, S. C. Nyburg, and R. J. O'Brien, *Chem. Comm.*, **18**, (1970), 1213.

31.   J. K. Fawcett, V. Kocman, and S. C. Nyburg, *Acta Crystallogr.*, **B30**, (1974), 1979–1982.

32.   E. J. Griffith, *Pure and Applied Chem.*, **44**, (1975), 173–200.

33.   H. Nariai, I. Motooka, and M. Tsuhako, *Bull. Chem. Soc. Jpn.*, **64**, (1991), 3205–3206.

34.   E. G. Griffith, *J. Am. Chem. Soc.*, **76**, (1954), 5862.

35.   E. G. Griffith, *J. Am. Chem. Soc.*, **78**, (1956), 3867–3870.

36.   G. W. Gryder, G. Donnay, and H. M. Ondik, *Acta Crystallogr.*, **11**, (1958), 38–40.

37.   O. H. Jarchow, *Acta Crystallogr.*, **17**, (1964), 1253–1262.

38.   M. T. Averbuch-Pouchot and A. Durif, *J. Solid State Chem.*, **60**, (1985), 13–19.

39.   E. Thilo and I. Grunze, *Z. anorg. allg. Chem.*, **290**, (1957), 209–222.

40.   M. Laügt, J. C. Guitel, I. Tordjman, and G. Bassi, *Acta Crystallogr.*, **B28**, (1972), 201–208.

41.   M. Beucher, *Dissertation*, (Univ. of Grenoble, France, 1968).

42.   M. Beucher and J. C. Grenier, *Mat. Res. Bull.*, **3**, (1968), 643–648.

43.   M. Laügt, A. Durif, and M. T. Averbuch-Pouchot, *Bull. Soc. fr. Minér. Cristallogr.*, **96**, (1973), 383–385.

44.   A. G. Nord and K. B. Lindberg, *Acta Chem. Scand.*, **A29**, (1975), 1–6.

45.   A. G. Nord, *Cryst. Struct. Comm.*, **11**, (1982), 1467–1474.

46.   A. G. Nord, *Acta Chem. Scand.*, **A37**, (1983), 539–543.

47.   E. A. Genkina, B. A. Maksimov, and O. K. Mel'nikov, *Sov. Phys. Crystallogr.*, **30**, (1985), 513–515.

48.   A. G. Nord, T. Ericsson, and P. E. Werner, *Z. Kristallogr.*, **192**, (1990), 83–90.

49.   A. G. Nord, *Mat. Res. Bull.*, **18**, (1983), 765–773.

50.   W. Gunsser, D. Fruehauf, K. Rohwer, A. Zimmermann, and A. Wiedenmann, *J. Solid State Chem.*, **82**, (1989), 43–51.

51.   M. Mohan Rao, M. Trojan, and L. Benes, *Mat. Res. Bull.*, **26**, (1991), 813–819.

52.   M. Trojan and L. Benes, *Mat. Res. Bull.*, **26**, (1991), 693–700.

53.   E. L. Krivovyazov, I. D. Sokolova, and N. K. Voskresenskaya, *Izv. Akad. Nauk SSSR, Neorg. Mater.*, **3**, (1967), 530–533.

54.   L. N. Shchegrov, N. M. Antraptseva, and I. G. Ponomareva, *Izv. Akad. Nauk SSSR, Neorg. Mater.*, **25**, (1989), 308–312.

55.   D. Z. Serazetdinov, E. V. Poletsev, and Yu. A. Kushnikov, *Russ. J. Inorg. Chem.*, **12**, (1967), 1599–1600.

56.   A. B. Bekturov, D. Z. Serazetdinov, Yu. A. Kushnikov, E. V. Poletaev, and S. M. Divnenko, *Russ. J. Inorg. Chem.*, **12**, (1967), 1242–1246.

57.   E. A. Genkina, N. S. Triodina, O. K. Mel'nikov, and B. A. Maksimov, *Izv. Akad. Nauk SSSR, Neorg. Mater.*, **24**, (1988), 1158–1162.

58.   M. Bagieu-Beucher, M. Gondrand, and M. Perroux, *J. Solid State Chem.*, **19**, (1976), 353–357.

59.   M. T. Averbuch-Pouchot, A. Durif, and M. Bagieu-Beucher, *Acta Crystallogr.*, **C39**, (1983), 25–26.

60.   R. C. Ropp and M. A. Aia, *Anal. Chem.*, **34**, (1962), 1288–1291.

61.   M. T. Averbuch-Pouchot, *Z. anorg. allg. Chem.*, **503**, (1983), 231–237.

62.   G. Foumakoye, *Thesis*, (Univ. of Liège, Belgium, 1986).

63.   M. Schneider, K. H. Jost, and H. Fichtner, *Z. anorg. allg. Chem.*, **500**, (1983), 117–122.

64.   U. Schülke, personal communication

65.   M. Schneider and K. H. Jost, *Z. anorg. allg. Chem.*, **580**, (1990), 175–180.

66.   M. Schneider and K. H. Jost, *Z. anorg. allg. Chem.*, **576**, (1989), 267–271.

67.   A. Durif and M. T. Averbuch-Pouchot, *Acta Crystallogr.*, **C42**, (1986), 927–928.

68.   H. Worzala, *Z. anorg. allg. Chem.*, **421**, (1976), 122–128.

69.   H. Worzala, *Z. anorg. allg. Chem.*, **445**, (1978), 27–35.

70.   B. Klinkert and M. Jansen, *Z. anorg. allg. Chem.*, **556**, (1988), 85–91.

71.   C. Cavero-Ghersi and A. Durif, *J. Appl. Crystallogr.*, **8**, (1975), 562–564.

72.   G. A. Bukhalova and I. A. Tokman, *Izv. Akad. Nauk SSSR, Neorg. Mater.*, **8**, (1972), 528–532.

73.   C. Martin, *Thesis*, (Univ. of Grenoble, France, 1972).

74.   A. Durif, C. Martin, I. Tordjman, and D. Tranqui, *Bull. Soc. fr. Minér. Cristallogr.*, **89**, (1966), 439–441.

75.   I. Tordjman, C. Martin, and A. Durif, *Bull. Soc. fr. Minér. Cristallogr.*, **90**, (1967), 293–298.

76.   A. Takenaka, I. Motooka, and H. Nariai, *Bull. Chem. Soc. Jpn.*, **62**, (1989), 2819–2823.

77.   J. B. Gill and R. M. Taylor, *J. Chem. Soc.*, (1964), 5905–5906.

78.   R. Andrieu and R. Diament, *C. R. Acad. Sci.*, **259**, (1964), 4708–4711.

79.   I. Mahama, M. Brunel-Laügt, and M. T. Averbuch-Pouchot, *C. R. Acad. Sci.*, Sér. C, **284**, (1977), 681–684.

80.   M. Schneider, *Z. anorg. allg. Chem.*, **503**, (1983), 238–240.

81.   M. T. Averbuch-Pouchot and A. Durif, *Acta Crystallogr.*, **C39**, (1983),

811–812.

82. M. T. Averbuch-Pouchot, to be published.

83. M. T. Averbuch-Pouchot and A. Durif, *J. Solid State Chem.*, **49**, (1983), 341–352.

84. M. T. Averbuch-Pouchot and A. Durif, *Acta Crystallogr.*, **C41**, (1985), 1557–1558.

85. M. T. Averbuch-Pouchot and A. Durif, *Acta Crystallogr.*, **C44**, (1988), 212–216.

86 A. Durif, M. T. Averbuch-Pouchot, and J. C. Guitel, *Acta Crystallogr.*, **C39**, (1983), 812–813.

87. A. Jouini, M. Dabbabi, and A. Durif, *J. Solid State Chem.*, **60**, (1985), 6–12.

88. A. Jouini, M. Soua, and M. Dabbabi, *J. Solid State Chem.*, **69**, (1987), 135–144.

89. C. Cavéro-Ghersi, *Thesis*, (Univ. of Grenoble, France, 1975).

90. I. Tordjman, R. Masse, and J. C. Guitel, *Acta Crystallogr.*, **B32**, (1976), 1643–1645.

91. S. B. Hendricks and R. W. G. Wyckoff, *Amer. J. Sci.*, **13**, (1927), 491–496.

92. F. d'Yvoire, *Bull. Soc. Chim.*, (1962),1237–1243.

93. P. Rémy and A. Boullé, *C. R. Acad. Sci.*, **258**, (1964), 927–929.

94. F. Liebau and H. P. Williams, *Angew. Chem.*, **76**, (1964), 303–304.

95. M. Bagieu-Beucher, *J. Appl. Crystallogr.*, **9**, (1976), 368–369.

96. M. Bagieu-Beucher and J. C. Guitel, *Acta Crystallogr.*, **B34**, (1978), 1439–1442.

97. G. Wappler, *Diplom Arbeit,* (Humbolt Univ., Berlin, Germany, 1958).

98. L. P. Mezentseva, A. I. Domanskii, and I. A. Bondar, *Russ. J. Inorg. Chem.* **22**, (1977), 43–45.

99. N. N. Chudinova, *Izv. Akad. Nauk SSSR, Neorg. Mater.*, **15**, (1979), 833–837.

100 M. Beucher, *Les Eléments des Terres Rares*, (Intern. Meeting, Paris-Grenoble, 1969).

101 R. Masse, J. C. Guitel, and A. Durif, *Acta Crystallogr.*, **B33**, (1977), 630–632.

102. M. Rzaigui and N. K. Ariguib, *J. Solid State Chem.*, **49**, (1983), 391–

398.

103.   M. A. Vaivada and Z. A. Konstant, *Dokl. Akad. Nauk SSSR, Neorg. Mater.*, **15**, (1979), 824–827.

104.   M. A. Vaivada and Z. A. Konstant, *Izv. Akad. Nauk SSSR, Neorg. Mater.*, **16**, (1980), 1810–1814.

105.   O. S. Tarasenkova, G. I. Dorokhova, N. N. Chudinova, B. N. Litvin, and N. V. Vinogradova, *Izv. Akad. Nauk SSSR, Neorg. Mater.*, **21**, (1985), 452–458.

106.   B. N. Litvin, G. I. Dorokhova, and O. S. Filipenko, *Sov. Phys. Dokl.*, **26**, (1981), 717–720.

107.   K. K. Palkina, V. G. Kuznetsov, N. N. Chudinova, and N. T. Chibiskova, *Izv. Akad. Nauk SSSR, Neorg. Mater.*, **12**, (1976), 730–734.

108.   N. N. Chudinova, N. V. Vinogradova, G. M. Balagina, and K. K. Palkina, *Izv. Akad. Nauk SSSR, Neorg. Mater.*, **13**, (1977), 1494–1499.

109.   H. Koizumi and J. Nakano, *Acta Crystallogr.*, **B33**, (1977), 2680–2684.

110.   K. K. Palkina, S. I. Maksimova, and N. T. Chibiskova, *Izv. Akad. Nauk SSSR, Neorg. Mater.*, **17**, (1981), 1248–1252.

111.   K. K. Palkina, S. I. Maksimova, and N. T. Chibiskova, *Sov. Phys. Dokl.*, **26**, (1981), 254–256.

112.   M. Rzaigui, M. T. Averbuch-Pouchot, and A. Durif, *Acta Crystallogr.*, **C39**, (1983), 1612–1613.

113.   G. I. Dorokhova, O. S. Filipenko, L. O. Atovmyan, and B. N. Litvin, *Russ. J. Inorg. Chem.*, **33**, (1988), 1581–1585.

114.   K. Byrappa and B. N. Litvin, *J. Mater. Sci.*, **18**, (1983), 2056–2062.

115.   K. Byrappa, I. I. Plyusnina, and G. I. Dorokhova, *J. Mater. Sci.*, **17**, (1982), 1847–1855.

116.   K. K. Palkina, A. K. Mustaev, S. I. Maksimova, R. Yu. Khusainova, and N. T. Chibiskova, *Russ. J. Inorg. Chem.*, **34**, (1989), 1533–1534.

117.   M. T. Averbuch-Pouchot, A. Durif, and J. C. Guitel, *Acta Crystallogr.*, **C44**, (1988), 888–890.

118.   M. T. Averbuch-Pouchot, A. Durif, and J. C. Guitel, *Acta Crystallogr.*, **C44**, (1988), 1416–1418.

119.   M. T. Averbuch-Pouchot, A. Durif, and J. C. Guitel, *Acta Crystallogr.*, **C45**, (1989), 428–430.

120.   M. Bdiri and A. Jouini, *C. R. Acad. Sci.*, Sér. 2, **309**, (1989), 881–885.

121. M. Bdiri and A. Jouini, *Eur. J. Solid State Inorg. Chem.*, **26**, (1989), 585–592.

122. M. Bdiri and A. Jouini, *C. R. Acad. Sci.*, Sér. 2, **308**, (1989), 1345–1348.

123. A. Jouini, *Acta Crystallogr.*, **C45**, (1989), 1877–1879.

124. M. Bdiri and A. Jouini, *J. Solid State Chem.*, **83**, (1989), 350–360.

125. M. T. Averbuch-Pouchot and A. Durif, *Acta Crystallogr.*, **C45**, (1989), 46–49.

126. M. T. Averbuch-Pouchot, A. Durif, and J. C. Guitel, *Acta Crystallogr.*, **C44**, (1988), 1189–1191.

127. M. Bagieu-Beucher, A. Durif, and J. C. Guitel, *Acta Crystallogr.*, **C44**, (1988), 2063–2065.

5.4.5. Cyclopentaphosphates

There is little to be mentioned about this category of phosphates. One can only report a very restricted number of examples of cyclopentaphosphates. Thilo and Schülke [1] showed that sodium phosphate glasses with a P/Na ratio close to one contain various cyclophosphates $Na_nP_nO_{3n}$ with n value varying from 3 to 8. After fractionating various solutions of these salts by acetone and hexamminecobalt(III) trichloride, they were able to isolate $Na_5P_5O_{15}.4H_2O$ and to prepare the corresponding barium and silver salts $Ba_5(P_5O_{15})_2.10H_2O$ and $Ag_5P_5O_{15}.2.6H_2O$.

The unique structural determination of a cyclopentaphosphate atomic arrangement was performed by Jost [2] with $Na_4NH_4P_5O_{15}.4H_2O$ obtained during the above experiments. This salt is triclinic, $P\bar{1}$, with the following bimolecular unit cell:

$$a = 8.73(2), \quad b = 15.66(3), \quad c = 6.81(2) \text{ Å}$$
$$\alpha = 93.9(4), \quad \beta = 106.1(4), \quad \gamma = 95.1(3)°$$

5.4.6. Cyclohexaphosphates

5.4.6.1. Introduction

The chemistry of cyclohexaphosphates made very rapid progresses during the last ten years. For a long time, the lack of a reliable process to produce large amounts of a starting material can probably explain this relatively late development. Nevertheless, before 1985, some cyclohexaphosphates,

$Na_6P_6O_{18}.6H_2O$, $Cu_2Li_2P_6O_{18}$, $Cr_2P_6O_{18}$ or $Cs_2(UO_2)_2P_6O_{18}$, for instance, were discovered by various methods and clearly characterized. The relatively low development in this domain is difficult to explain for as early as 1965, Griffith and Buxton [3] described a reproducible process for the preparation of useful amounts of sodium or lithium cyclohexaphosphates. This process was greatly improved in 1985 when Schülke and Kayser [4] performed a detailed investigation of the thermal behavior of LiH_2PO_4 and proposed a convenient procedure to prepare large amounts of pure $Li_6P_6O_{18}$. Still more recently, Averbuch-Pouchot [5] described the preparation and an accurate structural characterization of $Ag_6P_6O_{18}.H_2O$, opening the way to metathesis reactions deriving, from the Boullé's one [6], for the preparation of water-soluble cyclohexaphosphates.

5.4.6.2. Present state of the cyclohexaphosphate chemistry

If a relatively great number of cyclohexaphosphates are well characterized to date, mainly at a structural point of view, the relatively small amount of information that we have of their chemical behavior cannot authorize any valuable discussion aimed at comparing their basic chemical properties with those of the other kinds of cyclophosphates.

The *thermal behavior* of monovalent-cation cyclohexaphosphates is similar to what was observed in the two previously examined classes of cyclophosphates. When heated, they all transform into long-chain polyphosphates. We report above some examples of these transformations:

$$K_6P_6O_{18}.3H_2O \xrightarrow{363\,K} K_6P_6O_{18} \xrightarrow{553\,K} KPO_3 \ [7]$$

$$Rb_6P_6O_{18}.6H_2O \xrightarrow{323\,K} Rb_6P_6O_{18}.nH_2O \xrightarrow{373\,K} Rb_6P_6O_{18} \xrightarrow{573\,K} RbPO_3 \ [7-8]$$

$$Cs_6P_6O_{18}.6H_2O \xrightarrow{363\,K} Cs_6P_6O_{18}.nH_2O \xrightarrow{413\,K} Cs_6P_6O_{18} \xrightarrow{553\,K} CsPO_3 \ [7-8]$$

Only two divalent-cation derivatives were studied, $Cu_3P_6O_{18}.14H_2O$ [9–11] and $Cd_3P_6O_{18}.6H_2O$ [12]. Upon heating, the first one transforms into the copper cyclotetraphosphate:

$$2Cu_3P_6O_{18}.14H_2O \longrightarrow 3Cu_2P_4O_{12} + 14H_2O$$

and the second one into the cadmium long-chain polyphosphate above 588 K:

$$Cd_3P_6O_{18}.6H_2O \longrightarrow Cd(PO_3)_2 + 6H_2O$$

The formation of cadmium cyclotetraphosphate was observed before the last step of the reorganization leading to the long-chain polyphosphate.

Very little is known about the trivalent-cation derivatives. The chromium salt is stable up to 1273 K, but transforms rapidly into the C-form of the long-chain polyphosphate [13] at 1373 K. Under high-pressure, the same transformation occurs, but at a lower temperature:

$$Cr_2P_6O_{18} \xrightarrow[40\ Kbar]{1073\ K} Cr(PO_3)_3\ (C)$$

At about the same range of temperature (1223–1273 K), the aluminum salt transforms into the cyclotetraphosphate [14]:

$$2Al_2P_6O_{18} \longrightarrow Al_4(P_4O_{12})_3$$

These examples show that, as in all classes of condensed phosphates, the thermal behavior is mainly dependent on the nature of the associated cations

Monovalent- and mixed-monovalent cation cyclohexaphosphates – Most of the monovalent-cation cyclohexaphosphates are now well characterized as well as some recently prepared mixed-monovalent cation derivatives. These last compounds correspond to two different types of stoichiometries for the associated cations:
– an 1/1 order leading to $A_3B_3P_6O_{18}.nH_2O$ formula
– an 1/2 order corresponding to $A_4B_2P_6O_{18}.nH_2O$ compounds.

Griffith and Buxton [3] were the first to describe the preparation of the anhydrous *lithium* salt. Later, Schülke and Kayser [4] improved this process and reported the existence of one monohydrate, two forms of tetrahydrate and two forms of the anhydrous salt. The crystal structure of a pentahydrate was performed by Trunov *et al.* [15] and that of a more hydrated salt, $Li_6P_6O_{18}.9H_2O$, by Bagieu-Beucher and Rzaigui [16]. The existence of the hexahydrate reported by several authors as the normal hydrate crystallizing at room temperature has not yet been confirmed. Given the anomalies observed for the water molecule distribution in the structure reported for the pentahydrate [15], the sample studied is probably the hexahydrate.

The hexahydrate of the *sodium* salt is the only well characterized sodium salt. Its crystal structure was described by Jost [17].

The monohydrate is the only well characterized *silver* cyclohexaphosphate up to now. It was first described by Averbuch-Pouchot [5] who reported a detail process for its chemical preparation and a complete determination of its atomic arrangement. The existence of $Ag_6P_6O_{18}.2.2H_2O$ was reported by Thilo and Schülke [18].

Two crystalline *potassium* cyclohexaphosphates are well-known, the anhydrous salt and the trihydrate. The crystal structure of the anhydrous salt was performed almost simultaneously by Averbuch-Pouchot [19] and by Kholodkovskaya *et al.* [20], whereas that of the trihydrate was determined by the same last authors [21].

The *ammonium* derivative crystallizes as a monohydrate. Its preparation and its crystal structure were reported by Averbuch-Pouchot [22]. Its atomic arrangement is closely related to that reported for the silver salt.

The *rubidium* and *cesium* derivatives both exist as hexahydrates. These two hydrates, prepared by Averbuch-Pouchot and Durif [23], are isotypic. The crystal structure determination was almost simultaneously performed by these authors and by Kholodkovskaya *et al.* [24].

Table 5.4.19 lists the mixed-monovalent cation cyclohexaphosphates and reports shortly the procedures used for their preparations. These compounds belong to five different structural types. All the atomic arrangements were

Table 5.4.19
Mixed monovalent cation cyclohexaphosphates.

Formula	Nature of the preparation	Ref.
$Li_3Na_3P_6O_{18}.12H_2O$	Evaporation of an aqueous solution of $Li_6P_6O_{18}$ and $Na_6P_6O_{18}.6H_2O$ in the stoichiometric ratio	[4, 25–26]
$Li_3K_3P_6O_{18}.H_2O$	id. with $Li_6P_6O_{18}$ and $K_6P_6O_{18}.3H_2O$	[25–26]
$Li_2K_4P_6O_{18}.4H_2O$	id.	[27]
$Li_2Ag_4P_6O_{18}.2H_2O$	Aqueous solution of $AgNO_3$ slowly added to a solution of $Li_6P_6O_{18}$ in the ratio: $4AgNO_3/Li_6P_6O_{18}$	[28]
$Li_2(NH_4)_4P_6O_{18}.4H_2O$	Evaporation of an aqueous solution of $(NH_4)_6P_6O_{18}.H_2O$ and $Li_6P_6O_{18}$ in the stoichiometric ratio	[29]
$Na_2(NH_4)_4P_6O_{18}.2H_2O$	Evaporation of an aqueous solution of the two monovalent cyclohexaphosphates in the stoichiometric ratio	[30]
$Na_4Rb_2P_6O_{18}.6H_2O$	id.	[30]
$Na_4Cs_2P_6O_{18}.6H_2O$	id.	[30]
$Na_2Tl_4P_6O_{18}.2H_2O$	Aqueous solution of $TlNO_3$ slowly added to an aqueous solution of $Na_6P_6O_{18}.H_2O$ in the ratio: $4TlNO_3/Na_6P_6O_{18}$	[30]
$Ag_3(NH_4)_3P_6O_{18}.H_2O$	Aqueous solution of $AgNO_3$ slowly added to an aqueous solution of $(NH_4)_6P_6O_{18}.H_2O$ in the ratio: $3AgNO_3/Li_6P_6O_{18}$	[26]

determined. Among the compounds listed in Table 5.4.19, some of them belong to the same structural type. $Li_3K_3P_6O_{18}.H_2O$ and $Ag_3(NH_4)_3P_6O_{18}.H_2O$, for instance, are isotypic as well as the pairs listed below:

$Li_2(NH_4)_4P_6O_{18}.4H_2O$ and $Li_2K_4P_6O_{18}.4H_2O$

$Na_2(NH_4)_4P_6O_{18}.2H_2O$ and $Na_2Tl_4P_6O_{18}.2H_2O$

$Na_4Rb_2P_6O_{18}.6H_2O$ and $Na_4Cs_2P_6O_{18}.6H_2O$

We illustrate this category of compounds by a description of the atomic arrangement of one of them, $Li_3Na_3P_6O_{18}.12H_2O$ [25].

– The atomic arrangement in $Li_3Na_3P_6O_{18}.12H_2O$. This salt is trigonal (rhombohedral), $R\bar{3}c$, with the following unit-cell dimensions:

$a = 10.474(8)$, $c = 41.68(5)$ Å, with $Z = 6$ for the hexagonal setting

$a = 15.152$ Å, $\alpha = 40.44°$,   with $Z = 2$ for the rhombohedral setting

In spite of its apparent complexity and the large size of the unit cell, the atomic arrangement in this salt can be easily described as a succession of planes perpendicular to the threefold axis. A first family of planes contains the P_6O_{18} ring anions. These planes are located approximately in $z = nc/6$ and thus are separated by a distance of about 7 Å. The second family of planes contains a

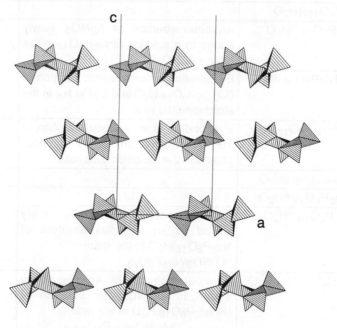

Figure 5.4.15. Planes of P_6O_{18} ring anions in $Li_3Na_3P_6O_{18}.12H_2O$ seen along the **b** axis.

two-dimensional network of corner-sharing LiO_4 tetrahedra and NaO_6 octahedra, with large voids centered around the $\bar{3}$ axes. Isolated $Na(H_2O)_6$ octahedra are located at the centers of these voids. The planes located approximately at $z = (2n+1)c/12$, alternate with those containing the phosphoric anions. The two independent P_6O_{18} ring anions have a $\bar{3}$ internal symmetry. In Figure 5.4.15 a projection along the **b** axis shows a group of planes of P_6O_{18} ring anions developing perpendicular to the **c** direction, whereas Figure 5.4.16 gives a representation of the two-dimensional network of LiO_4 and NaO_6 octahedra situated at $z = c/12$.

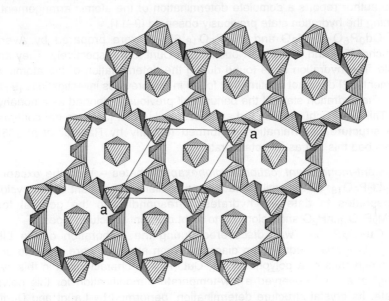

Figure 5.4.16. Projection, along the **c** axis, of the two-dimensional network of LiO_4 tetrahedra and NaO_6 octahedra at $z = c/12$ in the $Li_3Na_3P_6O_{18}.12H_2O$ atomic arrangement. The isolated octahedra represent the $Na(H_2O)_6$ groups.

Divalent-cation cyclohexaphosphates – Nine divalent-cation cyclohexaphosphates were described in chemical literature, but only four of them were investigated at a structural point of view to date.

The first investigation was performed by Lazarevski *et al.* [9–11] who reported the preparation and the thermal behavior of five cyclohexaphosphates, $Mn_3P_6O_{18}.9H_2O$, $Co_3P_6O_{18}.14H_2O$, $Ni_3P_6O_{18}.17H_2O$, $Cu_3P_6O_{18}.14H_2O$, and $Cd_3P_6O_{18}.16H_2O$. All were synthesized by the action of cyclohexaphosphoric acid on aqueous solutions of the corresponding perchlorates.

In the case of $Cu_3P_6O_{18}.14H_2O$, the P_6O_{18} ring is destroyed with the formation of the cyclotetraphosphate at high temperature:

$$2Cu_3P_6O_{18}.14H_2O \longrightarrow 3Cu_2P_4O_{12} + 14H_2O$$

Single crystals of this compound were later prepared by Averbuch-Pouchot [31] by adding water solutions of $CuCl_2$ or $Cu(NO_3)_2$ to an aqueous solution of $Li_6P_6O_{18}.6H_2O$ or by adding a water solution of copper nitrate to a solution of guanidinium cyclohexaphosphate in a stoichiometric ratio. After some days of evaporation of these different solutions at room temperature, crystals of $Cu_3P_6O_{18}.14H_2O$ appear as large turquoise calcite-like pseudo-rhombohedra. The author reports a complete determination of the atomic arrangement, confirming the hydration state previously observed [9–11].

$Cd_3P_6O_{18}.6H_2O$ and $Mn_3P_6O_{18}.6H_2O$ were prepared by Averbuch-Pouchot [32] and Averbuch-Pouchot and Durif [33] respectively. They are isotypic. The hydration state found during the determination of the atomic arrangement [32] does not confirm that found in the previous investigations [9–11].

The hydration state of the barium salt previously reported as a nonahydrate by Thilo and Schülke [18] and and Lazarevski *et al.* [34] was not confirmed by the structure determination performed recently by Rzaigui *et al.* [35] who described this salt as an octahydrate.

*Divalent-monovalent cation cyclohexaphos*phates – With the exception of $Cu_2Li_2P_6O_{18}$, all the well characterized bivalent-monovalent cation cyclohexaphosphates to date are hydrates corresponding to the general formula $M_2^{II}M_2^{I}P_6O_{18}.nH_2O$ and belonging to eight different structural types.

$Cu_2Li_2P_6O_{18}$ was discovered during the elaboration of the $LiPO_3-Cu_2P_4O_{12}$ phase-equilibrium diagram. When attempts were made to prepare single-crystals of a polyphosphate, $CuLi(PO_3)_3$, characterized in this system, Laügt *et al.* [36] observed a low-temperature modification for this polyphosphate. Its crystal structure determination, performed by Laügt and Durif [37], confirmed the results of the chromatographic analysis and proved the cyclic nature of the anion. According to the authors, this salt transforms into the polyphosphate at 773 K.

Most of the hydrates that we report now were prepared as polycrystalline samples by very classical methods of aqueous chemistry corresponding to the general reaction scheme:

$$M_6^{I}P_6O_{18} + 2M^{II}(NO_3)_2 \longrightarrow M_2^{I}M_2^{II}P_6O_{18}.nH_2O + 4M^{I}NO_3$$

All of them are sparingly water-soluble and, in some cases, the production of single crystals suitable for the structural investigations is tributary of non-orthodox processes. We report above a non exhaustive list of these numerous compounds that still are under systematic investigation and illustrate them by the description of the atomic arrangement in $Cu_2(NH_4)_2P_6O_{18}.8H_2O$.

$Zn_2Li_2P_6O_{18}.10H_2O$ and $Mn_2Li_2P_6O_{18}.10H_2O$ are isotypic. They were prepared and investigated by Averbuch-Pouchot [38].

$Ca_2Li_2P_6O_{18}.8H_2O$ and $Ca_2Na_2P_6O_{18}.8H_2O$ are also isotypic and were investigated by the same author [39]. They were both prepared as single crystals by adding solid gypsum in aqueous solutions of the alkali cyclohexaphosphates.

The cadmium–sodium salt, $Cd_2Na_2P_6O_{18}.14H_2O$, whose atomic arrangement was also described by Averbuch-Pouchot [40].

$Ca_2(NH_4)_2P_6O_{18}.6H_2O$ crystals were prepared using the procedure described for the calcium–lithium and calcium–sodium salts and its crystal structure performed by Averbuch-Pouchot [41].

The preparation and crystal structure of $Zn_2(NH_4)_2P_6O_{18}.8H_2O$ was described by Averbuch-Pouchot and Durif [42] and $Cd_2(NH_4)_2P_6O_{18}.9H_2O$ investigated by Averbuch-Pouchot *et al.* [43].

Suitable crystals of the copper–sodium salt, $Cu_2(NH_4)_2P_6O_{18}.8H_2O$, that we have selected to illustrate this family of compounds were difficult to elaborate. It is worth reporting this procedure with some details, as described by Averbuch-Pouchot and Durif [44], in order to illustrate the non-orthodox methods that we mentioned sometimes along this survey. An almost saturated aqueous solution of $(NH_4)_6P_6O_{18}.H_2O$ is added with the required amount of copper hydroxycarbonate, $CuCO_3.Cu(OH)_2$. In this slurry, drops of concentrated hydrochloric acid are added till the complete disparition of the hydroxycarbonate solid phase. Crystals of $Cu_2(NH_4)_2P_6O_{18}.8H_2O$ appear immediately as diamond-like plates. Polycrystalline samples of this very sparingly soluble salt can also be prepared by the classical methods of aque-ous chemistry.

– Structure of $Cu_2(NH_4)_2P_6O_{18}.8H_2O$. This salt is triclinic, $P\bar{1}$, with $Z = 1$ and the following unit-cell dimensions:

$$a = 7.413(3), \qquad b = 9.234(4), \qquad c = 9.634(4) \text{ Å}$$
$$\alpha = 116.23(5), \qquad \beta = 107.98(5), \qquad \gamma = 83.10(5)°$$

The P_6O_{18} ring anion, located around the inversion center at $(0,0,1/2)$ is centrosymmetric. The two crystallographically independent copper atoms are also located on inversion centers, the first one at $(0,0,0)$, the second one at $(1/2,1/2,0)$. Both are surrounded by distorted octahedra built by four oxygen atoms and two water molecules for the first one and two oxygen atoms and four water molecules for the second one. The atomic arrangement (Fig. 5.4.17) can be described as a succession of ribbons built by the P_6O_{18} phosphoric rings and one kind of CuO_6 octahedra extending parallel to the **c** direction. These ribbons are themselves interconnected along the **b** direction by the second kind of CuO_6 octahedra to form a layer parallel to the (b, c) plane. The connections thus established between the ring anions by the CuO_6 octahedra are not identical: one of the CuO_6 octahedron shares four of its oxygen atoms with the two adjacent P_6O_{18} groups, whereas the other one shares only two with two centrosymmetric P_6O_{18} rings. In both cases, all oxygen atoms are shared. The three-dimensional cohesion between these layers is established by the ammonium polyhedra and the hydrogen-bond network. One of the water molecule remains unbonded.

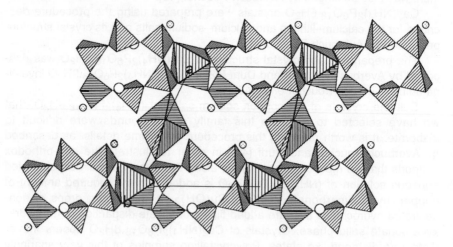

Figure 5.4.17. The atomic arrangement in $Cu_2(NH_4)_2P_6O_{18}.8H_2O$ in projection along the **a** direction. CuO_6 octahedra and P_6O_{18} ring anions are given by polyhedral representation. The smaller spotted circles represent the ammonium groups, the larger ones the non-bonded water molecules.

Trivalent-cation cyclohexaphosphates – The chromium, gallium, and aluminum cyclohexaphosphates crystallize as anhydrous compounds and are isotypic.

The chromium salt, $Cr_2P_6O_{18}$, was first obtained by Rémy and Boullé [45–46] during a systematic investigation of chromium phosphates.

The existence of $Ga_2P_6O_{18}$ was reported by Chudinova *et al.* [47] during a study of the Ga_2O_3–NH_4–P_2O_5–H_2O system. It was produced by heating at about 723 K, $GaNH_4HP_3O_{10}$, a triphosphate discovered during this investigation:

$$2GaNH_4HP_3O_{10} \longrightarrow Ga_2P_6O_{18} + 2H_2O + 2NH_3$$

The aluminum salt was obtained by Kanene *et al.* [14] by heating Al_2O_3 and $NH_4H_2PO_4$ at 823 K for 4 hours.

Bagieu-Beucher and Guitel [13] reported the preparation of single crystals and performed the determination of the atomic arrangement of the chromium salt. Table 5.4.20 gathers the main crystallographic data for these three salts.

Table 5.4.20
Crystal data for the anhydrous cyclohexaphosphates of trivalent cations.

Formula	a α	b β	c (Å) γ(°)	S. G.	Z	Ref.
$Cr_2P_6O_{18}$	8.311(4)	15.221(8) 105.85(5)	6.220(3)	$P2_1/a$	2	[13]
$Ga_2P_6O_{18}$	8.293	15.196 105.89	6.188	$P2_1/a$	2	[47]
$Al_2P_6O_{18}$	8.201(3)	15.062(3) 105.13(3)	6.101(2)	$P2_1/a$	2	[14]

Several hydrated trivalent-cation cyclohexaphosphates were recently described. They have various hydration states, generally high. In all of them, the determination of the atomic arrangements shows the existence of non-bonded water molecules. We simply report above a list of these hydrates with the corresponding references:

$Cr_2P_6O_{18}.21H_2O$, Rzaigui [48] and Bagieu-Beucher et al. [49]

$Y_2P_6O_{18}.18H_2O$, Lazarevski et al. [9]

$Ce_2P_6O_{18}.10H_2O$, Rzaigui [50] and Bagieu-Beucher and Rzaigui [51]

$Nd_2P_6O_{18}.12H_2O$, Trunov et al. [52]

$Yb_2P_6O_{18}.16H_2O$, Bagieu-Beucher and Rzaigui [53].

Table 5.4.21
List of some organic- and organo-metallic cation cyclohexaphosphates.

Formula	Name	Ref.
$(NH_3OH)_6P_6O_{18}.4H_2O$	hexakis(hydroxylammonium) cyclo-hexaphosphate tetrahydrate	[54]
$(C_2N_2H_{10})_3P_6O_{18}.2H_2O$	tris(ethylenediammonium) cyclohe-xaphosphate dihydrate	[55]
$(C_2H_5NH_3)_6P_6O_{18}.4H_2O$	hexakis(ethylammonium) cyclohe-xaphosphate tetrahydrate	[56]
$(N_2H_5)_2(N_2H_6)_2P_6O_{18}$	bis(monohydrazinium)-bis(dihydra-zinium) cyclohexaphosphate	[56]
$M^{II}(C_2N_2H_{10})_2P_6O_{18}.6H_2O$ M^{II} = Cu, Co, Ni, Mg, Zn, Fe	metal-bis(ethylenediammonium) cyclohexaphosphates hexahydrate	[55]

Organic- and organometallic cation cyclohexaphosphates – Following the characterization of the silver salt, $Ag_6P_6O_{18}.H_2O$, several organic- and organometallic cation cyclohexaphosphates were recently synthesized by the Boullé's process. In all these cases, the atomic arrangements were accurately determined. We list some of them in Table 5.4.21.

5.4.6.3. Adducts

The P_6O_{18} ring anions also appear in compounds like adducts with telluric acid or other salts as oxalates, nitrates and halides. This type of compounds will be described in the section devoted to adducts at the end of this chapter.

5.4.6.4. Chemical preparations of cyclohexaphosphates

For this very recently developed branch of the cyclophosphate chemistry, it seems useful to report the main chemical processes used for the preparation of cyclohexaphosphates to date in a detailed manner.

The Griffith-Buxton process – Griffith and Buxton [3] were the first to optimize the cyclization-condensation in the $Li_2O-P_2O_5$ system leading to the formation of lithium cyclohexaphosphate. A detailed procedure is described by the authors. For twenty years, this process was the only one permitting the preparation of appreciable amounts of cyclohexaphosphate, but was not extensively used by chemists. This lack of interest is probably explained by the multiple steps involved for the production of $Li_6P_6O_{18}$.

The Schülke and Kayser process – The Griffith-Buxton process was greatly improved recently and simplified by Schülke and Kayser [4]. Their process is based on a thermal dehydration-condensation of LiH_2PO_4, orientated toward the formation of $Li_6P_6O_{18}$ by a seeding of the initial product with a small quantity of $Li_6P_6O_{18}.6H_2O$. The experimental procedure is the following one. A mixture of 104 g of LiH_2PO_4 and 21 g of $Li_6P_6O_{18}.6H_2O$ is finely ground and heated at 623 K for 30 min. The resulting product is then ground and dissolved in 500 cm³ of water. Most of the time, three hours of mechanical stirring are necessary to perform this dissolution. The insoluble part, mainly $LiPO_3$ long-chain polyphosphate, is removed by filtration. 500 cm³ of methanol are then added to the solution leading to a complete precipitation of $Li_6P_6O_{18}.6H_2O$ within one day. The yield is almost the theoretical one. Schematically, the reaction is:

$$6LiH_2PO_4 + 0.2Li_6P_6O_{18}.6H_2O \xrightarrow[30\,mn]{623\,K} 1.2Li_6P_6O_{18} + 7.2H_2O$$

The Boullé's process – Since the characterization of the silver derivative,

$Ag_6P_6O_{18}.H_2O$, by Averbuch-Pouchot [5], a metathesis reaction deriving from the Boullé's process [6] can be used successfully for preparation of water-soluble cyclohexaphosphates. The reaction scheme is:

$$Ag_6P_6O_{18}.H_2O + 6M^ICl \longrightarrow M_6^IP_6O_{18} + 6AgCl$$

Many cyclohexaphosphates were prepared using this process and all the presently known organic-cation cyclohexaphosphates among them.

The use of fluorides – The insolubility of LiF can be used to prepare most of the alkali cyclohexaphosphates according to the following process:

$$6M^IF + Li_6P_6O_{18} \longrightarrow M_6^IP_6O_{18} + 6LiF$$

Other processes – Some other procedures, common to all classes of condensed phosphates, as *flux methods, ion-exchange,* and classical *aqueous chemistry* are also currently used for the preparation of cyclohexaphosphates.

5.4.6.5. Geometry of the P_6O_{18} rings

More than forty accurate determinations of cyclohexaphosphate atomic arrangements were performed authorizing the comparison of a good number of different P_6O_{18} ring-anions up to now.

Before examining the various numerical aspects of bond distances and angles in these rings briefly, it seems interesting to give a survey of their internal symmetry. Among all the presently known examples, one has no internal symmetry, but a strong pseudo-m conformation, one has m symmetry, one 2/m symmetry, two 3 symmetry, seven $\bar{3}$ and eighteen $\bar{1}$.

We limit this examination to what we consider as the main framework of a ring: the P–P distances and the P–O–P and P–P–P angles.

Among these three values, the examination of the rings in cyclohexaphosphates shows that two of them, the *P–P distances* and the *P–O–P angles* are quite comparable with what was previously measured in any kind of condensed phosphoric anions. For cyclohexaphosphates, the P–P distances, with a general average value of 2.927 Å and in range of 2.857–3.029 Å, as well as the P–O–P angles, ranging from 123.4 to 145.9° with an average of 132.4°, do not differ significantly from what we have reported for the other types of condensed anions that we have previously examined.

In the two categories of cyclophosphate anions that we have already analyzed in cyclotri- and cyclotetraphosphates, the P–P–P angles never depart significantly from their ideal values, 60±2° for cyclotriphosphates and 90±4° for cyclotetraphosphates. For the P_6O_{18} ring anions, if one examines first the most common ones ($\bar{1}$ internal symmetry), one observes very large deviations from the ideal value: 120°. The average angle is 111.9° with extrema of 85.9 and

142.8°. For higher symmetry rings (3 and 3̄) and so with more stresses, one could expect a more regular distribution of the P–P–P angles, but here the dispersion is also large. The average is 109.2°, with 87.5 and 131.7° as extreme values.

References

1.   E. Thilo and U. Schülke, *Z. anorg. allg. Chem.*, **344**, (1965), 293–307.

2.   K. H. Jost, *Acta Crystallogr.*, **B28**, (1972), 732–738.

3.   E. J. Griffith and R. L. Buxton, *Inorg. Chem.*, **4**, (1965), 549–551.

4.   U. Schülke and R. Kayser, *Z. anorg. allg. Chem.*, **531**, (1985), 167–175.

5.   M. T. Averbuch-Pouchot, *Z. Kristallogr.*, **189**, (1990), 17–23.

6.   A. Boullé, *C. R. Acad. Sci.*, **206**, (1938), 517–519.

7.   N. N. Chudinova, L. A. Borodina, U. Schülke, and K. H. Jost, *Izv. Akad. Nauk SSSR, Neorg. Mater.*, **25**, (1989), 459–465.

8.   N. N. Chudinova, L. A. Borodina, and U. Schülke, *Izv. Akad. Nauk SSSR, Neorg. Mater.*, **25**, (1989), 303–307.

9.   E. V. Lazarevski, L. V. Kubasova, N. N. Chudinova, and I. V. Tananaev, *Izv. Akad. Nauk SSSR, Neorg. Mater.*, **16**, (1980), 120–125.

10.   E. V. Lazarevski, L. V. Kubasova, N. N. Chudinova, and I. V. Tananaev, *Izv. Akad. Nauk SSSR, Neorg. Mater.*, **18**, (1982), 1544–1549.

11.   E. V. Lazarevski, L. V. Kubasova, N. N. Chudinova, and I. V. Tananaev, *Izv. Akad. Nauk SSSR, Neorg. Mater.*, **18**, (1982), 1550–1556.

12.   H. Nariai, I. Motooka, Y. Kanaji, and M. Tsuhako, *Phosphorus Res. Bull.*, **1**, (1991), 113–118.

13.   M. Bagieu-Beucher and J. C. Guitel, *Acta Crystallogr.*, **B33**, (1977), 2529–2533.

14.   Z. Y. Kanene, Z. A. Konstant, and V. V. Krasnikov, *Izv. Akad. Nauk SSSR, Neorg. Mater.*, **21**, (1985), 1552–1554.

15.   W. K. Trunov, L. N. Kholodkovskaya, L. A. Borodina, and N. N. Chudinova, *Kristallografiya*, **34**, (1989), 748–751.

16.   M. Bagieu-Beucher and M. Rzaigui, *J. Solid State Chem.*, **108**, (1994), 11–17.

17.   K. H. Jost, *Acta Crystallogr.*, **19**, (1965), 555–560.

18. E. Thilo and U. Schülke, *Z. anorg. allg. Chem.*, **341**, (1965), 293–307.

19. M. T. Averbuch-Pouchot, *Acta Crystallogr.*, **C45**, (1989), 1273–1275.

20. L. N. Kholodkovskaya, L. A. Borodina, W. K. Trunov, and N. N. Chudinova, *Izv. Akad. Nauk SSSR, Neorg. Mater.*, **25**, (1989), 466–469.

21. L. N. Kholodkovskaya, L. A. Borodina, W. K. Trunov, and N. N. Chudinova, *Izv. Akad. Nauk SSSR, Neorg. Mater.*, **25**, (1989), 454–458.

22. M. T. Averbuch-Pouchot, *Acta Crystallogr.*, **C45**, (1989), 539–540.

23. M. T. Averbuch-Pouchot and A. Durif, *C. R. Acad. Sci.*, Sér. 2, **308**, (1989), 1699–1702.

24. L. N. Kholodkovskaya, L. A. Borodina, W. K. Trunov, and N. N. Chudinova, *Izv. Akad. Nauk SSSR, Neorg. Mater.*, **25**, (1989), 470–476.

25. M. T. Averbuch-Pouchot, *Z. anorg. allg. Chem.*, **574**, (1989), 225–234.

26. M. T. Averbuch-Pouchot, *Acta Crystallogr.*, **C47**, (1991), 930–932.

27. M. T. Averbuch-Pouchot, personal communication.

28. M. T. Averbuch-Pouchot and A. Durif, *Acta Crystallogr.*, **C47**, (1991), 1150–1152.

29. M. Elmokhtar, M. Rzaigui, and A. Jouini, *Acta Crystallogr.*, **C49**, (1993), 435–437.

30. M. T. Averbuch-Pouchot and A. Durif, *Acta Crystallogr.*, **C47**, (1991), 932–936.

31. M. T. Averbuch-Pouchot, *Acta Crystallogr.*, **C45**, (1989), 1275–1277.

32. M. T. Averbuch-Pouchot, *Z. anorg. allg. Chem.*, **570**, (1989), 138–144.

33. M. T. Averbuch-Pouchot and A. Durif, *C. R. Acad. Sci.*, Sér. 2, **309**, (1989), 535–537.

34. E. V. Lazarevski, L. V. Kubasova, N. N. Chudinova, and I. V. Tananaev, *Izv. Akad. Nauk SSSR, Neorg. Mater.*, **17**, (1981), 486–491.

35. M. Rzaigui, M. T. Averbuch-Pouchot, and A. Durif, *Acta Crystallogr.*, **C48**, (1992), 241–243.

36. M. Laügt, A. Durif, and C. Martin, *J. Appl. Crystallogr.*, **7**, (1974), 448–449.

37. M. Laügt and A. Durif, *Acta Crystallogr.*, **B 30**, (1974), 2118–2121.

38. M. T. Averbuch-Pouchot, *Acta Crystallogr.*, **C45**, (1989), 1856–1858.

39. M. T. Averbuch-Pouchot and A. Durif, *Acta Crystallogr.*, **C46**, (1990), 968–970.

40. M. T. Averbuch-Pouchot, *Acta Crystallogr.*, **C46**, (1990), 10–13.

41.   M. T. Averbuch-Pouchot, *Acta Crystallogr.*, **C46**, (1990), 2005–2007.

42.   M. T. Averbuch-Pouchot and A. Durif, *Eur. J. Solid State Inorg. Chem.*, **30**, (1993), 573–581.

43.   I. Tordjman,. M. T. Averbuch-Pouchot, and A. Durif, to be published.

44.   M. T. Averbuch-Pouchot and A. Durif, *Acta Crystallogr.*, **C47**, (1991), 1148–1150.

45.   P. Rémy and A. Boullé, *C. R. Acad. Sci.*, **258**, (1964), 927–929.

46.   P. Rémy and A. Boullé, *Bull. Soc. Chim. Fr.*, (1972), 2213–2221.

47.   N. N. Chudinova, I. Grunze, and L. S. Guzeeva, *Izv. Akad. Nauk SSSR, Neorg. Mater.*, **23**, (1987), 616–621.

48.   M. Rzaigui, *J. Solid State Chem.*, **89**, (1990), 340–344.

49.   M. Bagieu-Beucher, M. T. Averbuch-Pouchot, and M. Rzaigui, *Acta Crystallogr.*, **C47**, (1991), 1364–1366.

50.   M. Rzaigui, *J. Solid State Chem.*, in press.

51.   M. Bagieu-Beucher and M. Rzaigui, *Acta Crystallogr.*, **C47**, (1991), 1789–1791.

52.   V. K. Trunov, N. N. Chudinova, and L. A. Borodina, *Dokl. Akad. Nauk SSSR*, **300**, (1988), 1375–1379.

53.   M. Bagieu-Beucher and M. Rzaigui, *Acta Crystallogr.*, **C48**, (1992), 244–246.

54.   A. Durif and M. T. Averbuch-Pouchot, *Acta Crystallogr.*, **C46**, (1990), 2026–2028.

55.   A. Durif and M. T. Averbuch-Pouchot, *Acta Crystallogr.*, **C45**, (1989), 1884–1887.

56.   M. T. Averbuch-Pouchot and A. Durif, *Acta Crystallogr.*, **C47**, (1991), 1579–1583.

5.4.7. Cyclooctaphosphates

5.4.7.1. Introduction

From paper-chromatography experiments, the existence of phosphoric ring anions larger than $[P_6O_{18}]^{6-}$ was clearly established more than forty years ago. In 1968 Schülke [1–2] reported the preparation of a crystalline cyclooctaphosphate, $Pb_4P_8O_{24}$, and the possibility to prepare the corresponding sodium salt. It is only since 1992 that a renewal of interest for this class of cyclophosphates appeared. Between these two dates, the great majority of the well characterized cyclooctaphosphates were discovered during the study of various phase-equilibrium diagrams and prepared as crystals by flux methods at relatively high temperature. The first structural characterization of a $[P_8O_{24}]^{8-}$ ring anion did not occur before 1975, when Laügt and Guitel [3] determined the crystal structure of the copper–ammonium salt, $Cu_3(NH_4)_2P_8O_{24}$.

The limited number of cyclooctaphosphates cannot authorize a general survey of their main physico-chemical properties. From the authors's experience, one can simply note that, under normal conditions, this large phosphoric ring appears to be very stable in aqueous solution.

5.4.7.2. Present state of cyclooctaphosphate chemistry

Monovalent-cation cyclooctaphosphates – The crystal structure of the *sodium salt*, $Na_8P_8O_{24}.6H_2O$, whose preparation will be examined further in this section, was performed by Schülke *et al.* [4].

Schülke and Chudinova [5] described the preparation and investigated the thermal behavior and the solubility of five hydrates, $M_8^IP_8O_{24}.6H_2O$, with $M^I =$ Li, K, Rb, Cs, and NH_4, all prepared through ion-exchange resins from the sodium salt. The solubility of the sodium salt is relatively small in compa-rison with the other alkali or ammonium derivatives.

In the structural determination of the *ammonium cyclooctaphosphate* made by Schülke *et al.* [6], this salt appears as a trihydrate, $(NH_4)_8P_8O_{24}.3H_2O$. The authors also report a study of its thermal behavior showing that its stability is very limited. It losses its crystallization water as low as 343 K and at 413 K transforms into the form (I) of ammonium long-chain polyphosphate through an exothermic reaction:

$$(NH_4)_8P_8O_{24}.3H_2O \xrightarrow{\ 343\,K\ } (NH_4)_8P_8O_{24} + 3H_2O \xrightarrow{\ 413\,K\ } 8NH_4PO_3$$

Brühne and Jansen [7] prepared crystals of an hydrated *cesium derivative*, $Cs_8P_8O_{24}.8H_2O$, by using a gel technique and performed its crystal structure determination.

These two latter structural investigations did not confirm the uniform state of hydration reported by Schülke and Chudinova [5] for these salts.

*Divalent- and divalent-monovalent cation c*yclooctaphosphates – Among the very limited number of *divalent-cation cyclooctaphosphates*, we must emphasize the lead salt which is the key for the preparation of any kind of starting material up to now. $Pb_4P_8O_{24}$ was discovered by Schülke [1–2] during a study of the thermal behavior of lead cyclotetraphosphate tetrahydrate. He observed that this salt can be converted into anhydrous lead cyclooctaphosphate under appropriate conditions. According to this author, the best yields are obtained when $Pb_2P_4O_{12}.4H_2O$ is first heated at 383 K for 30 min and then at 623 K again for 30 min. This transformation can be schematized by the following two steps:

$$2Pb_2P_4O_{12}.4H_2O \longrightarrow 2Pb_2H_4(PO_4)(P_3O_{10}) + 4H_2O \longrightarrow Pb_4P_8O_{24} + 8H_2O$$

In fact, the process is more complicated and can be decomposed into four distinct steps:

- (a) formation of crystalline $Pb_2P_4O_{12}.2H_2O$ by partial dehydration
- (b) hydrolysis of the cyclotetraphosphate anion by the residual crystal water leading to a complex mixture of mono-, di-, and triphosphates whose main component seems to be $Pb_2H_4(PO_4)(P_3O_{10})$
- (c) condensation of these various anions into 30% of the polyphosphate, $Pb(PO_3)_2$, and 70% of the anhydrous cyclotetraphosphate, $Pb_2P_4O_{12}$
- (d) conversion of $Pb_2P_4O_{12}$ into $Pb_4P_8O_{24}$

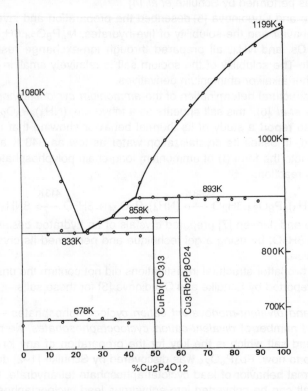

Figure 5.4.18. The $Cu_2P_4O_{12}$–$RbPO_3$ phase-equilibrium diagram.

Steps (c) and (d) occur simultaneously. The obtained $Pb_4P_8O_{24}$ can be used for the preparation of the sodium cyclooctaphosphate hexahydrate by the action of sodium sulphide according to the reaction:

$$Pb_4P_8O_{24} + 4Na_2S \longrightarrow Na_8P_8O_{24} + 4PbS$$

Several other divalent-cation derivatives were reported: $Ca_4P_8O_{24}.16H_2O$, by Schülke [1–2], $Ni_4P_8O_{24}.19H_2O$, $Mn_4P_8O_{24}.17H_2O$, and $Cd_4P_8O_{24}.12H_2O$ by Lavrov et al. [8].

The field of the *divalent-monovalent cation cyclooctaphosphates* is mainly dominated by a series of four isotypic $Cu_3M_2^IP_8O_{24}$ compounds. The rubidium, cesium, and thallium salts were characterized by Laügt [9] during a systematic study of the $Cu_2P_4O_{12}-M^IPO_3$ phase-equilibrium diagrams. All of them are incongruent melting compounds decomposing at 893, 939, and 909 K respectively. Figure 5.4.18 represents the $Cu_2P_4O_{12}-RbPO_3$ phase-equilibrium diagram. The ammonium derivative, $Cu_3(NH_4)_2P_8O_{24}$, was characterized during an investigation of the $Cu_2O-(NH_4)_2O-P_2O_5-H_2O$ system. Its crystal structure determination by Laügt and Guitel [3] provided the first geometric data of a P_8O_{24} group. Table 5.4.22 gathers the main crystallographic data for these four cyclooctaphosphates.

Table 5.4.22
Main crystallographic data for the four isotypic $Cu_3M_2^IP_8O_{24}$ cyclooctaphosphates.
The space group is $P\bar{1}$ and $Z = 1$.

Formula	a	b	c (Å)	Ref.
	α	β	γ (°)	
$Cu_3(NH_4)_2P_8O_{24}$	9.846(2)	7.962(2)	7.261(2)	[3, 9]
	80.98(3)	110.79(3)	110.61(3)	
$Cu_3Rb_2P_8O_{24}$	9.797(4)	8.035(3)	7.256(3)	[9]
	80.93(3)	110.35(3)	110.48(3)	
$Cu_3Cs_2P_8O_{24}$	9.913(2)	7.998(1)	7.298(1)	[9]
	81.55(2)	109.20(2)	109.14(2)	
$Cu_3Tl_2P_8O_{24}$	9.862(2)	7.922(2)	7.273(2)	[9]
	81.62(3)	110.43(3)	109.88(3)	

Trivalent- and monovalent-trivalent cation cyclooctaphosphates – Many compounds of general formula $A_2B_2P_8O_{24}$, with A^{III} = Al, Ga, Fe, and V and B^I = K, Rb, and NH_4 were described in chemical literature. Most of them were characterized during investigations of various $A_2O_3-B_2O-P_2O_5-H_2O$ systems by flux methods. We summarize these investigations in Table 5.4.23. During these various investigations, it was frequently observed that the $M^{III}M^I(H_2P_2O_7)_2$ diphosphates were the most stable compounds in these systems. When heated, these diphosphates lead to the $M_2^{III}B_2^{II}P_8O_{24}$ with very good yields in many cases. For instance, in the case of the gallium–potassium salt:

$$2GaK(H_2P_2O_7)_2 \xrightarrow{603 K} Ga_2K_2P_8O_{24} + 4H_2O \ [11]$$

In spite of some deficiencies in the crystallographic characterizations of some of them, it can be assumed that all the $M_2^{III}B_2^{II}P_8O_{24}$ are isotypic. The structure was determined by Palkina *et al.* [19] with the gallium–potassium salt: $Ga_2K_2P_8O_{24}$.

Various other trivalent- or trivalent-monovalent cation cyclooctaphosphates were investigated:

– $[Ga(OH)_2]_8P_8O_{24}.16H_2O$, $GaNa_5P_8O_{24}$ and $Ga_8(P_8O_{24})_3$ by Lazarevski *et al.* [20]

– $Y_8(P_8O_{24})_3.nH_2O$ (n ~ 30) and $YK_5P_8O_{24}.10H_2O$ by Lazarevski *et al.* [21]

but, no crystallographic investigation was performed for any of these salts.

Table 5.4.23
Summary of the investigations leading to the preparation of $A_2^{III}B_2^IP_8O_{24}$ cyclooctaphosphates.

Formula	Nature of the investigation	Ref.
$Ga_2K_2P_8O_{24}$	Investigation by flux methods at 573–623 K and at 423–773 K of the Ga_2O_3–K_2O–P_2O_5–H_2O system with starting molar ratio P/K/Ga = 15/5/1 by Chudinova *et al.* (mp = 1003 K)	[10–11]
$Ga_2Rb_2P_8O_{24}$	Same type of investigation with very similar conditions by Chudinova *et al.*	[12]
$Ga_2(NH_4)_2P_8O_{24}$	Same type of investigation between 423 and 623 K by Chudinova *et al.*	[13]
$Cr_2M_2^IP_8O_{24}$ (M^I = K, Rb, NH$_4$)	Same type of investigation with very similar conditions (473–673 K) by Chudinova *et al.*	[14]
$M_2^{III}K_2P_8O_{24}$ (M^{III} = Al, Ga, Fe) $Al_2Rb_2P_8O_{24}$	Thermal decomposition at about 650 K of the $M^{III}M^I(H_2P_2O_7)_2$ diphosphates by Grunze and co-workers	[15–17]
$V_2K_2P_8O_{24}$	Obtained by Lavrov *et al.* during the investigation of the V_2O_3–K_2O–P_2O_5–H_2O system and by firing $VK(H_2P_2O_7)_2$ at 620 K for 10–12h	[18]

Organic-cation cyclooctaphosphates – Very recently, Averbuch-Pouchot *et al.* [22–23] synthesized two organic-cation derivatives: $[C(NH_3)_3]_8P_8O_{24}.2H_2O$, the guanidinium cyclooctaphosphate dihydrate, and the ethylenediammonium cyclooctaphosphate hexahydrate $[NH_3(CH_2)_2NH_3]_4P_8O_{24}.6H_2O$.

Adducts – Cyclooctaphosphate anions were also observed in two kinds of adducts, cyclooctaphosphate-nitrates and cyclooctaphosphate-tellurates. Such compounds will be reported in the section devoted to adducts.

5.4.7.3. The atomic arrangement in $(NH_4)_8P_8O_{24}.3H_2O$

We give as structural illustration the atomic arrangement in the ammonium cyclooctaphosphate trihydrate, $(NH_4)_8P_8O_{24}.3H_2O$, that we think is well representative of this family of phosphates.

$(NH_4)_8P_8O_{24}.3H_2O$ [6] is monoclinic, Cc, with $Z = 4$ and the following unit-cell dimensions:

$$a = 24.27(1), \quad b = 6.700(3), \quad c = 20.59(1) \text{ Å}, \quad \beta = 112.06(6)°$$

The structure built by the stacking of the P_8O_{24} and NH_4 groups, is pseudo-centrosymmetric. The only clear departure from centrosymmetry is observed for the water molecules. Figure 5.4.19 is a projection of this structure along the **b** axis. For the representation of this atomic arrangement, the arbitrary origin was fixed so as to locate the ring anion around the origin of the unit cell. The phosphoric ring anion itself has a strong pseudo-centrosymmetric configuration.

The eight independent ammonium groups have various types of oxygen coordination. Within a range of 3.50 Å, three of them have eight neighbors, three have seven, one has six, and the last one has only five. Within these various NO_n polyhedra, the N–O distances range between 2.772 and 3.483 Å.

The water molecules are not, as is most frequently seen, dispersed within

Table 5.4.24
Main features of the P_8O_{24} group in $(NH_4)_8P_8O_{24}.3H_2O$.

P–P–P (°)	P–O(L)–P (°)	P–P (Å)
128.5	126.7	2.887
118.2	130.5	2.915
100.6	129.7	2.910
109.9	127.8	2.876
123.1	128.2	2.887
121.7	128.7	2.890
100.4	131.5	2.930
112.1	126.2	2.874

the arrangement. They are assembled in groups of three, building an irregular O(W1)–O(W2)–O(W3) triangle with the following edge lengths and angles:

O(W1)–O(W2)	3.479(8) Å	O(W2)–O(W1)–O(W3)	49.8(2)°
O(W1)–O(W3)	3.205(6)	O(W1)–O(W2)–O(W3)	60.1(2)
O(W2)–O(W3)	2.824(9)	O(W1)–O(W3)–O(W2)	70.2(2)

Inside this triangle, the only hydrogen bond is established between O(W2) and O(W3). All the other hydrogen atoms of these water molecules are involved in H–bonds with external oxygen atoms of the adjacent phosphoric groups.

We report, in Table 5.4.24, the main numerical values in the P_8O_{24} group observed in this compound.

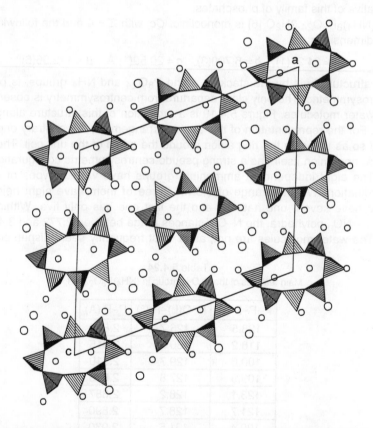

Figure 5.4.19. Projection, along the **c** axis, of the atomic arrangement in $(NH_4)P_8O_{24}.3H_2O$. The smaller circles represent ammonium groups and the larger ones the water molecules.

Table 5.4.25

Main features of the P_8O_{24} groups

Formula	P–P–P (°)	P–O–P (°)	P–P (Å)	Sym.	Ref.
$Na_8P_8O_{24}.6H_2O$	123.9	138.4	3.011	$\bar{1}$	[5]
	121.0	126.3	2.880		
	147.8	127.3	2.891		
	146.7	128.5	2.902		
$Cs_8P_8O_{24}.8H_2O$	123.7	137.2	2.997	$\bar{1}$	[7]
	121.8	128.2	2.900		
	112.7	126.0	2.882		
	117.6	127.5	2.886		
$Cu_3(NH_4)_2P_8O_{24}$	119.9	129.1	2.888	$\bar{1}$	[3]
	92.1	134.8	2.928		
	112.2	146.3	2.930		
	123.3	134.9	3.018		
$Ga_2K_2P_8O_{24}$	131.4	123.1	2.818	$2/m$	[19]
	138.0	134.4	2.933		
		136.0	2.947		
$Ag_9NaP_8O_{24}(NO_3)_2.4H_2O$	102.5	133.9	2.958	$2/m$	[25]
	108.7	129.5	2.889		
		128.8	2.900		
$(NH_4)_8P_8O_{24}.Te(OH)_6.2H_2O$	110.3	130.0	2.884	$\bar{1}$	[26]
	118.7	137.0	2.936		
	105.7	134.6	2.946		
	106.8	136.4	2.965		
	106.8	136.5	2.960	$\bar{1}$	
	103.7	134.9	2.952		
	114.6	126.3	2.868		
	119.9	131.0	2.908		
$K_8P_8O_{24}.Te(OH)_6.2H_2O$	107.4	132.4	2.946	$\bar{1}$	[27]
	105.1	132.6	2.939		
	117.5	132.2	2.925		
	101.1	133.3	2.953		
$(eda)_4P_8O_{24}.Te(OH)_6.2H_2O$	109.1	133.1	2.941	$\bar{1}$	[28]
	118.8	127.5	2.875		
	118.4	128.4	2.895		
	123.8	127.3	2.871		

– eda = ethylenediammonium or $[NH_3–(CH_2)_2–NH_3]^{2+}$

– the numerical values corresponding to the ammonium salt were given in the description of its atomic arrangement

– in $Te(OH)_6.(NH_4)_8P_8O_{24}.2H_2O$, there are two crystallographic independent ring anions

5.4.7.4. Geometry of the cyclooctaphosphate anion

The main geometrical features of the ring framework, P–P distances, P–O– P and P–P–P angles (Table 5.4.25) are those observed in nine of the presently known P_8O_{24} rings. In this table are also included the data for the rings found in various adducts. The data dealing with the ammonium salt, $(NH_4)_8P_8O_{24}.3H_2O$, already reported in Table 5.4.24, are omitted. In view of so few examples, any kind of discussion similar to what was possible with a larger number of examples for other smaller rings, P_6O_{18} for instance [24], seems fruitless here. Nevertheless, one can note that, among the ten rings presently investigated, two have a 2/m internal symmetry, seven are centrosymmetric and one has no internal symmetry. Among the numerical data reported in Tables 5.4.24 and 5.4.25, the P–P distances ranging from 2.818 to 3.018 Å and the P–O–P angles varying from 123.1 to 146.3° are within the ranges commonly observed in all the other classes of cyclophosphates.

The values reported for the P–P–P angles can appear as very dispersed varying from 92.1 to 146.7°, but are in fact quite comparable to the range of values observed in cyclohexaphosphates (85.9 to 142.8°) [24].

The numerical data analyzed here shortly are limited to values involved in the condensation phenomenon. In no way we dicuss the features of the main building unit, the PO_4 tetrahedron. This is presently the matter of a long study involving the investigation of several hundreds of tetrahedra.

References

1.   U. Schülke, *Z. anorg. allg. Chem.*, **360**, (1968), 231–246.

2.   U. Schülke, *Angew. Chem., Int. Ed. Engl.*, **7**, (1968), 71.

3.   M. Laügt and J. C. Guitel, *Z. Kristallogr.*, **141**, (1975), 203–216.

4.   U. Schülke, M. T. Averbuch-Pouchot, and A. Durif, *J. Solid State Chem.*, **98**, (1992), 213–218.

5.   U. Schülke and N. N. Chudinova, *Izv. Akad. Nauk SSSR, Neorg. Mater.*, **10**, (1974), 1697–1703.

6.   U. Schülke, M. T. Averbuch-Pouchot, and A. Durif, *Z. anorg. allg. Chem.*, **619**, (1993), 374–380.

7.   B. Brühne and M. Jansen, *Z. anorg. allg. Chem.*, **619**, (1993), 1633– 1638.

8.   A. V. Lavrov, T. A. Bykanova, and N. N. Chudinova, *Izv. Akad. Nauk*

SSSR, Neorg. Mater., **13**, (1977), 334–338.

9.   M. Laügt, *Thesis*, (Univ. of Grenoble, France, 1974).

10.   N. N. Chudinova, I. V. Tananaev, and M. A. Avaliani, *Izv. Akad. Nauk SSSR, Neorg. Mater.*, **13**, (1977), 2234–2237.

11.   N. N. Chudinova, M. A. Avaliani, L. S. Guzeeva, and I. V. Tananaev, *Izv. Akad. Nauk SSSR, Neorg. Mater.*, **14**, (1978), 2054–2057.

12.   N. N. Chudinova, M. A. Avaliani, L. S. Guzeeva, and I. V. Tananaev, *Izv. Akad. Nauk SSSR, Neorg. Mater.*, **15**, (1979), 2176–2179.

13.   N. N. Chudinova, I. Grunze, and L. S. Guzeeva, *Izv. Akad. Nauk SSSR, Neorg. Mater.*, **23**, (1987), 616–621.

14.   I. Grunze and N. N. Chudinova, *Izv. Akad. Nauk SSSR, Neorg. Mater.*, **24**, (1988), 988–993.

15.   J. Grunze, W. Hilmer, N. N. Chudinova, and H. Grunze, *Izv. Akad. Nauk SSSR, Neorg. Mater.*, **20**, (1984), 287–291.

16.   I. Grunze and H. Grunze, *Z. anorg. allg. Chem.*, **512**, (1984), 39–47.

17.   I. Grunze, N. N. Chudinova, and K. K. Palkina, *Izv. Akad. Nauk SSSR, Neorg. Mater.*, **19**, (1983), 1943–1945.

18.   A. V. Lavrov, M. Ya. Voitenko, and E. G. Tselebrovskaya, *Izv. Akad. Nauk SSSR, Neorg. Mater.*, **17**, (1981), 99–103.

19.   K. K. Palkina, S. I. Maksimova, V. G. Kusznetsov, and N. N. Chudinova, *Dokl. Akad. Nauk SSSR*, **245**, (1979), 1386–1389.

20.   E. V. Lazarevski, L. V. Kubasova, N. N. Chudinova, and I. V. Tananaev, *Izv. Akad. Nauk SSSR, Neorg. Mater.*, **15**, (1979), 2180–2184.

21.   E. V. Lazarevski, L. V. Kubasova, and N. N. Chudinova, *Izv. Akad. Nauk SSSR, Neorg. Mater.*, **19**, (1983), 498–499.

22.   M. T. Averbuch-Pouchot, A. Durif, and U. Schülke, *Eur. J. Solid State Inorg. Chem.*, **30**, (1993), 741–750.

23.   M. T. Averbuch-Pouchot, A. Durif, and U. Schülke, *Eur. J. Solid State Inorg. Chem.*, **30**, (1993), 557–563.

24.   M. T. Averbuch-Pouchot and A. Durif, *Eur. J. Solid State Inorg. Chem.*, **28**, (1991), 9–22.

25.   M. T. Averbuch-Pouchot and A. Durif, *Acta Crystallogr.*, **C48**, (1992), 1173–1176.

26.   M. T. Averbuch-Pouchot and A. Durif, *Acta Crystallogr.*, **C48**, (1993), 361–363.

27.   U. Schülke, M. T. Averbuch-Pouchot, and A. Durif, *Z. Kristallogr.*, **204**,

(1993), 143–152.

28.	M. T. Averbuch-Pouchot and U. Schülke, *Z. anorg. allg. Chem.*, in press.

5.4.8. Cyclononaphosphates

In the section on trivalent-cation long-chain polyphosphates, we mentioned that the $M^{III}(PO_3)_3$ compounds with small trivalent cations as M^{III} = Al, Cr, Ga, V, Mn, Fe, etc... can crystallize in six different structural types. Only three representatives of the hexagonal E-form are presently known. Little is known about these compounds because single crystals suitable for structural investigations could not be obtained. Their hexagonal unit cells (Table 5.4.26) were measured with very small crystals and refined from powder data by Bagieu [1].

Table 5.4.26
Hexagonal unit-cell dimensions of the three $M^{III}(PO_3)$ compounds belonging to the E-form.

Formula	a	c (Å)
Al(PO$_3$)$_3$	10.940(8)	9.192(8)
Cr(PO$_3$)$_3$	11.017(6)	9.381(7)
Fe(PO$_3$)$_3$	11.11(2)	9.54(2)

Very recently the authors obtained larger crystals of very poor quality for a new mixed-cation compound $(Fe,V)(PO_3)_3$. One of these crystals was nevertheless sufficient to obtain, with a very low accuracy, a model of the atomic arrangement which revealed that these compounds are in fact cyclononaphosphates and therefore must be formulated as $M^{III}_3P_9O_{27}$.

The material used for this study was prepared by a flux method involving a large excess of phosphoric acid at 583 K. The crystals obtained appear as lemon-yellow hexagonal platlets and are very often lenticular.

The crystallographic investigation showed that the unit-cell dimensions are:

$$a = 11.11, \quad c = 9.553 \text{ Å}$$

and the proper space to be $P6_3cm$ with $Z = 2$.

We give a short description of the structural results obtained in this unique investigation of a cyclononaphosphate. Figure 5.4.20 is a projection along the **c** direction of this atomic arrangement. The phosphoric ring anions are located around the 6_3 axes and have 3*mmm* internal symmetry. There are two kinds of $(V, Fe)O_6$ octahedra in the arrangement. The first ones are built around the 6_3

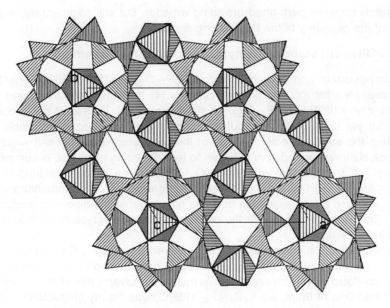

Figure 5.4.20. Projection, along the **c** direction, of the atomic arrangement in $(V,Fe)_3P_9O_{27}$.

axes, the second ones around the internal threefold axes. Along the 6_3 axis, $(V, Fe)O_6$ octahedra alternate with the phosphoric anions. The octahedra located on the 6_3 axes, half way between the ring anions, share their six oxygen atoms with the two adjacent P_9O_{27} groups, whereas those located on the threefold axes share their oxygen atoms with six different ring anions providing so the connection between the $-P_9O_{27}-MO_6-P_9O_{27}-MO_6-$ rows located around the 6_3 axes in the **a** and **b** directions.

Due to the inaccuracy of this investigation, we do not report any numerical values.

5.4.9. Cyclodecaphosphates

5.4.9.1. Introduction

The first cyclodecaphosphate, $Ba_2Zn_3P_{10}O_{30}$, was characterized in 1982 by Bagieu-Beucher et al. [2–3]. Cyclodecaphosphates are still relatively rare compounds in spite of a process recently described by Schülke [4] for the production of the alkali derivatives. Since this discovery, several new structural investigations were performed providing a better, but still fragmentary, knowledge of the geometry of the $P_{10}O_{30}$ ring anions.

5.4.9.2. Present state of the cyclodecaphosphate chemistry

The potassium and ammonium salts, $K_{10}P_{10}O_{30}.4H_2O$ and $(NH_4)_{10}P_{10}O_{30}$, were prepared for the first time by Schülke [4] through a process using the barium–zinc salt, $Ba_2Zn_3P_{10}O_{30}$, as starting material. Details of this process have not yet been published, but its outlines are described in an article [5] reporting the structure determination of the potassium salt. This salt which is very soluble in water and very resistant to hydrolytic ring cleavage, is converted at very low temperature (433–443 K) to crystalline potassium long-chain polyphosphate, KPO_3, through an exothermic reaction. The ammonium derivative prepared by ion exchange from the potassium salt is also very soluble in water. This was a handicap for the preparation of single crystals suitable for a structural investigation.

Attempts were made to reproduce the *silver* salt, $Ag_{10}P_{10}O_{30}$, reported by Schülke [5], but they were unsuccessful up to now. During these experiments, Averbuch-Pouchot [6] observed the formation of several mixed silver–potassium salts. One of them, $Ag_4K_6P_{10}O_{30}.10H_2O$, was clearly characterized and its atomic arrangement determined by Averbuch-Pouchot *et al.* [7].

$Mn_4K_2P_{10}O_{30}.18H_2O$ was prepared by Schülke and Averbuch-Pouchot [8] who described its atomic arrangement.

$Ba_2Zn_3P_{10}O_{30}$ was characterized during experiments performed by Bagieu-Beucher and El Horr [9]. In order to optimize a flux method to prepare $Ba_2Zn(PO_3)_6$, a long-chain polyphosphate appearing in the $BaO–ZnO–P_2O_5$ system, these authors obtained a new species that was identified as a cyclodecaphosphate when its crystal structure was solved [2–3].

Very recently, the first organic-cation derivative, the guanidinium cyclodecaphosphate tetrahydrate, $[C(NH_3)_3]_{10}P_{10}O_{30}.4H_2O$, was prepared and structurally investigated by Averbuch-Pouchot and Schülke [10].

5.4.9.3. Chemical preparation of cyclodecaphosphates

For this very new section of cyclophosphate chemistry, it seems interesting to report the chemical preparations of $Ba_2Zn_3P_{10}O_{30}$ and $K_{10}P_{10}O_{30}.4H_2O$ as recently optimized by Schülke [4–5] in detail.

Single crystals of $Ba_2Zn_3P_{10}O_{30}$ were first prepared by Bagieu-Beucher *et al.* [3]. The method used by these authors was considerably improved recently by Schülke [4–5]. A mixture of $BaCO_3$, $ZnCO_3$, and H_3PO_4 in the exact molar ratio and seeded with a few crystals of $Ba_2Zn_3P_{10}O_{30}$ is heated at 773 K in a platinum dish for two hours. The yield is 100%. The crystals used as seeds are prepared by melting a stoichiometric mixture of $BaCO_3$, $ZnCO_3$, and H_3PO_4 at 1023 K for ten min. The melt is then tempered at 773 K for 12 hours.

The process used by Schülke [4–5] for the preparation of the potassium salt, $K_{10}P_{10}O_{30}.4H_2O$, can be schematized by the following three steps:

$$Ba_2Zn_3P_{10}O_{30} + 2K_2SO_4 \longrightarrow 2BaSO_4 + Zn_3K_4P_{10}O_{30} \text{ (amorphous)}$$

$$Zn_3K_4P_{10}O_{30} + 3Na_2S \longrightarrow 3ZnS + K_4Na_6P_{10}O_{30} \text{ (solution)}$$

$$\overset{\text{K}^+ \text{ cation exchanger}}{K_4Na_6P_{10}O_{30} \longrightarrow K_{10}P_{10}O_{30} \text{ (solution)}}$$

The following experimental conditions are used: 18.9 g of $Ba_2Zn_3P_{10}O_{30}$, 5.2 g of K_2SO_4, and 40 ml of H_2O are intensively grinded in an agate swing mill for 8 hours, then under vigorous stirring a solution of 10.5 g of $Na_2S.9H_2O$ in 400 ml of H_2O is added to the suspension. The ZnS and $BaSO_4$ formed during the first two steps are then eliminated by filtration and an aqueous solution of $K_4Na_6P_{10}O_{30}$ is obtained. This last salt can be precipitated by the addition of 400 ml of methanol. The pure potassium salt, $K_{10}P_{10}O_{30}.4H_2O$, is then prepared from $K_4Na_6P_{10}O_{30}.aq.$ by ion exchange with a strong acidic cationic exchanger in its K^+ form.

5.4.9.4. The $[C(NH_2)_3]_{10}P_{10}O_{30}.4H_2O$ atomic arrangement

The atomic arrangement in the guanidinium cyclodecaphosphate tetrahydrate was selected to illustrate this category of compounds. It was prepared by the action of guanidinium carbonate on an aqueous solution of cyclodecaphosphoric acid prepared by ion-exchange from a solution of the potassium salt. The slow evaporation of the resulting solution leads to stout triclinic prisms at room temperature.

$[C(NH_2)_3]_{10}P_{10}O_{30}.4H_2O$ is triclinic, $P\bar{1}$, with $Z = 1$, and the following unit-cell dimensions:

$$a = 12.192(8), \quad b = 14.083(9), \quad c = 9.317(6) \text{ Å}$$
$$\alpha = 91.25(3), \quad \beta = 103.61(3), \quad \gamma = 71.22(3)°$$

The present arrangement is, like in all the organic-cation cyclophosphates, a stacking of ring anions and organic groups, whose cohesion is performed by a network of hydrogen bonds interconnecting, in the present case, the hydrogen atoms of the NH_2 groups to the external oxygen atoms of the phosphoric anion. Here, the $P_{10}O_{30}$ group located around the inversion center at $(0,0,0)$ is centrosymmetric. The examination of its geometrical features does not exhibit special features when compared to what is commonly observed in higher phosphoric ring anions. In this arrangement, the five crystallographically independent guanidinium groups have various orientations as shown in Figure 5.4.21. In each guanidinium group, the sum of the three N–C–N angles is close to

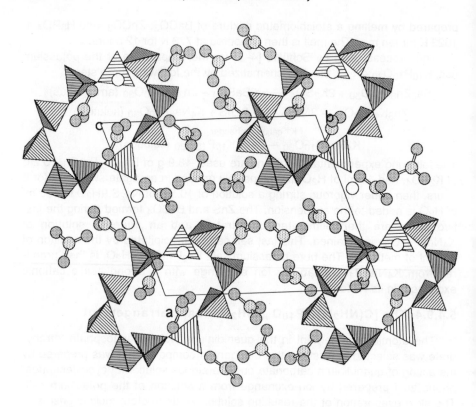

Figure 5.4.21. Projection, along the **c** axis, of the structure of $[C(NH_2)_3]_{10}P_{10}O_{30}.4H_2O$. The phosphoric anions are given by polyhedral representation, the spotted circles are the carbon atoms, the grey ones the nitrogen atoms and the open ones the water molecules. Hydrogen atoms are omitted.

360°, confirming the planarity generally observed in these groups. As sometimes observed in NH_2 or NH_3 rich compounds, some of hydrogen atoms of the NH_2 groups here could not find acceptors. The examination of the neighboring oxygen of the fifteen NH_2 radicals explains this observation. Thirteen of them have two or three oxygen neighbors within a range of 3.20 Å, two of them have only one.

5.4.9.5. Geometry of the $P_{10}O_{30}$ ring anions

As we have mentioned earlier, the $P_{10}O_{30}$ ring anions are very rare and one can only report five examples up to now. Therefore, in view of so few data any kind of discussion seems fruitless, like in the P_8O_{24} groups. Nevertheless, we report, in Table 5.4.27, the main geometrical features of the five $P_{10}O_{30}$ rings presently known. Among the numerical values reported in this table, the

Table 5.4.27
Main geometrical features of the five $P_{10}O_{30}$ ring anions presently investigated.

Formula	P–P–P (°)	P–O–P (°)	P–P (Å)	Sym.	Ref.
$Ba_2Zn_3P_{10}O_{30}$	112.9	130.0	2.915	2	[2–3]
	102.1	133.8	2.915		
	131.9	144.3	3.002		
	99.7	123.3	2.838		
	126.6	128.6	2.870		
$K_{10}P_{10}O_{30}\cdot4H_2O$	86.6	137.1	3.011	2	[5]
	111.5	133.1	2.950		
	133.3	128.9	2.911		
	107.1	131.3	2.936		
	90.5	134.9	2.963		
		138.2	3.006		
$Ag_4K_6P_{10}O_{30}\cdot10H_2O$	99.8	131.5	2.986	$\bar{1}$	[7]
	96.0	130.1	2.925		
	136.9	127.9	2.911		
	116.3	134.2	2.891		
	99.4	137.2	2.945		
$Mn_4K_2P_{10}O_{30}\cdot18H_2O$	114.6	135.9	2.938	$\bar{1}$	[8]
	138.8	129.4	2.928		
	85.5	133.5	2.890		
	87.9	134.3	2.944		
	98.0	127.1	2.869		
$(gua)_{10}P_{10}O_{30}\cdot4H_2O$	138.7	130.9	2.925	$\bar{1}$	[10]
	144.1	125.6	2.867		
	138.8	135.5	2.958		
	121.4	127.8	2.881		
	107.6	129.2	2.904		

gua = $[C(NH_2)_3]^+$ or guanidinium

P–P distances varying from 2.870 to 3.011 Å and the P–O–P angles from 123.3 and 144.3° are inside the ranges commonly observed in the crystal chemistry of any kind of cyclophosphates. The P–P–P angles varying from 86.6 to 136.9° P–P angles are in conformity with all that was previously observed in higher phosphoric rings [11]. This wide range of values may appear surprising.

Two of the five rings reported in Table 5.4.27 have two-fold internal symmetry but fundamentally different geometries. In $Ba_2Zn_3P_{10}O_{30}$ the two-fold axis is perpendicular to the mean plane of the phosphorus atoms whereas in $K_{10}P_{10}O_{30}.4H_2O$ the two-fold axis is parallel to this mean plane since it passes through two centrosymmetric bonding oxygen atoms of the ring.

References

1.   M. Bagieu, *Thesis*, (Univ. of Grenoble, France, 1980)

2.   M. Bagieu-Beucher, A. Durif, and J. C. Guitel, *J. Solid State Chem.*, **45**, (1982), 159–163.

3.   M. Bagieu-Beucher, A. Durif, and J. C. Guitel, *J. Solid State Chem.*, **40**, (1981), 248.

4.   U. Schülke, not published.

5.   U. Schülke, M. T. Averbuch-Pouchot and A. Durif, *Z. anorg. allg. Chem.*, **612**, (1992), 107–112.

6.   M. T. Averbuch-Pouchot, not published.

7.   M. T. Averbuch-Pouchot, A. Durif, and U. Schülke, *J. Solid State Chem.*, **97**, (1992), 299–304.

8.   U. Schülke and M. T. Averbuch-Pouchot, *Z. anorg. allg. Chem.*, **620**, (1994), 545–550.

9.   M. Bagieu-Beucher and N. El Horr, personal communication.

10. M. T. Averbuch-Pouchot and U. Schülke, *Z. Kristallogr.*, **210**, (1995), 129–132.

11. M. T. Averbuch-Pouchot and A. Durif, *Eur. J. Solid State Inorg. Chem.*, **28**, (1991), 9–22.

5.4.10. Cyclododecaphosphates

5.4.10.1. Present state of the cyclododecaphosphate chemistry

Very little information is available on cyclododecaphosphates. The first to be characterized with certainty was $V_3Cs_3P_{12}O_{36}$ discovered by Lavrov *et al.* [1] during an investigation of various $V_2O_3–M_2O–P_2O_5–H_2O$ systems. These authors also showed that the corresponding iron–cesium compound is isotypic and performed the crystal structure determination of $Cs_3V_3P_{12}O_{36}$ [2].

The vanadium–cesium salt is stable in argon up to 1173 K and at this temperature decomposes into the C-form of $V(PO_3)_3$ and $CsPO_3$:

$$V_3Cs_3P_{12}O_{36} \longrightarrow 3CsPO_3 + 3V(PO_3)_3$$

Grunze *et al* (3) obtained similar compounds when investigating the thermal behavior of various gallium-caesium phosphates. For instance, they observed the formation of $Ga_3Cs_3P_{12}O_{36}$ during the thermal decomposition of $GaCsHP_3O_{10}$:

$$6GaCsHP_3O_{10} \longrightarrow Ga_3Cs_3P_{12}O_{36} + 3GaCsP_2O_7 + 3H_2O$$

$GaCsHP_3O_{10}$ is trimorphic and its temperature of decomposition, function of the form used, can vary from 823 to 853 K. These authors performed the crystal structure of $Ga_3Cs_3P_{12}O_{36}$.

In addition, $Cr_3(NH_4)_3P_{12}O_{36}$, $Cr_3Rb_3P_{12}O_{36}$, and $Cr_3Cs_3P_{12}O_{36}$ were isolated by Grunze and Chudinova [4] during an investigation of various Cr_2O_3–M_2O–P_2O_5–H_2O systems between 473 and 673 K.

All the compounds described above are isotypic and crystallize in cubic unit cells close to 14.5 Å. The space group is $Pa3$ and $Z = 4$. The unit cells of two of them were accurately measured:

$$a = 14.543 \text{ Å for } V_3Cs_3P_{12}O_{36} \text{ [1, 4]}$$
$$a = 14.374 \text{ Å for } Ga_3Cs_3P_{12}O_{36} \text{ [3]}$$

The atomic arrangement is highly symmetric and is rather difficult to illustrate clearly. The $P_{12}O_{36}$ ring anions built around the ternary axes have in fact a $\bar{3}$ internal symmetry and are centered by a vanadium atom building its octahedral

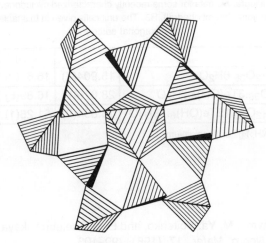

Figure 5.4.22. Projection along the ternary axis of a $P_{12}O_{36}$ group centered by a VO_6 octahedron in $V_3Cs_3P_{12}O_{36}$.

coordination with six of the external atoms of the ring. This arrangement, represented in projection along a threefold axis in Figure 5.4.22, is the unique example of a phosphoric ring centred by an associated cation polyhedron. We list above the main geometrical features of the $P_{12}O_{36}$ ring observed in $V_3Cs_3P_{12}O_{36}$:

P–P–P (°)	P–O–P (°)	P–P (Å)
85.8	133.8	2.945
111.3	136.3	2.952

They are not significantly different from what was observed in other large phosphoric rings.

Very recently, Schülke [5] successfully performed the synthesis of some alkali cyclododecaphosphates by using $Fe_3Cs_3P_{12}O_{36}$ as starting material. Some experiments carried out by Averbuch-Pouchot [6] with these alkali derivatives showed that they are extremely water soluble and hence very difficult to crystallize. From these materials, Schülke and Averbuch-Pouchot were, nevertheless, able to produce crystals of some other cyclododecaphosphates, guanidinium cyclododecaphosphate hexahydrate and two adducts with telluric acid, $Cs_{12}P_{12}O_{36} \cdot 3Te(OH)_6 \cdot 9H_2O$ and $(gua)_{12}P_{12}O_{36} \cdot Te(OH)_6 \cdot H_2O$. Crystal structure of the first one was recently performed [7], the last two are still under investigation. We summarize in Table 5.4.28 the main crystallographic data for these new compounds.

Table 5.4.28
Main crystallographic data for some recently characterized cyclododecaphosphates.
All of them are trigonal (rhombohedral), R$\bar{3}$. The unit cells given in this table correspond to the hexagonal setting.

Formula	a	c (Å)	Z	Ref.
$(gua)_{12}P_{12}O_{36} \cdot 6H_2O$	15.904(7)	16.67(2)	2	[7]
$Cs_{12}P_{12}O_{36} \cdot 3Te(OH)_6 \cdot 9H_2O$	28.07(4)	16.89(6)	6	[6]
$(gua)_{12}P_{12}O_{36} \cdot 12Te(OH)_6 \cdot 18H_2O$	15.854(9)	51.26(1)	3	[6]

gua = $[C(NH_2)_3]^+$ or guanidinium

References

1. A. V. Lavrov, M. Ya. Voitenko, and E. G. Tselebrovskaya, *Izv. Akad. Nauk SSSR, Neorg. Mater.*, **17**, (1981), 99–103.

2. A. V. Lavrov, V. P. Nikolaev, G. G. Sadikov, and M. Y. Voitenko, *Dokl.*

Akad. Nauk SSSR, **259**, (1981), 103–106.

3.   I. Grunze, K. K. Palkina, N. N. Chudinova, L. S. Guzeeva, M. A. Avaliani, and S. I. Maksimova, *Izv. Akad. Nauk SSSR, Neorg. Mater.*, **23**, (1987), 610–615.

4.   I. Grunze and N. N. Chudinova, *Izv. Akad. Nauk SSSR, Neorg. Mater.*, **24**, (1988), 988–993.

5.   U. Schülke, to be published.

6.   U. Schülke and M. T. Averbuch-Pouchot, to be published.

7.   U. Schülke and M. T. Averbuch-Pouchot, *Z. anorg. allg. Chem.*, in press.

5.4.11. Thiocyclophosphates

5.4.11.1. Introduction

The occurance of sulfur-substituted anions in cyclophosphate chemistry is very recent. Wolf and Meisel [1] described the synthesis of several phosphorus oxide-sulfides $P_4O_{(10-n)}S_n$ by reaction of P_4S_{10} with P_4O_{10} some years ago. According to the ratio P_4S_{10}/P_4O_{10} in the starting mixture, various phosphorus oxide-sulfides can be obtained. Besides the well-known $P_4O_6S_4$, the following compounds were obtained by the authors for the first time: $P_4O_5S_5$, $P_4O_4S_6$, $P_4O_3S_7$, $P_4O_2S_8$, and P_4OS_9. They are separated by fractional distillation or crystallization. A detailed study of these phosphorus oxide-sulfides was reported by Meisel [2].

Some of these compounds were used as starting materials for the preparation of thiocyclophosphates by a process similar to that described for the production of the cyclotetraphosphoric acid hydrolysis of P_4O_{10} at low temperature. For instance, the reaction in the case of $P_4O_6S_4$ is:

$$P_4O_6S_4 + 2H_2O \longrightarrow H_4P_4O_8S_4$$

leading to the formation of the tetrathiocyclotetraphosphoric acid.

In the corresponding salts the observed cyclic anions have a geometry similar to that of cyclotri- or cyclotetraphosphoric rings, but with one or two external oxygen atoms substituted by sulphur atoms.

To date two families of thiocyclophosphates are well established, the *thiocyclotriphosphates* corresponding to the $[P_3O_6S_3]^{3-}$ and $[P_3O_3S_6]^{3-}$ anions, and the *thiocyclotetraphosphates* corresponding to the $[P_4O_8S_4]^{4-}$ anion.

We will examine these two families separately. For this very recent class of compounds, we report the chemical preparations with more details as usual.

5.4.11.2. Thiocyclotriphosphates

As mentioned before, two types of anionic groups are observed in this category of compounds. In the first one, corresponding the $[P_3O_6S_3]^{3-}$ anion, only one external oxygen atom of each tetrahedron is substituted by a sulphur atom, whereas, in the second type corresponding to the $[P_3O_3S_6]^{3-}$ anion, all the external oxygen atoms are substituted.

The first type has only one representative: $(NH_4)_3P_3O_6S_3$. It was obtained by Meisel *et al.* [3] by fluorhydrolysis of $P_4O_6S_4$ in glacial acetic acid. Finely ground, $P_4O_6S_4$ is slowly added to an ice-cooled saturated solution of ammonium fluoride with acetic acid. Crystals of $(NH_4)_3P_3O_6S_3$ precipitate after two hours of stirring. The reaction scheme is:

$$3NH_4F + P_4O_6S_4 + 2AcOH \longrightarrow (NH_4)_3P_3O_6S_3 + H(POSF_2) + Ac_2O + HF$$

The crystal structure determination was performed by Meisel *et al.* [3].

The second type of thiocyclotriphosphates has more representatives. The sodium and thallium salts, $Na_3P_3O_3S_6$ and $Tl_3P_3O_3S_6$, were first prepared by Wolf and Meisel [4]. The sodium salt was prepared according to the following scheme at 283 K:

$$P_4O_3S_6 + 4NaHCO_3 \longrightarrow Na_3P_3O_3S_6 + NaH(PO_3H) + 4CO_2 + H_2O$$

Recently, the crystal structures of the potassium and cesium salts were performed by Palkina *et al.* [5].

5.4.11.3. Thiocyclotetraphosphates

Kuvshinova *et al.* [6] prepared a series of monovalent-cation thiocyclotetraphosphates: $Li_4P_4O_8S_4.9H_2O$, $Na_4P_4O_8S_4.6H_2O$, $K_4P_4O_8S_4.2H_2O$, $Rb_4P_4O_8S_4.2H_2O$, $Cs_4P_4O_8S_4$, $Tl_4P_4O_8S_4$, $(NH_4)_4P_4O_8S_4.2H_2O$, and some organic salts: $(C_5H_5NH)_4P_4O_8S_4$, $[C(NH_2)_3]_4P_4O_8S_4$, $(C_9H_7NH)_4P_4O_8S_4$, and $(C_2H_5NH_3)_4P_4O_8S_4.2H_2O$ — $[C_5H_5NH]^+$ = pyridinium, $[C(NH_2)_3]^+$ = guanidinium, $[C_9H_7NH]^+$ = quinolinium, and $[C_2H_5NH_3]^+$ = ethylammonium. The properties of these salts, mainly their thermal behavior, were also investigated by the authors. We report some of their observations.

Upon heating, the *lithium salt*, $Li_4P_4O_8S_4.9H_2O$, after several steps of dehydration (8, 6, 4, $2H_2O$), transforms first, at about 723 K, into $Li_4P_4O_{12}$ and then, at about 823 K, into a mixture of $LiPO_3$ (HT) and $Li_6P_6O_{18}$.

In the case of the *sodium derivative*, $Na_4P_4O_8S_4.6H_2O$, the tetrahydrate and the dihydrate are observed successively between room temperature and 423 K. Finally, this salt transforms into $Na_3P_3O_9$ above 700 K.

The *potassium and rubidium salts* become anhydrous above 423 K and at higher temperature transform into the corresponding long-chain polyphosphates.

The melting point of the *cesium salt* is observed at about 650 K. Above this temperature, it transforms into $CsPO_3$.

We have more details about the chemical preparations for some of these salts. A number of them were prepared as single crystals and the atomic arrangements determined.

$Na_4P_4O_8S_4.6H_2O$ was obtained by Ilyukhin *et al.* [7] by slowly adding ground $P_4O_6S_4$ into a cooled aqueous solution of $NaHCO_3$ (an excess of about 20% of carbonate with respect to the stoichiometry of the reaction is used). At the end of the reaction, the excess carbonate is destroyed by acetic acid and ethanol is added to precipitate the sodium salt which is very difficult to crystallize and is not very stable in air. Nevertheless, the authors could perform the crystal structure determination.

The anhydrous cesium salt was prepared by Ilyukhin *et al.* [8] using a similar procedure. The atomic arrangement is described by the authors.

Nikolaev and Kuvshinova [9] synthesized $Ba_2P_4O_8S_4.10H_2O$ using the same procedure. Crystals were obtained by recrystallization in acetic acid. This salt is stable for years in the mother liquor, but transforms into the hexahydrate within some minutes in air.

The thermal transformation of Ba and Sr tetrathiocyclotetraphosphate hexahydrates were studied by Prodan *et al.* [10] in vacuum and by Kubshinova *et al.* [11] in air and the influence of water vapor on their topochemical transformations by Petrovskaya *et al.* [12].

The anhydrous guanidinium salt was prepared by Kuvshinova *et al.* [6] through a two-step process:

$$P_4O_6S_4 + 2H_2O \xrightarrow{\text{273--278 K}} H_4P_4O_8S_4$$

$$H_4P_4O_8S_4 + 2[C(NH_2)_3]_2CO_3 \longrightarrow [C(NH_2)_3]_4P_4O_8S_4 + 2CO_2 + 2H_2O$$

This salt which is stable in air at room temperature, is sparingly water soluble. Its atomic arrangement was described by Meisel *et al.* [13].

5.4.11.4. The ammonium trithiocyclotriphosphate structure

We will illustrate this type of componds by a description of the atomic arrangement in the ammonium trithiocyclotriphosphate.

$(NH_4)_3P_3O_6S_3$ [3] is orthorhombic, *Pnma*, with $Z = 4$, and the following unit-cell dimensions:

$$a = 12.450(8), \quad b = 12.755(8), \quad c = 8.154(6) \text{ Å}$$

Figure 5.4.23 shows a projection of the atomic arrangement along the **a** direction. The ring anion has mirror symmetry with one phosphorus atom, one sulfur atom, and two oxygen atoms located on the mirror plane. One of the two crystallographic independent ammonium group is also located on this mirror plane.

This ammonium group and the $P_3O_6S_3$ groups build layers centred by the mirror planes at $y = 1/4$ and $3/4$. The ammonium groups, in general crystallographic position, form corrugated layers between the mirror planes. The ring anion is built by two crystallographically independent strongly distorted tetrahedra including a P–S bond length longer than 1.9 Å and three normal P–O distances. The NH_4 located on the mirror plane has a sixfold coordination of O atoms whereas the second one in a general position, has sevenfold including three sulfur atoms. Within these two polyhedra the N–O distances range between 2.744 and 3.273 Å and the N–S distances between 3.395 and 3.497 Å.

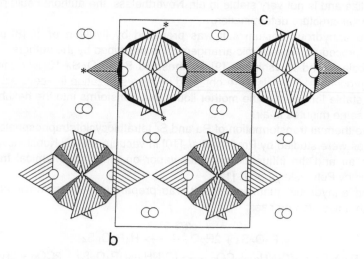

Figure 5.4.23. The atomic arrangement in $(NH_4)_3P_3O_6S_3$ in projection along the **a** direction. The starred atoms of the PO_3S tetrahedra are the S atoms and the open circles the NH_4 groups.

5.4.11.5. A cyclotriphosphate with a totally sulfur substituted anion

Among the thiocyclophosphates that we have examined above, one or two external oxygen atoms of the ring anions were substituted by sulphur atoms. A compound described by Wolf and Meisel [14] corresponds very probably to a complete substitution, $(NH_4)_3P_3S_9$. Unfortunately, no structural study ascertains the ring nature of the anion suggested by the chemical formula of this salt.

This compound was prepared by the reaction of P_4S_{10} or P_4S_9 in liquid ammonia below 240 K:

$$P_4S_9 + 6NH_3 \longrightarrow (NH_4)_3P_3S_9 + P(NH_2)_3$$

$$P_4S_{10} + 6NH_3 \longrightarrow (NH_4)_3P_3S_9 + SP(NH_2)_3$$

This salt crystallizes on cooling down to 195 K. It reacts with PCl_3 rebuilding the adamantine-like structure of P_4S_9:

$$(NH_4)_3P_3S_9 + PCl_3 \longrightarrow P_4S_9 + 3NH_4Cl$$

When heated for 100 hours in vacuum at 513 K it decomposes according to:

$$(NH_4)_3P_3S_9 \longrightarrow 3PNS + 6H_2S$$

5.4.11.6. The thiocyclophosphate rings

To date the structural examples are insufficient to justify any meaningful comparisons. Only one trithiocyclotriphosphate is presently well investigated. Nevertheless, one can note that the five $P_4O_8S_4$ rings, all centrosymmetric, have P–P–P angles departing more significantly from the ideal value than in the P_4O_{12} rings. Thus, for these five $P_4O_8S_4$ rings the P–P–P angles range from 97.7 to 95.6° (Table 5.4.29).

Table 5.4.29
Main geometrical features of $P_3O_6S_3$ and $P_4O_8S_4$ anions.

Formula	P–P (Å)	P–O–P (°)	P–P–P (°)	Sym.	Ref
$(NH_4)_3P_3O_6S_3$	2.938	131.3	2x60.4	m	[3]
	2x2.977	2x136.0	59.1		
$Na_4P_4O_8S_4.6H_2O$	2.951	132.0	97.7	$\bar{1}$	[6–7]
	2.993	133.0	82.3		
	2.957	133.0	83.7	$\bar{1}$	
	3.015	141.5	96.3		
$Cs_4P_4O_8S_4$	2.973	133.7	82.8	$\bar{1}$	[6, 8]
	2.981	134.2	97.2		
$Ba_2P_4O_8S_4.10H_2O$	2.935	131.1	82.6	$\bar{1}$	[9]
	2.953	132.0	97.4		
$(gua)_4P_4O_8S_4$	2.981	135.3	95.6	$\bar{1}$	[6, 13]
	2.977	135.1	84.4		

gua = $[C(NH_2)_3]^+$ or guanidinium

The most important difference between P_3O_9 or P_4O_{12} phosphoric rings and the present thio-rings lies in the constitution of the tetrahedron. The presence of a sulfur atom leading to an average P–S distance longer than 1.93 Å induces a strong deformation for this polyhedron. Table 5.4.30 reports the geometry of such a PO_3S tetrahedron as observed in $[C(NH_2)_3]_4P_4O_8S_4$.

If compared to the normal non-sustituted PO_4 tetrahedra observed in other types of condensed phosphoric anions, the main differences in the PO_3S and PO_2S_2 tetrahedra arise from the P–S bonds, significantly longer than the P–O distances. Therefore, two or three edges of the tetrahedron corresponding to the S–O distances are much longer (2.876 < S–O < 2.976 Å) than the remaining ones corresponding to the O–O distances (2.478 < O–O < 2.560 Å).

Table 5.4.30

Main geometrical features of a PSO_3 tetrahedron of the $P_4S_4O_8$ ring anion observed in $[C(NH_2)_3]_4P_4O_8S_4$.

P	S	O(E)	O(L1)	O(L2)
S	1.9427(8)	2.976(2)	2.876(2)	2.926(2)
O(E)	119.37(8)	1.495(2)	2.560(2)	2.478(2)
O(L1)	107.84(6)	111.17(9)	1.608(2)	2.504(2)
O(L2)	110.04(7)	105.29(8)	101.70(8)	1.621(1)

Values reported in the diagonal correspond to P–O or P–S distances. The O–P–O(S) angles are given in the lower left triangle and the O–O(S) distances in the upper right triangle. O(E) is an external oxygen of the ring whereas O(L) atoms belong to the ring.

On the contrary, the O–P–O(S) angles are in a remarkable accordance with all was found in P_nO_{3n} ring anions. The value of 119.37° observed for the S–P–O(E) is quite comparable with the O(E)–P–O(E) values observed in any type of phosphoric rings and the average value of O–P–O(S) in this tetrahedron (109.23°) is similar to what is commonly encountered in any kind of condensed phosphoric anions.

References

1.   G. U. Wolf and M. Meisel, *Z. anorg. allg. Chem.*, **509**, (1984), 101–110

2.   M. Meisel, *Z. Chem.*, **23**, (1983), 117–125.

3.   M. Meisel, G. U. Wolf, and M. T. Averbuch–Pouchot, *Acta Crystallogr.*, **C47**, (1991), 1368–1370.

4.   G. U. Wolf and M. Meisel, *Z. Chem.*, **20**, (1980), 452.

5.   K. K. Palkina, S. I. Maksimova, N. T. Chibiskova, M. S. Pautina, G. U. Wolf, and T. B. Kuvshinova, *Izv. Akad. Nauk SSSR, Neorg. Mater.*, **28**, (1992), 1486–1490.

6.   T. B. Kuvshinova, G. U. Wolf, and M. Meisel, *Izv. Akad. Nauk SSSR*,

Neorg. Mater., **20**, (1984), 1056–1062.

7.   V. V. Ilyukhin, V. R. Kalinin, T. B. Kuvshinova, and I. V. Tananaev, *Dokl. Akad. Nauk SSSR*, **266**, (1981), 1387–1391.

8.   V. V. Ilyukhin, V. R. Kalinin, T. B. Kuvshinova, and I. V. Tananaev, *Dokl. Akad. Nauk SSSR*, **267**, (1982), 85–88.

9.   V. P. Nikolaev and T. B. Kuvshinova, *Izv. Akad. Nauk SSSR, Neorg. Mater.*, **23**, (1987), 622–625.

10.   E. A. Prodan, L. I. Petrovskaya, and T. B. Kuvshinova, *Izv. Akad. Nauk SSSR, Neorg. Mater.*, **25**, (1989), 1339–1343.

11.   T. S. Kubshinova, G. U. Wolf, T. N. Kuzmina, and M. S. Pautina, *Izv. Akad. Nauk SSSR, Neorg. Mater.*, **25**, (1989), 1349–1355.

12.   L. I. Petrovskaya, M. S. Pautina, E. A. Prodan, T. B. Kuvshinova, and S. I. Pytlev, *Izv. Akad. Nauk SSSR, Neorg. Mater.*, **27**, (1991), 1023–1027.

13.   M. Meisel, G. U. Wolf, and M. T. Averbuch–Pouchot, *Acta Crystallogr.*, **C46**, (1990), 2239–2241.

14.   G. U. Wolf and M. Meisel, *Z. anorg. allg. Chem.*, **494**, (1982), 49–54.

5.5. ULTRAPHOSPHATES

5.5.1. Introduction

At the beginning of this book, in the chapter devoted to the classification of phosphates, we have reported the situation of ultraphosphates inside phosphate chemistry and gave their chemical definition and a few details on the basic geometric properties of the corresponding anions. Let us simply remember that the general formulation for the presently well characterized ultraphosphate anions is given by:

$$[P_{n+2}O_{3n+5}]^{n-} \quad \text{with } n = 2, 3, 4, 5, \text{ and } 6$$

In Chapter 7, we shall come back on some non-satisfactory aspects of the nomenclature.

Kroll [1–3] was credited for their discovery for a long time, but in fact, as early as 1885, Clève [4] characterized the samarium ultraphosphate, SmP_5O_{14}, and some years later, Johnsson [5] prepared some LnP_5O_{14} compounds and performed detailed morphological measurements on single crystals of the lanthanum derivative in 1889. Both Clève and Johnsson called these compounds "anhydro-metaphosphates". Twenty years later, Kroll [1–3] discovered some new P_2O_5-rich compounds, but he ignored the previous investigations and named them "ultraphosphates". Then for more than thirty years, the literature about ultraphosphates was purely speculative and it is only in 1944 that Hill *et al.* [6–7] characterized two calcium ultraphosphates, CaP_4O_{11} and $Ca_2P_6O_{17}$ clearly during their investigation of a part of the $CaO–P_2O_5$ system. Nevertheless, it must be said that Kroll [1–3] was the first to report the existence of $Ca_2P_6O_{17}$. The phase diagram as elaborated by Hill *et al.* [6–7] is given in Figure 5.5.1.

Soon after the systematic crystallographic investigation of the LnP_5O_{14} compounds performed by Beucher [8] during the late sixties, these materials were recognized as efficient laser materials and this provoked a renewed interest in ultraphosphates. In fact, they were the first examples of stoichiometric laser crystals. During the seventies an explosive development of the scientific and technical literature occurred that was devoted not only to these ultraphosphates (mainly the Nd salt) but to any kind of rare-earth bearing phosphates.

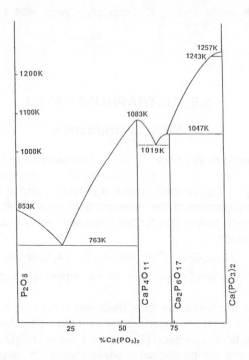

Figure 5.5.1. The $Ca(PO_3)_2$–P_2O_5 phase-equilibrium diagram as elaborated by Hill *et al.* [6–7].

This trend was beneficial, inducing a great development of the crystal chemistry of rare-earth condensed phosphates which were rather neglected before. Another beneficial result of the Beucher's study [8] was to prove that ultraphosphates, or at least some of them, can exist as stable compounds. It provoked a new interest in these salts, mainly in the former Soviet Union and in Germany, where a number of new ultraphosphates were synthesized and clearly characterized during the last twenty five years.

A review of crystal chemistry of ultraphosphates was recently published by the authors [9].

5.5.2. General Properties

The minor interest in ultraphosphates for a long time must probably be attributed to what was commonly and erroneously believed about their properties. Ultraphosphates were said to hydrolyse very rapidly when in contact with water vapor present in the air, this instability was attributed to the existence of the triply linked PO_4 tetrahedra present in the anionic framework. The "Antibranching Rule" introduced by Van Wazer and Holst in 1950 [10] summarizes the commonly accepted opinion at this time: *"In any environment in which reactions involving the degradation of condensed phosphates are possible, it is to be expected that assemblies in which three of the four oxygens of a PO_4 tetrahedron are shared with other PO_4 tetrahedra will be exceedingly unstable and will degrade much more rapidly as compared to those in which one or two oxygens are shared"*. These words can probably be explained by the quasi-purely speculative state of the knowledge of ultraphosphates which was based upon a very limited number of experiments at this time.

The still restricted number of ultraphosphates cannot yet authorize a valuable survey of their general behavior. The LnP_5O_{14} were the subject of a good number of investigations and their properties are presently rather well-known, but all papers dealing with the other kinds of ultraphosphates contain an accurate structural determination generally with some lines devoted to the chemical preparation and very few words on their basic properties.

5.5.2.1. Thermal behavior

There is an evident lack of accurate data concerning the thermal behavior of ultraphosphates, nevertheless we know that all the LnP_5O_{14} decompose on prolonged heating according to the following scheme:

$$LnP_5O_{14} \longrightarrow LnPO_4 + 2P_2O_5$$

This mechanism of decomposition was investigated for both the rare-earth and bismuth ultraphosphates by Dudko *et al.* [11].

According to Lavrov *et al.* [12], the nickel ultraphosphate, NiP_4O_{11}, decomposes into the cyclotetraphosphate when heated in air:

$$2NiP_4O_{11} \longrightarrow Ni_2P_4O_{12} + 2P_2O_5$$

5.5.2.2. Behavior towards water

Most of the earlier work dealing with the action of water on ultraphosphates was performed with more or less glassy samples because crystalline ultraphosphates were not yet known. All the observations report a very slow solubility. Some of these ultraphosphate compositions decrepitate violently with a screaming noise when in contact with water. Small particles can spring out from the surface and sometimes be thrown to a distance of 30 cm.

The behavior of ultraphosphates towards water is variable. The LnP_5O_{14} single crystals usually prepared by flux processes are extracted from the flux by washing with water without any alteration. At room temperature, they can be kept in contact with water for several days with no apparent degradation.

At room temperature, $Ca_2P_6O_{17}$ is not stable in a wet atmosphere, but $YCaP_7O_{20}$ crystals are extracted from the excess of flux by boiling water.

Tests were performed on the solubility of the two calcium ultraphosphates in order to check their possible usage as fertilizers. They have very low solubility in water and hydrolyse as soon as they dissolve.

5.5.3. The Present State of Ultraphosphate Chemistry

No example of crystalline monovalent-cation ultraphosphates were reported in chemical literature up to now. Nevertheless, from the behavior of some phosphoric fluxes currently used in crystal growth experiments, evidence exists for ultraphosphate anions in presence of alkali metals.

5.5.3.1. Divalent-cation ultraphosphates

The presently well characterized divalent-cation ultraphosphates correspond to two different chemical formulas, $M^{II}P_4O_{11}$ and $M_2^{II}P_6O_{17}$.

$M^{II}P_4O_{11}$ ultraphosphates are known for M^{II} = Mg [13–14], Zn [15], Co [16], Ni [12, 32], Mn [17], and Ca [6–7, 18–20] and $M_2^{II}P_6O_{17}$ ultraphosphates for M^{II} = Ca [6–7, 21], Sr [22–23], and Cd [22–24]. The calcium and cadmium derivatives seem to be isotypic, but in the two structure determinations the authors encountered some unusual features, as for instance, the location of a bonding oxygen atom on an inversion center implying the existence of a P–O–P bond angle of 180°. In our opinion, it seems that this structure type needs a deeper investigation in spite of the accuracy of the result.

As can be seen from Table 5.5.1 which gathers the main crystallographic features of these salts, some of them (Mg, Ca, Mn) are polymorphic.

Table 5.5.1

Main crystallographic features for divalent-cation ultraphosphates.

Formula	a α	b β	c (Å) γ (°)	S. G.	Z	Ref.
MgP_4O_{11}	5.343(1)	22.228(5) 110.89(1)	7.451(2)	$P2_1/c$	4	[13]
MgP_4O_{11}	9.670(5)	35.40(4)	14.521(8)	$Pmc2_1$	24	[14]
CaP_4O_{11}	8.856(4)	12.72(1) 134.26(5)	12.148(8)	$P2_1/c$	4	[18–19]
CaP_4O_{11} at 358 K	12.683	12.090	12.627	$Aba2$	8	[20]
MnP_4O_{11}	8.608(2)	8.597(4)	12.464(5) 97.30(4)	$P2_1/n$	4	[17]
MnP_4O_{11}	5.452(1)	22.407(8) 109.60(1)	7.501(1)	$P2_1/c$	4	[32]
NiP_4O_{11}	9.375(1)	8.019(1) 100.73(1)	11.135(2)	$P2_1/n$	4	[12, 32]
CuP_4O_{11}	9.193(2) 69.74(2)	9.248(2) 70.79(2)	10.516(2) 89.73(2)	$P\bar{1}$	2	[32]
ZnP_4O_{11}	5.302(1)	22.242(4) 110.13(1)	7.412(1)	$P2_1/c$	4	[15]
$Ca_2P_6O_{17}$	5.753(2)	18.265(5) 111.12(1)	7.625(2)	$P2_1/c$	2	[21]
$Sr_2P_6O_{17}$	7.158(2)	13.02(2) 105.26(2)	7.172(1)	$P2_1$	2	[22–23]
$Cd_2P_6O_{17}$	7.566(4)	5.486(4)	18.082(5) 111.22(4)	$P2_1/n$	2	[22–24]

5.5.3.2. Miscellaneous ultraphosphates

Before reviewing the most important family of ultraphosphates, the lanthanide derivatives, we report the existence of various types of ultraphosphates.

Chudinova et al. [25] identified $FeNa_3P_8O_{23}$ during an investigation of the Fe_2O_3–Na_2O–P_2O_5 system. Later, Palkina et al. [26] characterized two isomorphous derivatives, $AlNa_3P_8O_{23}$ and $GaNa_3P_8O_{23}$. The crystal structures of these three compounds were determined and provided the first example of a finite ultraphosphate anionic group.

Hamady and Jouini [27] prepared and determined the atomic arrangement of $YCaP_7O_{20}$ very recently.

$(TaO_2)_4P_6O_{17}$ and $(UO_2)_2P_6O_{17}$ were characterized by Chernorukov *et al.* [28] and Lavrov [29] respectively. The atomic arrangement in the second salt was determined by Gorbunova *et al.* [30].

5.5.3.3. Yttrium, rare-earth, and bismuth ultraphosphates

In this section, we cannot report the great amount of literature devoted to these compounds since 1969, so much the more that two previous and recent reviews [10, 31] of this domain already presented this literature in the form of annotated bibliographies. Thus, we selected a restricted number of studies in order to illustrate the main types of anionic frameworks existing in this family of

Table 5.5.2

Main crystallographic data for bismuth and rare-earth ultraphosphates belonging to the first structure type. The space group is $P2_1/a$ and $Z = 4$.

Formula	a α	b β	c (Å) $\gamma(°)$	Ref.
BiP_5O_{14}	13.06(1)	9.02(1) 90.6(1)	8.77(1)	[36]
LaP_5O_{14}	13.18(1)	9.112(3) 90.38(5)	8.820(3)	[8]
CeP_5O_{14}	13.11(1)	9.063(3) 90.45(5)	8.790(3)	[8]
PrP_5O_{14}	13.08(1)	9.041(3) 90.42(5)	8.787(3)	[8]
NdP_5O_{14}	13.03(1)	9.001(3) 90.48(5)	8.768(3)	[8, 37–38]
SmP_5O_{14}	12.99(1)	8.944(3) 90.41(5)	8.757(3)	[8, 33]
EuP_5O_{14}	12.93(1)	8.930(3) 90.45(5)	8.751(3)	[8, 39]
GdP_5O_{14}	12.93(1)	8.904(3) 90.49(5)	8.743(3)	[8]
HoP_5O_{14} (a)	12.855	8.822 90.54	8.695	[8]
TbP_5O_{14}	12.91(1)	8.887(3) 90.49(5)	8.728(3)	[8]

a: high-pressure form

compounds. All the compounds described in this section, have a common formula, TP_5O_{14}, (T = Y, Bi or Ln) and belong to four structure types. We have already mentioned that some of them were characterized at the end of the last century by Clève [4] and Johnsson [5], but the fundamental pioneering work was performed by Beucher [9] who prepared as single crystals YP_5O_{14} and all the LnP_5O_{14} ultraphosphates and have shown by a careful crystallographic investigation that these compounds belonged to three different structural types, some of them being polymorphic. The crystal structures of these three structural types were established by Bagieu–Beucher and co-workers:

- Type I with SmP_5O_{14} by Tranqui *et al.* [33]
- Type II with HoP_5O_{14} by Bagieu-Beucher *et al.* [34]
- Type III with HoP_5O_{14} by Tranqui *et al.* [35]

Later, a fourth structural type was described by Rzaigui *et al.* [40] for the cerium salt. Unit-cell dimensions of this variety of CeP_5O_{14} are:

$$a = 9.227(5), \quad b = 8.890(5), \quad c = 7.219(4) \text{ Å}$$
$$\alpha = 110.12(5), \quad \beta = 102.68(5), \quad \gamma = 82.13(5)^\circ$$

Table 5.5.3
Main crystallographic data for yttrium and rare-earth ultraphosphates belonging to the second structure type. The space group is *C2/c* and $Z = 8$.

Formula	a α	b β	c (Å) $\gamma(^\circ)$	Ref.
TbP_5O_{14}	12.91(1)	12.80(1) 91.31(5)	12.48(1)	[8]
DyP_5O_{14}	12.90(1)	12.79(1) 91.30(5)	12.46(1)	[8]
HoP_5O_{14}	12.88(1)	12.77(1) 91.34(5)	12.42(1)	[8, 34]
ErP_5O_{14}	12.85(1)	12.74(1) 91.28(5)	12.40(1)	[8, 42]
TmP_5O_{14}	12.84(1)	12.73(1) 91.26(5)	12.38(1)	[8, 40]
YbP_5O_{14}	12.84(1)	12.72(1) 91.25(5)	12.37(1)	[8, 43]
LuP_5O_{14}	12.81(1)	12.71(1) 91.24(5)	12.34(1)	[8]
YP_5O_{14}	12.87(1)	12.76(1) 91.28(5)	12.43(1)	[8]

The space group is $P1$ and $Z = 2$. This compound is the only one crystallizing with this structural type up to now.

We report in Table 5.5.2, 5.5.3 and 5.5.4, the main crystal data for the (Ln, Bi)P_5O_{14} compounds belonging to these three forms.

It is worth reporting that almost simultaneously with the Bagieu-Beucher's investigations [33–35], Jaulmes [41] prepared single crystals of the lanthanum derivative. Soon after these studies, BiP_5O_{14} was recognized as an isotype of the lanthanum salt by Chudinova and Jost [36].

Some short surveys dealing with these TP$_5O_{14}$ compounds were published by Bagieu–Beucher and Tranqui [44], Bagieu [45], Durif [46] and Bondar *et al.* [47]. Bagieu [45] examined carefully the geometries of the associated cation coordination polyhedra in the first three structural types.

In addition to these basic investigations, the discovery of the optical properties of rare-earth ultraphosphates induced a good number of studies in various domains of optics, electronics, and chemistry along which the crystal structures of the LnP_5O_{14} ultraphosphates as well as those of many other rare-earth bearing phosphates were determined repeatedly.

Table 5.5.4

Main crystallographic data for yttrium, bismuth, and rare-earth ultraphosphates belonging to the third structure type. The space group is *Pnma* and $Z = 4$.

Formula	a	b	c (Å)	Ref.
	α	β	$\gamma(°)$	
DyP_5O_{14}	8.726(3)	12.75(1)	8.950(3)	[8]
HoP_5O_{14}	8.720(3)	12.71(1)	8.926(3)	[8]
ErP_5O_{14}	8.712(3)	12.68(1)	8.919(3)	[8]
YP_5O_{14}	8.718(3)	12.73(1)	8.939(3)	[8]
BiP_5O_{14} (a)	8.726(12)	12.950(20)	8.958(18)	[48]

a: high-temperature form

5.5.4. The Ultraphosphate Anions

We will not illustrate the present section with the description of atomic arrangements, but with a survey of the various geometries adopted by the ultraphosphate anions. We have adopted this approach for two reasons:

– the atomic arrangements in ultraphosphates are generally very intricate and difficult to describe

– various geometries could exist for a given anionic formula. Thus, the $[P_5O_{14}]^{3-}$ anion can build ribbons in forms I and III of the LnP_5O_{14} derivatives or a three-dimensional network in form II of the same compounds and the $[P_6O_{17}]^{4-}$ anion exists as infinite layers in $Ca_2P_6O_{17}$, $Cd_2P_6O_{17}$, and $Sr_2P_6O_{17}$ or as a three-dimensional network in $(UO_2)_2P_6O_{17}$.

The ultraphosphate anions can adopt various geometries: finite groups, infinite ribbons, infinite layers or three-dimensional networks.

5.5.4.1. Finite anionic group

The unique ultraphosphate with a finite anionic group, $[P_8O_{23}]^{6-}$, was first observed in $FeNa_3P_8O_{23}$ [25] and later in the isotypic aluminium and gallium derivatives [26]. Figure 5.5.2 gives a perspective view of this anion. The latter has a ternary internal symmetry and can be simply described as being built by two tetrahedra, which are located on the ternary axis and represented on the left and right extremities of the drawing, interconnected by three P_2O_7 groups. Another way to describe this anion is to consider it as a ring of six PO_4 tetrahedra in which two opposite tetrahedra (those located on the ternary axis), are linked by a P_2O_7 group.

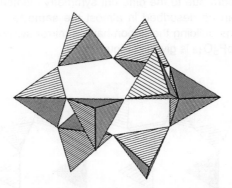

Figure 5.5.2. A perspective view of the eight tetrahedra building the P_8O_{23} group.

5.5.4.2. Infinite ribbon anions

Three types of infinite ribbon anions were observed in forms I and III of the rare-earth and bismuth ultraphosphates and in $YCaP_7O_{20}$ [27].

The infinite ultraphosphate anion observed in form I can be described in several ways. One can consider it as built by two infinite $(PO_3)_n$ chains interconnected by PO_4 groups. Another way to describe this anion is to see it as constructed by a succession of centrosymmetric eight-member rings interconnected through the ternary tetrahedra. A projection of this anion as observed in

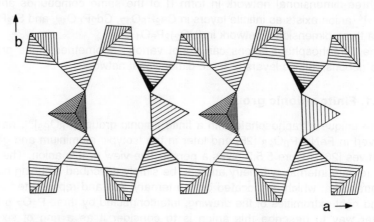

Figure 5.5.3. The P_5O_{14} infinite ribbon as observed in SmP_5O_{14} (form I).

SmP_5O_{14} is given in Figure 5.5.3. With some slight differences in the orientation of the tetrahedra due to the different symmetry elements, the anion observed in form III can be described in almost the same terms. In this form, the eight-member rings building the ribbon have a mirror symmetry. Such an anion as observed in HoP_5O_{14} is given in Figure 5.5.4.

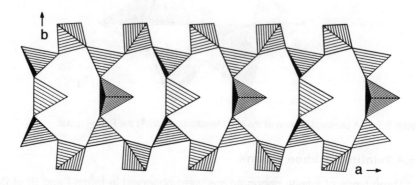

Figure 5.5.4. The P_5O_{14} infinite ribbon as observed in HoP_5O_{14} (form III).

In $YCaP_7O_{20}$, the ultraphosphate anion is a ribbon that can be described as built by a succession of centrosymmetric $P_{10}O_{30}$ rings, each ring sharing two of its PO_4 tetrahedra with the adjacent one. A representation of this anion is given in Figure 5.5.5.

Figure 5.5.5. The P_7O_{20} infinite ribbon as observed in $YCaP_7O_{20}$.

5.5.4.3. Infinite two-dimensional anions

Several types of infinite two-dimensional anions were found in various ultraphosphates, CaP_4O_{11}, $Sr_2P_6O_{17}$, $Ca_2P_6O_{17}$, $Cd_2P_6O_{17}$, and in form IV of rare-earth ultraphosphates.

The $[P_4O_{11}]^{2-}$ anion in CaP_4O_{11} – The atomic arrangement in CaP_4O_{11} [19] is made by a succession of layers of the infinite anion and layers of calcium atoms alternating perpendicular to the **a** direction. The internal arrangement of the

Figure 5.5.6. A projection, along the **a** direction, of the infinite ultraphosphate anion in CaP_4O_{11}.

phosphoric layer is represented in Figure 5.5.6, a projection along the **a** axis. Like in the previous examples several possibilities are offered here also to describe this network. One can consider it to be built up by a set of infinite $(PO_3)_n$ chains interconnected by PO_4 tetrahedra or as a tiling of P_8O_{24} and $P_{12}O_{36}$ rings.

The $[P_6O_{17}]^{4-}$ anion in $Cd_2P_6O_{17}$ – In $Cd_2P_6O_{17}$ [22], the atomic arrangement can be described as built by a succession of two kinds of layers parallel to the (110) planes. The first type of layer containing the infinite phosphoric anion alternate with planes of cadmium atoms. The internal arrangement of the phosphoric layer is shown in projection along the **b** axis in Figure 5.5.7. This phosphoric anion can be depicted as built by a tiling of centrosymmetric rings of fourteen tetrahedra interconnected in a two-dimensional way through their ternary PO_4 tetrahedra. Each ring is connected to six of its neighbors.

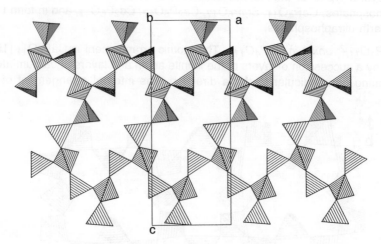

Figure 5.5.7. A projection, along the **b** direction, of the infinite P_6O_{17} ultraphosphate anion in $Cd_2P_6O_{17}$.

The $[P_6O_{17}]^{4-}$ anion in $Sr_2P_6O_{17}$ – The atomic arrangement in $Sr_2P_6O_{17}$ [23] is also a layer arrangement. Two kinds of very corrugated layers alternate perpendicular to the **a** direction. The first type of layer contains the strontium atoms, the second one the phosphoric anion. The details of the anion framework are given in Figure 5.5.8. As in the case of the cadmium ultraphosphate, the infinite anion can be described as a tiling of fourteen member rings interconnected through their branching tetrahedra.

Figure 5.5.8. A projection, along the **a** direction, of the infinite P_6O_{17} ultraphosphate anion in $Sr_2P_6O_{17}$.

5.5.4.4. Three-dimensional anionic networks

They were observed in the form II of rare-earth ultraphosphates and in the uranyl ultraphosphate. As can be expected for three-dimensional networks of tetrahedra, these anions are difficult to represent clearly.

The $[P_6O_{17}]^{4-}$ anion in $(UO_2)_2P_6O_{17}$ – In $(UO_2)_2P_6O_{17}$ [29], the framework of the PO_4 tetrahedra building the anion can be described in several ways. The most appropriate is to consider it to be built by a set of $P_{14}O_{42}$ rings interconnected in a three-dimensional way by sharing some tetrahedra. In Figure 5.5.9 we have simply depicted the main building unit of this arrangement, an isolated $P_{14}O_{42}$ ring.

The $[P_5O_{14}]^{3-}$ anion in the form II of LnP_5O_{14} – In LnP_5O_{14} (form II), the anionic framework exhibits some unexpected features. In the first step, one can consider it to be built up by three different types of interconnected thick layers. Layers of the first type are approximately centered by the planes $x = 1/4$ and 3/4, those of the second type are located around the planes $y = 1/4$ and 3/4 and the third ones around the planes $z = 0$ and 1/2. This stacking of layers creates channels in which are located the associated cations.

It is interesting to examine the internal arrangement of each type of layer separately through a projection perpendicular to its plane. One can then notice

(Figure 5.5.10) that the layers of the first type are built by a succession of very corrugated independent phosphoric chains running along the **b** axis, whereas those of the second type (Figure 5.5.11) are constituted of a tiling of independent P_8O_{24} rings. In the third type of layers (Figure 5.5.12), independent slightly corrugated infinite chains develop along the $[1\bar{1}0]$ direction.

All the drawings used to illustrate this atomic arrangement were performed using the data reported for HoP_5O_{14} [34].

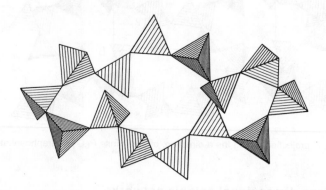

Figure 5.5.9. An isolated $P_{14}O_{42}$ ring as found in the ultraphosphate anion of $(UO_2)_2P_6O_{17}$.

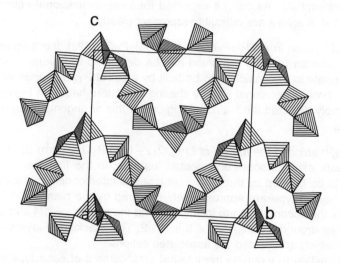

Figure 5.5.10. A projection, along the **a** direction, of an isolated anionic layer of the first type in form-II of a LnP_5O_{14} ultraphosphate.

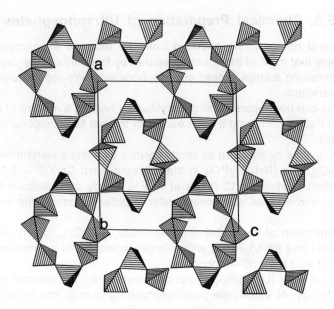

Figure 5.5.11. A projection, along the **b** direction, of an isolated anionic layer of the second type in form-II of a LnP$_5$O$_{14}$ ultraphosphate.

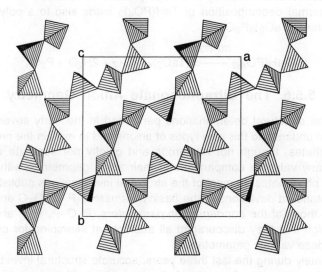

Figure 5.5.12. A projection, along the **c** direction, of an isolated anionic layer of the third type in form-II of a LnP$_5$O$_{14}$ ultraphosphate.

5.5.5. Chemical Preparation of Ultraphosphates

No general rules can be deduced from the reported experiments. One notices simply that most of them were prepared by flux methods by using initial mixtures containing a large excess of phosphoric acid. We report below some typical preparations.

CaP_4O_{11} can be prepare as single crystals by heating a mixture of $CaCO_3$, H_3PO_4, and P_4O_{10} containing a large excess of the last two components at 773 K for ten days.

$YCaP_7O_{20}$ was synthesized as single crystals by using a starting mixture of $CaCO_3$, Y_2O_3, and $(NH_4)H_2PO_4$ in the following ratio: Ca/Y/P = 1/2/7. This mixture is then heated for 12 hours at 723 K, slowly cooled down to room temperature, and washed with boiling water in order to remove the excess of flux.

The preparation of the nickel ultraphosphate, NiP_4O_{11}, is very similar. A mixture of NiO and $(NH_4)_2HPO_4$ with a large excess of the ammonium salt is heated at 893–933 K for 30–40 hours.

The formation of $(UO_2)_2P_6O_{17}$ was observed during the thermal evolution of $(UO_2)H(PO_3)_3$. At 973 K, the polyphosphate transforms into the ultraphosphate:

$$2(UO_2)H(PO_3)_3 \longrightarrow (UO_2)_2P_6O_{17} + H_2O$$

Such a process leads evidently to a polycrystalline compound.

The thermal decomposition of $TaH(PO_4)_2$ leads also to a polycrystalline ultraphosphate $(TaO_2)_4P_6O_{17}$:

$$4TaH(PO_4)_2 \xrightarrow{723-773\,K} (TaO_2)_4P_6O_{17} + 2H_2O + P_2O_5$$

5.5.6. The Ultraphosphate Anion Geometry

The first structural determinations performed in the early seventies were sufficient to understand this new types of anions and to explain the properties of ultraphosphates, though not numerous and chiefly not accurate enough to authorize any valuable comparisons of their anion geometries with the other classes of phosphates. In most of the structural investigations published at this time, the standard deviations of the *basic parameters* (P–O, O–O and O–P–O) as well as those of the *condensation parameters* (P–O–P, P–P and P–P–P) were so high that they discouraged all attempts at searching for correlations between these various parameters.

Fortunately during the last three years, accurate structural investigations of ultraphosphates were performed including new structural types or redeterminations of the previous ones. In spite of being restricted in number, these new data are nevertheless sufficient to clarify this subject in a first step. They have been included in a wide investigation of the phosphoric anion geometries presently undertaken by the authors. From a first rapid survey of their analysis, it appears that the geometry of ultraphosphate anions does not differ significantly of what can be expected in its main lines in a condensed phosphoric anion.

In an ultraphosphate anion, two kinds of tetrahedron coexist: the ternary tetrahedron whose three oxygen atoms, denoted O(L), are involved in P–O–P bonds and the internal tetrahedron in which only two are involved in such bonds.

If one examines the *condensation parameters* first, one oberves that like in all other condensed phosphoric anions, the P–P–P angles can vary in a wide range (85–145°), the P–P distances and the P–O–P angles are in their usual range (2.850–3.020 Å) and (126–145°).

The *basic parameters* inside the *internal tetrahedra* are similar to those observed in any kinds of condensed phosphoric anion.

The differences observed for the basic parameters in the tetrahedron corresponding to the *ternary phosphorus* are slight but significant. Here, the P–O distances corresponding to the three oxygen atoms involved in the P–O–P bonds are significantly shorter than those observed in an internal tetrahedron. The complete analysis of the ultraphosphate anions is far from being terminated so we have simply reported two numerical examples corresponding to two compounds of similar formula, MgP_4O_{11} and NiP_4O_{11}, but belonging to two different structural types. In these two anionic frameworks, there are two internal tetrahedra and two ternary tetrahedra. In Table 5.5.5, we compare the values of the P–O(L) distances in the two kinds of tetrahedra:

As a consequence of that, the Baur's distortion coefficient [49] for the P–O values, DI(PO), is very significantly smaller for the ternary tetrahedra than for

Table 5.5.5

For both compounds the left-hand side column corresponds to the P–O(L) distances in the ternary tetrahedra, the right-hand side column to the P–O(L) distances in the internal tetrahedra.

MgP_4O_{11}

P–O (Å)	P–O (Å)
1.557	1.628
1.569	1.632
1.575	
1.571	1.615
1.561	1.626
1.554	

NiP_4O_{11}

P–O (Å)	P–O (Å)
1.545	1.629
1.559	1.621
1.582	
1.562	1.636
1.550	1.630
1.587	

the internal tetrahedra. For instance, an average value of 0.027 in CuP_4O_{11} is observed for the ternary tetrahedra against 0.045 for the internal tetrahedra.

A similar observation can be made about the $O(L)-P-O(L)$ angles in the two kinds of tetrahedra. These angles whose average value is generally smaller than 100° in the internal tetrahedra are, with rare exceptions, significantly larger in the ternary tetrahedra. In the two examples reported above, their average values are 105.0° for the magnesium salt and 104.2° for the nickel derivative.

References

1. A. V. Kroll, Z. anorg. Chem., 76, (1912), 387–418.

2. A. V. Kroll, Z. anorg. Chem., 77, (1912), 1–40.

3. A. V. Kroll, Z. anorg. Chem., 78, (1912), 95–133.

4. P. T. Clève, Bull. Soc. Chim., 43, (1885), 169.

5. K. R. Johnsson, Ber. dtch. Chem. Ges., 22, (1889), 976–980.

6. W. H. Hill, G. T. Faust, and D. S. Reynolds, Am. J. Sci., 242, (1944), 457–477.

7. W. H. Hill, G. T. Faust, and D. S. Reynolds, Am. J. Sci., 242, (1944), 542–562.

8. M. Beucher, "Les Ultraphosphates", (Les Eléments des Terres Rares, Intern. Meeting), 1969, Paris–Grenoble, France.

9. M. T. Averbuch–Pouchot and A. Durif, Z. Kristallogr., 201, 1992, 69–92.

10. J. R. Van Wazer and K. A. Holst, J. Am. Chem. Soc., 72, (1950), 639–644.

11. G. D. Dudko, V. V. Fedorov, and R. S. Shevelevich, Izv. Akad. Nauk SSSR, Neorg. Mater., 22, (1986), 456–459.

12. A. V. Lavrov, T. A. Bykanova, and Yu. M. Kessler, Izv. Akad. Nauk SSSR, Neorg. Mater., 13, (1977), 329–333.

13. D. Stachel, H. Paulus, C. Guenter, and H. Fuess, Z. Kristallogr., 199, 275–276 (1992).

14. O. V. Yakubovich, O. V. Dimitrova, and A. I. Vidrevich, Kristallografiya, 38, (1993), 77–85.

15. C. Baez–Doelle, D. Stachel, I. Svoboda, and H. Fuess, Z. Kristallogr., 203, (1993), 282–283.

16.   A. Olbertz, D. Stachel, I. Svoboda, and H. Fuess, *Z. Kristallogr.*, in press.

17.   L. Kh. Minacheva, M. A. Porai–Koshits, A. S. Antsyshkina, V. G. Ivanova, and A. V. Lavrov, *Koord. Khim.*, **1**, (1975), 421–428.

18.   M. Beucher, *Mat. Res. Bull.*, **4**, (1969), 15–18.

19.   I. Tordjman, M. Bagieu–Beucher, and R. Zilber, *Z. Kristallogr.*, **140**, (1974), 145–153.

20.   M. Schneider, J. Buschmann, and P. Luger, *Z. anorg. allg. Chem.*, **620**, (1994), 766–770.

21.   D. Stachel, H. Paulus, I. Svoboda, and H. Fuess, *Z. Kristallogr.*, **202**, (1982), 117–118.

22.   A. S. Antsyshkina, M. A. Porai–Koshits, L. Kh. Minacheva, V. G. Ivanova, and A. V. Lavrov, *Koord. Khim.*, **4**, (1978), 448–454.

23.   A. S. Antsyshkina, M. A. Porai–Koshits, L. Kh. Minacheva, A. V. Lavrov, and V. G. Ivanova, *Dokl. Akad. Nauk SSSR*, **229**, (1976), 896–897.

24.   S. Jaulmes, *C. R. Acad. Sci.*, Sér. C, **268**, (1969), 935–937.

25.   N. N. Chudinova, K. K. Palkina, N. B. Komarovskaya, S. I. Maksimova, and N. T. Chibiskova, *Dokl. Akad. Nauk SSSR*, **306**, (1989), 635–638.

26.   K. K. Palkina, S. I. Maksimova, N. T. Chibiskova, N. N. Chudinova, and N. B. Karmanovskaya, *Neorg. Mater.*, **29**, (1993), 119–120.

27.   A. Hamady and T. Jouini, *J. Solid State Chem.*, **111**, (1994), 442–446.

28.   N. G. Chernorukov, N. P. Egorov, and T. A. Galanova, *Izv. Akad. Nauk SSSR, Neorg. Mater.*, **17**, (1981), 328–332.

29.   A. V. Lavrov, *Izv. Akad. Nauk SSSR, Neorg. Mater.*, **15**, (1979), 942–946.

30.   Yu. E. Gorbunova, S. A. Linde, and A. V. Lavrov, *Russ. J. Inorg. Chem.*, **26**, (1981), 383–386.

31.   A. Durif, Crystal Chemistry of Condensed Phosphates, (Plenum Press, New York, 1995).

32.   A. Olbertz, D. Stachel, I. Svoboda, and H. Fuess, private communication.

33.   D. Tranqui, M. Bagieu, and A. Durif, *Acta Crystallogr.*, **B30**, (1974), 1751–1755.

34.   M. Bagieu, I. Tordjman, A. Durif, and G. Bassi, *Cryst. Struct. Comm.*, **3**, (1973), 387–390.

35.   D. Tranqui, M. Bagieu–Beucher, and A. Durif, *Bull. Soc. fr. Minér. Cristallogr.*, **95**, (1972), 437–440.

36.   N. N. Chudinova and K. H. Jost, *Z. anorg. allg. Chem.*, **400**, (1973), 185–

188.

37.   H. Y-P. Hong, *Acta Crystallogr.*, **B 30**, (1974), 468–474.

38.   K. R. Albrand, R. Attig, J. Fenner, J. P. Jeser, and D. Mootz, *Mat. Res. Bull.*, **9**, (1974), 129–140.

39.   R. Parrot, C. Barthou, B. Canny, B. Blanzat, and G. Collin, *Phys. Rev.*, **B11**, (1975), 1001–1012.

40.   M. Rzaigui, N. Kbir Ariguib, M. T. Averbuch–Pouchot, and A. Durif, *J. Solid State Chem.*, **52**, (1984), 61–65.

41.   S. Jaulmes, *C. R. Acad. Sci.*, Sér. C, **268**, (1969), 935–937.

42.   B. Jezowska-Trzebiatowska, Z. Mazurak, and T. Lis, *Acta Crystallogr.*, **B36**, (1980), 1639–1641.

43.   H. Y-P. Hong and J. W. Pierce, *Mat. Res. Bull.*, **9**, (1974), 179–190.

44.   M. Bagieu-Beucher and D. Tranqui, *Bull. Soc. fr. Minéral. Cristallogr.*, **93**, (1970), 505–508.

45.   M. Bagieu, *Thesis*, (Univ. of Grenoble, France, 1980).

46.   A. Durif, *Bull. Soc. fr. Minér. Cristallogr.*, **94**, (1971), 314–318.

47.   I. A. Bondar, L. P. Mezentseva, A. I. Domanskii, and M. M. Piryutko, *Russ. J. Inorg. Chem.*, **20**, (1975), 1448–1452.

48.   K. K. Palkina, N. N. Chudinova, B. N. Litvin, and N. V. Vinogradova, *Izv. Akad. Nauk SSSR, Neorg. Mater.*, **17**, (1981), 1501–1503.

49.   W. H. Baur, *Acta Crystallogr.*, **B30**, (1974), 1195–1215.

5.6. ADDUCTS AND HETEROPOLYPHOSPHATES

5.6.1. Introduction

Phosphoric groups are frequently observed in several kinds of compounds which cannot be classified as phosphates. We shall come back to the problems generated in the domains of classification and nomenclature by these compounds in a further section, but let us explain briefly at the beginning of this chapter that this type of compounds includes two categories of salts: the *adducts* and derivatives including *heteropolyanions*.

If in a given compound, a phosphoric anion coexists with another type of phosphoric or non-phosphoric anion and if these two entities do not share any oxygen atom then such a compound must be called an *adduct*. On the contrary, if one or more oxygen atoms are shared between the two anionic entities (as observed, for instance, in a series of condensed anions including both PO_4 and CrO_4 tetrahedra interconnected by P–O–Cr bonds), then we are dealing with another class of salts whose anions should be called *heteropolyanions*.

5.6.2. Adducts

Adducts involving phosphates as components include not only well-known phosphate families as apatites, but also more and more numerous new compounds. Thus, we will not report an exhaustive review of these adducts. We will simply try to show the great variety of these compounds illustrating this survey by some examples selected among the recent developments in this field. We will begin with the description of the adducts between phosphates and telluric acid, $Te(OH)_6$, which constitute the most represented family in our survey.

5.6.2.1. Adducts with telluric acid

At the beginning of the present century, Weinland et Prause [1] reported the existence of a series of compounds that they have formulated:
- $2Na_2O.2TeO_3.P_2O_5.9H_2O$
- $2(NH_4)_2O.TeO_3.P_2O_5.4H_2O$
- $4(NH_4)_2O.2TeO_3.3P_2O_5.11H_2O$

With such a formulation, it was not possible to know if these compounds including both PO_4 tetrahedra and TeO_6 octahedra were adducts or heteropolyanion-containing salts. The first structural investigations, performed in the early eighties, demonstrated clearly that all these arrangements include both PO_4 tetrahedra and TeO_6 octahedra coexisting as independent units and must in fact be written as:
- $Te(OH)_6.Na_2HPO_4.H_2O$
- $Te(OH)_6.2(NH_4)_2HPO_4$
- $Te(OH)_6.2NH_4H_2PO_4.(NH_4)_2HPO_4$

The investigations were then enlarged systematically and today one can say that most of the water soluble monovalent- or organic-cation phosphates form adducts with telluric acid whether they are condensed or not. Lithium salts are the only exception up to now.

Most of these adducts can be obtained by evaporation of aqueous solutions prepared with the required stoichiometries, at room temperature.

In all the atomic arrangements, the three-dimensional cohesion is performed by a network of hydrogen bonds connecting the OH radicals of telluric acid to the external oxygen atoms of the phosphoric anions and to the water molecules when they are present.

One can simply classify telluric acid-phosphate adducts as a function of the nature of the phosphoric anion.

The telluric acid-monophosphate adducts – Telluric acid-alkali, silver, thallium, or ammonium monophosphate adducts were systematically investigated by Averbuch-Pouchot and co-workers between 1980 and 1985. We have listed some of them below:

- $Te(OH)_6.Na_2HPO_4.H_2O$ [2]
- $Te(OH)_6.Na_2HPO_4.NaH_2PO_4$ [3]
- $Te(OH)_6.Rb_2HPO_4.RbH_2PO_4$ [4]
- $Te(OH)_6.2(NH_4)_2HPO_4$ [2]
- $Te(OH)_6.2TlH_2PO_4.Tl_2HPO_4$ [5]
- $Te(OH)_6.2TlH_2PO_4$ [5]

One of these adducts, $Te(OH)_6.2NH_4H_2PO_4.(NH_4)_2HPO_4$, also known as TAAP, proved to be a very efficient pyroelectric and ferroelectric material [6].

The telluric acid-oligophosphate adducts – Such adducts were not deeply investigated till now. One can, nevertheless, report the existence of some telluric acid-diphosphate adducts:

- $Te(OH)_6.K_3HP_2O_7.H_2O$ [7]
- $Te(OH)_6.2Rb_2H_2P_2O_7$ [8]
- $Te(OH)_6.2(NH_4)_2H_2P_2O_7$ [8]
- $Te(OH)_6.Cs_2H_2P_2O_7$ [9]

It is to be noted that no example of similar adducts with higher oligophosphates, as tri- ou tetraphosphates, was prepared to date. Owing to their insolubility, long-chain polyphosphates cannot form adducts with telluric acid.

The telluric acid-cyclophosphate adducts – First, a systematic investigation of telluric acid-cyclotriphosphate adducts was done by Boudjada [10]. During this study, six compounds were characterized. Later, $Te(OH)_6.Na_3P_3O_9.K_3P_3O_9$ [11] and $Te(OH)_6.Cs_3P_3O_9.H_2O$ [12] were prepared by Averbuch-Pouchot. The main crystallographic data for telluric acid-cyclotriphosphate adducts are gathered in Table 5.6.1.

After this first investigation, telluric acid-cyclophosphate adducts were characterized with almost all families of cyclophosphates. Thus for cyclotetraphosphates for example:

- $2Te(OH)_6.(NH_4)_4P_4O_{12}.2H_2O$ [17]
- $Te(OH)_6.K_4P_4O_{12}.2H_2O$ [18–19]
- $Te(OH)_6.Rb_4P_4O_{12}.2H_2O$ [19]

A number of telluric acid-cyclohexa-, cycloocta-, cyclodeca-, and cyclododecaphosphates are also known.

Table 5.6.1
Main crystallographic data for telluric acid-cyclotriphosphate adducts.

Formula	a α	b β	c (Å) γ (°)	S. G.	Z	Ref.
T A.2Na$_3$P$_3$O$_9$.6H$_2$O	11.67(1)	11.67(1)	12.12(1)	$P6_3/m$	2	[10, 13]
T A.K$_3$P$_3$O$_9$.2H$_2$O	15.57(2)	7.438(6) 107.2(1)	14.85(1)	$P2_1/c$	4	[10, 14]
T A.Rb$_3$P$_3$O$_9$.H$_2$O	15.56(1)	8.376(4) 113.33(2)	13.705(4)	$P2_1/a$	4	[10, 15]
T A.2(NH$_4$)$_3$P$_3$O$_9$	11.16(1)	11.16(1)	17.86(1)	$R\bar{3}$	3	[10, 16]
T A.Cs$_3$P$_3$O$_9$.H$_2$O	7.279(2)	13.984(8) 90.42(2)	17.071(4)	$P2_1/c$	4	[12]
T A.2Tl$_3$P$_3$O$_9$	11.168(1)	11.168(1)	11.733(3)	$P6_3/m$	2	[10]
T A.Cs$_2$NaP$_3$O$_9$	12.946(1)	9.174(1) 107.60(1)	13.406(1)	$C2/c$	4	[10]
T A.Na$_3$P$_3$O$_9$.K$_3$P$_3$O$_9$	18.42(1)	10.644(5) 119.76(5)	12.348(8)	$C2/c$	4	[11]

T A = telluric acid or Te(OH)$_6$

Recently, similar adducts involving simple organic cations were also characterized. For instance, Averbuch-Pouchot and Durif [20] described 2Te(OH)$_6$.(C$_2$N$_2$H$_{10}$)$_3$P$_6$O$_{18}$.2H$_2$O, an adduct between telluric acid and the ethylenediammonium cyclohexaphosphate.

The atomic arrangement in Te(OH)$_6$.2Na$_3$P$_3$O$_9$.6H$_2$O – We illustrate this kind of adducts with the description of the structure of Te(OH)$_6$.2Na$_3$P$_3$O$_9$.6H$_2$O [13]. This salt is hexagonal, $P6_3/m$, with a unit cell containing two formula units and having the following parameters:

$$a = 11.67(1), \quad b = 12.12(1) \text{ Å}$$

A projection of this arrangement, along the **c** direction, is represented in Figure 5.6.1. The structure can be explained as being built by a succession of two kinds of layers alternating perpendicular to the direction of projection. In the first type of layers centered around the planes $z = 1/4$ and $3/4$ are located the sodium atoms and the water molecules. Around planes $z = 0$ and $1/2$ are located the second type of layers containing the telluric groups and the phosphoric anions. The Te(OH)$_6$ group with its central tellurium atom located at the origin has a $\bar{6}$ internal symmetry. The P$_3$O$_9$ phosphoric ring built around the internal $\bar{6}$ axis has ternary symmetry. As in all telluric acid adducts, the Te(OH)$_6$

group is very regular, with the present arrangement of six Te–O distances of 1.923 Å.

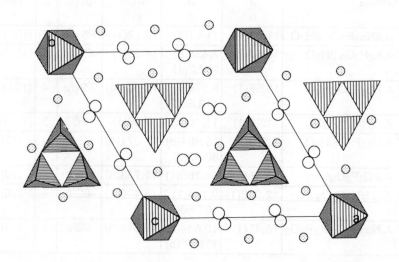

Figure 5.6.1. Projection of Te(OH)$_6$.2Na$_3$P$_3$O$_9$.6H$_2$O along the **c** axis. Owing to the presence of mirror planes in z = 1/4 and 3/4, two Te(OH)$_6$ and two P$_3$O$_9$ groups are superimposed in projection. The spotted circles represent the sodium atoms and the open the water molecules.

5.6.2.2. Adducts with phosphoric acid

As telluric acid, the monophosphoric acid also appears as a component of some adducts with phosphates.

Two isotypic compounds, Zn(H$_2$PO$_4$)$_2$.2H$_3$PO$_4$ reported by Salmon and Terrey [21] and also by Komskra and Satava [22] and Co(H$_2$PO$_4$)$_2$.2H$_3$PO$_4$ described by Herak *et al.* [23] are well-known. The corresponding cadmium arsenate, Cd(H$_2$AsO$_4$)$_2$.2H$_3$AsO$_4$, investigated by Boudjada *et al.* [24] is isotypic with these two phosphate adducts.

The same type of adducts was also and more frequently observed in the domain of organic-cation phosphates. Some of them will be be examined in Chapter 6 which is devoted to the anionic networks.

5.6.2.3. Adducts with oxides and hydroxides

Such compounds are the most numerous among the phosphate adducts. They are also very often called oxyphosphates and hydroxyphosphates.

Among them, the very important *apatite* family, including both oxy- and hydroxyphosphates, is the most common. In spite of the importance of this class of compounds for both biology and industry, we do not report any details about it since Elliot [25] published a very complete analysis of this family of compounds recently.

Many other systems reveal the existence of phosphate-oxides. We have examined some of these compounds.

$FePO_4.Fe_2O_3$, $Fe_3(PO_4)_2.FeO$, and $3FePO_4.Fe(OH)_3$ were characterized in the $Fe_2O_3-FeO-P_2O_5-H_2O$ system by Modaressi *et al.* [26], Bouchdoug *et al.* [27] and Ijjali *et al.* [28] respectively.

Merker and Wondratschek [29] reported the existence of $Pb_3(PO_4)_2.5PbO$ and $Pb_3(PO_4)_2.PbO$ during their investigation of the $PbO-P_2O_5$ system. Later, Brixner and Foris [30] published a process of crystal growth and crystal data for these two compounds.

The $M^{III}PO_4-M_2^{III}O_3$ systems, first investigated by Tananaev *et al.* [31] for M^{III} = Al, Cr, and Y, show the existence of several very refractory phosphate-oxides. $AlPO_4.Al_2O_3$, $CrPO_4.2Cr_2O_3$, $2YPO_4.Y_2O_3$, and $YPO_4.2Y_2O_3$ melt congruently at 2303, 2463, 2303, and 2383 K, respectively. In addition, in the $CrPO_4-Cr_2O_3$ system, an uncongruent melting compound corresponding probably to $CrPO_4.Cr_2O_3$, decomposes at about 1973 K. Following these results, a systematic investigation of the $LnPO_4-Ln_2O_3$ systems was performed by Serra [32] and Serra *et al.* [33] using a solar furnace. They obtained four series of new phosphate-oxides:

- $Ln_7P_3O_{18}$ or $3LnPO_4.2Ln_2O_3$
- Ln_3PO_7 or $LnPO_4.Ln_2O_3$
- $Ln_8P_2O_{17}$ or $2LnPO_4.3Ln_2O_3$
- $Ln_{12}P_2O_{23}$ or $2LnPO_4.5Ln_2O_3$

From X-ray diffraction measurements, they concluded that inside a given series all the derivatives were isotypic. Later, Rouanet *et al.* [34] tried to elaborate the shape of some of the $LnPO_4-Ln_2O_3$ phase-equilibrium diagrams from these results.

5.6.2.4. Adducts with halides

The monophosphate-halides – This family of compounds provides a good number of examples. Among them, many are well-known natural compounds. We have listed some of them below:

- the two isotypic classical minerals: the *wagnerite*, $Mg_3(PO_4)_2.MgF_2$, often formulated as $Mg_2(PO_4)F$, and the *triplite*, $(Mn,Fe)_2(PO_4)F$
- the *isokite*, $CaMg(PO_4)F$ said to be a *sphene* analogue
- the *fluorapatites* and *chlorapatites*, $Ca_5(PO_4)_3F$ and $Ca_5(PO_4)_3Cl$,

which are abundant and well-known species.

In spite of the great amount of literature dealing with the monophosphate-halides, the structural situation in this family is still confused. Nevertheless, some recent and accurate crystal structure determinations bring a beginning of clarification in this field. Among them, we have selected the investigations of $Mn_2(PO_4)F$ by Rea and Kostiner [35], $Mg_2(PO_4)Cl$ [36] and $Mn_2(PO_4)Cl$ [37] by the same authors, $Fe_2(PO_4)F$ by Yakubovich et al. [38], $KFe(PO_4)F$ by Matvienko et al. [39], and $Ca_2(PO_4)Cl$ by Greenblatt et al. [40].

The diphosphate-halides – They are less common salts. Three of them were recently investigated, $Fe_2K_2(P_2O_7)F_2$ and $Mn_2K_2(P_2O_7)F_2$ by Yakubovich et al. [41–42] and $Cd_3(NH_4)Na_2(P_2O_7)_2Cl$ by Ivanov et al. [43]. All of them were obtained by hydrothermal processes during the investigation of various systems:

- $FeO–(NH_4)_2HPO_4–KHF_2–H_2O$ for $Fe_2K_2(P_2O_7)F_2$
- $MnO–(NH_4)_2HPO_4–KF–H_2O$ (723 K, 98,000 kPa) for $Mn_2K_2(P_2O_7)F_2$
- $CdO–(NH_4)_2HPO_4–NaCl–H_2O$ (773 K, 80,000–140,000 kPa) in the case of $Cd_3(NH_4)Na_2(P_2O_7)_2Cl$

Their atomic arrangements were all accurately determined.

The cyclophosphate-halides – The first observation of such compounds was done by Averbuch–Pouchot and Durif [44] during the final stage of crystallization of aqueous solutions of ammonium cyclohexaphosphate monohydrate, $(NH_4)_6P_6O_{18}.H_2O$, prepared by a metathesis reaction using $Ag_6P_6O_{18}.H_2O$ and NH_4Cl as starting materials. This adduct formation was attributed to the use of an impure silver cyclohexaphosphate or to a small excess of ammonium chloride during the metathesis reaction. The few crystals obtained were of poor quality. Similar compounds were further observed in the cases of the bromide and iodide derivatives. Reproducible preparations leading to good quality

Table 5.6.2
Main crystallographic features of the $(NH_4)_6P_6O_{18}.NH_4X.H_2O$ (X = Cl, Br, I) compounds.

Formula	a α	b β	c (Å) γ (°)	S. G.	Z	Ref.
$(NH_4)_6P_6O_{18}.NH_4Cl.H_2O$	6.783(3) 101.48(2)	10.101(8) 90.84(3)	19.33(1) 107.31(2)	$P\bar{1}$	2	[44]
$(NH_4)_6P_6O_{18}.NH_4Br.H_2O$	6.678(5) 100.58(5)	10.11(1) 91.34(5)	19.30(2) 107.23(5)	$P\bar{1}$	2	[44]
$(NH_4)_6P_6O_{18}.NH_4I.H_2O$	14.96(1)	24.82(1) 91.86(5)	6.710(6)	$P2_1/n$	2	[44]

crystals were then elaborated and the crystal structure determinations made possible. The chemical formula common to this series of salts is $(NH_4)_6P_6O_{18}.NH_4X.H_2O$ (X = Cl, Br, I). As shown by Table 5.6.2 which reports their main crystallographic features, the chloride and bromide derivatives are isotypic. The atomic arrangement of the iodide derivative, also determined by the same authors [44], shows great similarities with those of the first two salts.

5.6.2.5. Adducts with oxosalts

The phosphate-borates – They are not common compounds. The coexistence as independent units of BO_3 groups and phosphoric anions in an atomic arrangement has been observed in only one series of compounds, investigated by Palkina et al. [45] and formulated as $Ln_7O_6(BO_3)(PO_4)_2$ (Ln = La $\longrightarrow$ Dy). In this particular case and because of the presence of some oxygen atoms in the atomic arrangement belonging neither to the phosphoric group nor to the borate group, the present compounds are in fact phosphate-borate-oxides.

$Na_3PO_4.NaBO_2.18H_2O$ was reported by Harris [46], but no structural data are available.

The phosphate-carbonates – Most of these adducts are natural compounds. The bradleyite-group has a good number of representatives including for instance: the *sidorenkite* $MnNa_3(PO_4)(CO_3)$, the *bradleyite* $MgNa_3(PO_4)(CO_3)$, the *bonshtedtite* $FeNa_3(PO_4)(CO_3)$, and an unnamed mineral of formula $SrNa_3(PO_4)(CO_3)$. Several of these salts were the subject of careful structural investigations as *sidorenkite* by Kurova et al. [47] and $SrNa_3(PO_4)(CO_3)$ by Sokolova and Khomyakov [48].

The phosphate-nitrates – In this class of adducts, various phosphoric anions can coexist with the NO_3 groups.

Two *monophosphate-nitrates* are presently well characterized. The first, $CaH_2(NO_3)(PO_4).H_2O$, was prepared by Frazier and Lehr [49] during a careful revision of the $CaO-N_2O_5-P_2O_5-H_2O$ system. Probably because its stability region is limited, this salt was not observed by the first investigators of the system [50–53]. No structural data are reported for this compound. The second, $Hg_4(NO_3)(PO_4).H_2O$, precipitates as tufts of yellowish needles from an aqueous solution of $Na_3PO_4.12H_2O$, $HgNO_3.H_2O$, and HNO_3, at room temperature. Its crystal structure was accurately determined [54].

Several *cyclophosphate-nitrates* have also been described:

– $K_3P_3O_9.KNO_3$ and $(NH_4)_3P_3O_9.NH_4NO_3$, these two *cyclotriphosphate-nitrates* were observed by Schülke [55] during various experiments dealing with ammonium or alkali cyclotriphosphates.

– $M_6^IP_6O_{18}.M^INO_3.H_2O$ (M^I = K, NH_4, Rb), this series of three isotypic *cyclohexaphosphate-nitrates* was recently reported by Averbuch–Pouchot and Durif [56]. These salts are prepared by slow evaporation of an aqueous solution of the two components in stoichiometric amounts, at room temperature. The atomic arrangement of this type of compounds was accurately determined by the authors.

– $Ag_9NaP_8O_{24}(NO_3)_2.4H_2O$, the unique example of *cyclooctaphosphate-nitrate*, was characterized by Averbuch–Pouchot and Durif [57] during attempts to elaborate a silver cyclooctaphosphate in order to generalize the Boullé's metathesis reaction for the preparation of water soluble cyclooctaphosphates. This orthorhombic compound with:

$$a = 17.254, \quad b = 7.543, \quad c = 23.465 \text{ Å}, \quad Cmcm, \quad Z = 4,$$

exhibits a very regular arrangement of phosphoric anions and NO_3 groups that we have represented in Figure 5.6.2.

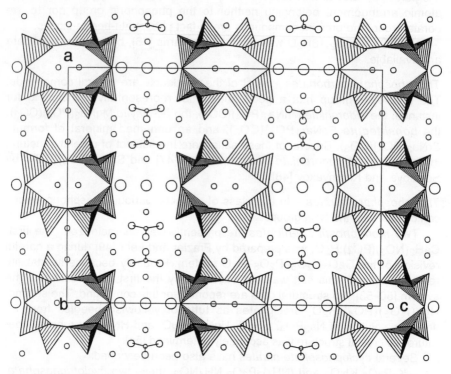

Figure 5.6.2. Projection of the atomic arrangement of $Ag_9NaP_8O_{24}(NO_3)_2.4H_2O$ along the **b** direction. In order of decreasing size, open circles represent water molecules, O atoms of the NO_3 groups, Ag, and N atoms. Phosphoric rings are shown by polyhedral representation.

The phosphate-sulfates and -chromates – A number of natural compounds contain both PO_4 and SO_4 or CrO_4 groups. Among them, we can report $Pb_2CuOH(PO_4)(CrO_4)$, the *vauquelinite* [58], or $Pb_2CuOH(PO_4)(SO_4)$, the *tsumebite* [59].

A series of synthetic phosphate-sulfates, $M^{II}SO_4.M^{II}_3(PO_4)_2$ (M^{II} = Pb, Sr...) isotypic with the mineral *eulytite*, $Bi_4(SiO_4)_3$, was described by Durif [60–61].

The scattering factors of S and P have very similar values so that, in many investigations, it was difficult and often impossible to decide between an ordered or a statistical distribution of the SO_4 and PO_4 groups. In more recent studies, as for instance the investigation of a series of synthetic $M^IH_2PO_4.M^IHSO_4$ compounds with M^I = NH_4 or K, the existence of an order was clearly established.

The phosphate-silicates – *The phosinaite*, $Na_{12}Ca_3Ce_{0.67}(Si_4O_{12})(PO_4)_4$, a typical example of phosphate-silicate, was investigated by Krutik *et al.* [62]. This compound is orthorhombic, $P2_12_12$, $Z = 2$, with the following unit-cell dimensions:

$$a = 7.234, \quad b = 14.670, \quad c = 12.231 \text{ Å}$$

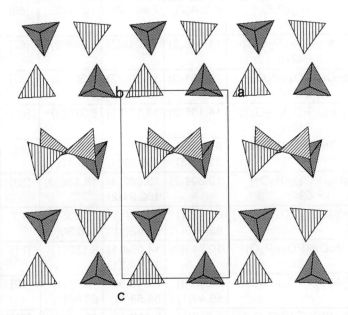

Figure 5.6.3. A projection along the **b** direction of an anionic layer of PO_4 tetrahedra and Si_4O_{12} rings in phosinaite. The associated cations are omitted.

In this very intricate atomic arrangement, the anionic part built by independent PO_4 groups and Si_4O_{12} rings forms thick layers spreading around the planes y = 0 and 1/2 respectively. A projection along the **b** direction of such a layer is shown in Figure 5.6.3.

The phosphate-oxalates – These derivatives are rather rare and one can only report the existence of the two isotypic salts, $Mg_2(NH_4)_4P_6O_{18}(C_2O_4).6H_2O$ and $Mn_2(NH_4)_4P_6O_{18}(C_2O_4).6H_2O$. The preparation of these two cyclohexa-phosphate-oxalates characterized by Averbuch–Pouchot and Durif [63] is diffi-

Table 5.6.3
Main crystallographic data for monophosphate-diphosphates.

Formula S. G. Z	a α	b β	c (Å) γ (°)	Ref.
$Na_5H_2(PO_4)(P_2O_7)$ $P\bar{1}$ 2	10.862(2) 128.27(4)	8.426(3) 99.21(4)	7.001(2) 88.84(3)	[64]
$K_2H_5(PO_4)(P_2O_7)$ $C2/c$ 8	31.272(7) 99.85(1)	7.428(1)	9.253(1)	[65]
$K_2H_8(PO_4)_2(P_2O_7)$ $Pca2_1$ 4	9.364(2)	7.458(2)	19.560(2)	[66]
$Na_7Al_4(PO_4)(P_2O_7)_4$ $P42_1c$ 2	14.046(3)	14.046(3)	6.169(2)	[67]
$Na_7Cr_4(PO_4)(P_2O_7)_4$ $P42_1c$ 2	14.061(3)	14.061(3)	6.306(2)	[67]
$Na_7Fe_4(PO_4)(P_2O_7)_4$ $P42_1c$ 2	14.105(3)	14.105(3)	6.378(2)	[67]
$Na_3Th(PO_4)(P_2O_7)$ $P\bar{1}$ 2	8.734(2) 93.33(3)	8.931(2) 108.29(4)	6.468(1) 110.10(4)	[68]
$K_2Ni_4(PO_4)_2(P_2O_7)$ $C2/c$ 8	10.304(2)	13.682(3) 102.91(2)	18.139(3)	[69]
$Cs(MoO)_2(PO_4)(P_2O_7)$ $P\bar{1}$ 2	6.342(1) 83.083(7)	9.676(1) 97.42(1)	10.035(1) 108.30(1)	[70]
$K(MoO)_2(PO_4)(P_2O_7)$ $C2/c$ 8	10.743(2)	14.084(1) 126.42(1)	8.852(1)	[71]
$K(MoO)_2(PO_4)(P_2O_7)$ $P\bar{1}$ 1	8.846(8) 56.49(1)	8.846(9) 55.59(1)	10.01(1) 68.87(1)	[72]
$KMoWO(PO_4)(P_2O_7)$ $Pbcm$ 4	8.818(1)	9.157(1)	12.384(1)	[73]

cult and requires several months in some cases. A very diluted aqueous solution of the bivalent-cation oxalate is added drop by drop (some drops a day) to a solution of the alkali phosphate kept at room temperature. After several months, very sparingly water-soluble crystals of these phosphate-oxalates appear in the solution. The crystal structure of this type of compounds was solved by using the manganese salt.

The mixed-anion phosphates – Before ending this survey of adducts between phosphates and oxosalts, one must not forget the wide category of compounds denominated as mixed-anion phosphates till now. They are compounds containing two kinds of phosphoric groups. The number of such salts increases rapidly.

Except in the case of the very few of them that have been prepared by classical reactions of aqueous chemistry, such as $Na_5H_2(PO_4)(P_2O_7)$ [64], $K_2H_5(PO_4)(P_2O_7)$ [65], and $K_2H_8(PO_4)_2(P_2O_7)$ [66], there is no rule for the che-

Table 5.6.4
Main crystallographic data for mixed-anion phosphates containing anions having higher degrees of condensation.

Formula S. G. Z	a α	b β	c (Å) γ (°)	Ref.
$CsTa_2(PO_4)_2(P_3O_{10})$ $P\bar{1}$ 2	5.1355(6) 82.54(1)	10.900(1) 87.13(1)	14.392(1) 85.05(1)	[74]
$Ta_2Rb_2H(PO_4)_2(P_5O_{16})$ $P2_1/m$ 2	5.173(1)	18.458(5) 95.63(2)	10.803(4)	[75]
$(NH_4)Cd_6(P_2O_7)_2(P_3O_{10})$ Pm 4	6.785(3)	5.494(2) 107.28(4)	27.199(5)	[76]
$Cs_2Mo_5O_2(P_2O_7)_3(P_3O_{10})$ $P\bar{1}$ 2	14.395(1) 98.316(6)	15.600(1) 90.824(7)	6.4571(6) 90.073(6)	[76]
$CaNb_2O(P_2O_7)(P_4O_{13})$ $C2/m$ 4	13.264(9)	10.577(8) 96.09(1)	12.393(9)	[77]
$KTa(P_2O_7)(PO_3)_2$ $P2_12_12_1$ 4	7.045(1)	8.402(1)	17.494(4)	[78]
$KAl_2(H_2P_3O_{10})(P_4O_{12})$ $C2/c$ 4	11.864(3)	8.332(3) 99.67(2)	17.317(4)	[79]
$Pb_2Cs_3(P_4O_{12})(PO_3)_3$ $P\bar{1}$ 2	6.808(5) 86.23(1)	7.875(6) 96.96(1)	22.12(1) 113.98(1)	[80]
$Sr_2Cs_3(P_4O_{12})(PO_3)_3$ $P\bar{1}$ 2	6.922(5) 86.91(5)	8.055(5) 97.99(5)	21.97(5) 115.15(5)	[80]

mical preparation of these adducts. The vast majority of them have been obtained in various flux-method experiments as unexpected products that could not be identified until their crystal structures had been solved. A selection of crystallographic data for monophosphate-diphosphates is gathered in Table 5.6.3 and that for mixed-anion phosphates containing anions with higher degrees of condensation in Table 5.6.4.

We will illustrate this important class of materials with the description of the unique example of an adduct including both P_4O_{12} ring anions and infinite $(PO_3)_n$ chains, $Pb_2Cs_3(P_4O_{12})(PO_3)_3$.

The $Pb_2Cs_3(P_4O_{12})(PO_3)_3$ atomic arrangement – Lead-cesium cyclotetraphosphate-long-chain polyphosphate and the isotypic strontium-cesium salt were characterized by Averbuch-Pouchot [80] during various investigations of the PbO–Cs$_2$O–P$_2$O$_5$ and SrO–Cs$_2$O–P$_2$O$_5$ systems by flux method. The determination of the atomic arrangement was made with the lead-cesium derivative. The crystal data for these two compounds are reported in Table 5.6.4.

If one considers the stacking of the anions only, this structure can be considered as a layer arrangement. As can be seen in Figure 5.6.4, all the P_4O_{12} anions are located in the planes $z = 0$ and $1/2$, whereas the infinite $(PO_3)_n$ chains spread in the planes $z = 1/4$ and $3/4$. Two crystallographically independent ring anions, both centrosymmetric, coexist in the structure. The infinite $(PO_3)_n$ anion has a period of three tetrahedra. The general feature of such a

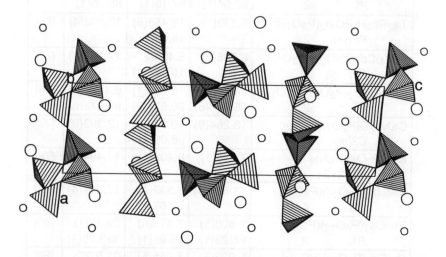

Figure 5.6.4. Projection, along the **b** axis, of $Pb_2Cs_3(P_4O_{12})(PO_3)_3$. The large open circles represent the cesium atoms and the smaller ones the lead atoms.

chain as observed in a layer centered on the plane $z = 0.25$ is shown in Figure 5.6.5. Both the two independent lead atoms have sevenfold coordination, whereas within a range of 3.5 Å two of the cesium atoms have eight neighbors and the third has nine.

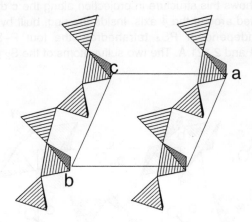

Figure 5.6.5. Projection, along the **c** axis, of the infinite phosphoric chain sited in the anionic layer located around $z = 0.25$.

5.6.2.6. Various adducts

Various other phosphate-adducts were reported. A *guanidinium monophosphate-chloride*, $[C(NH_2)_3]_3PO_4.[C(NH_2)_3]Cl.2H_2O$ was described by Averbuch-Pouchot and Durif [81]. This compound forms when there is an excess of guanidinium chloride in the metathesis reactions involving Ag_3PO_4 as starting material.

A unique example of *phosphate-adduct with an organic molecular compound* is provided by $Cu_2(C_2N_2H_{10})(HP_2O_7)_2.(C_2N_2H_8)_2.3H_2O$, the ethylenediammonium bis[copper(II) monohydrogendiphosphate ethylenediamine] trihydrate, whose structure, recently investigated by Gharbi *et al.* [82], contains both ethylenediammonium cations and ethylenediamine molecules.

Some unexpected compounds are sometimes encountered along phosphate chemistry. Thus, a *thio-compound* simply reported as P_2NbS_8 by Evain *et al.* [83] is in fact an adduct between a totally substituted cyclotetraphosphate and sulphur and consequently must be written $Nb_2P_4S_{12}.2S_2$. We will describe this unusual derivative now.

The $Nb_2P_4S_{12}.2S_2$ atomic arrangement – Its space group is tetragonal, $P\bar{4}n2$, with $Z = 4$ and its unit-cell dimensions are:

$$a = 12.048, \quad c = 7.207 \text{ Å}$$

The atomic arrangement of this niobium (IV) compound includes a P_4S_{12} totally sulfur-substituted cyclotetraphosphate cycle and pairs of sulfur atoms. Figure 5.6.6 shows this structure in projection along the **c** direction. The P_4S_{12} rings are located around the $\bar{4}$ axis. Inside the ring, built by only one crystallographically independent PS_4 tetrahedron, the four P–S distances range between 1.981 and 2.121 Å. The two sulfur atoms of the S_2 pairs are separated by 2.014 Å.

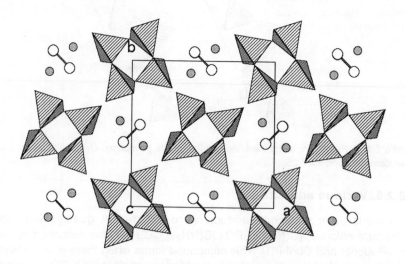

Figure 5.6.6. Projection of NbP_2S_8 along the **c** direction. The P_4S_{12} rings are shown by polyhedral representation. The grey circles are the Nb atoms and the open the S atoms of the S_2 pairs.

5.6.3. Heteropolyphosphates

At the beginning of this section, we defined a phosphoric heteropolyanion as one in which a phosphoric anion shares one or more oxygens with another anionic entity.

In many cases, there is no ambiguity. This is true, for instance, for finite heteropolyanions and for most of the linear ones. But, probably due to a lack of knowledge about the true nature of the chemical bonds, it is in many cases difficult to establish clearly the limits of these complex anions mainly for two- or three-dimensional infinite entities. Hence one must be very circumspect when discussing heteropoyanions. For example, shall we consider $Be_2NH_4P_3O_{10}$ as containing a three-dimensional $[Be_2P_3O_{10}]^-$ anion as did the first authors or simply as an ammonium-beryllium triphosphate as we did in the section of this book devoted to oligophosphates? For such a compound that is now well characterized, the question may appear as purely academic, but, nevertheless, the problem remains illustrating either the misuse of some words or a lack of knowledge or of definite rules.

5.6.3.1. Finite heteropolyanions

The phosphomolybdates and phosphotungstates – We will not deal here with the finite heteropolyanions found in the well-known phosphomolybdates and phosphotungstates, widely used in analytical chemistry since the last century, because they are described in all classical textbooks.

The phosphosulphates – Phosphosulphates are not so widely known. Thilo and Blumenthal [84] were the first to investigate this field. By melting mixtures of sodium long-chain polyphosphate or cyclophosphates and sodium disulphate with starting mole ratios such that $(P_2O_5 + SO_3)$: $Na_2O > 1$, they obtained various mixed condensed anions built by a linear linkage of alternating corner-sharing PO_4 and SO_4 tetrahedra. The authors also studied the rate of hydrolysis of these salts.

Some years later, Lampe [85–86] resumed these investigations and succeeded in the preparation of some potassium and sodium derivatives reported as $Na_3PS_2O_{10}.x\,H_2O$, $K_3PS_2O_{10}$, and $K_4P_2S_2O_{13}$. Unfortunately, in spite of the fact that these compounds are well crystallized, no structural information is available.

The phosphochromates – About ten years ago, a systematic investigation of various $CrO_3–M_2O–P_2O_5–H_2O$ and $CrO_3–MO–P_2O_5–H_2O$ systems was performed, leading to the characterization of a new family of heteropolyanions of the general formula:

$$[PCr_nO_{3n+4}]^{3-}\quad \text{with } n = 1, 2, 3, 4$$

We have reported below some examples of these salts:
- $CuK_2H_2(PCrO_7)_2$ [87] for n = 1
- $K_2HCr_2PO_{10}$ [88], $BaHCr_2PO_{10}.3H_2O$, and $BaHCr_2PO_{10}.H_2O$ [89] for n = 2
- $Na_3PCr_3O_{13}.3H_2O$ [90] and $(NH_4)_2HCr_3AsO_{13}$ [91] for n = 3
- $Li_3Cr_4PO_{16}.3H_2O$ [92], $(NH_4)_3PCr_4O_{16}$ [93], $K_3PCr_4O_{16}$ [94], and $Rb_3PCr_4O_{16}$ [95] for n = 4

For n = 1 and 2, the anions have geometries similar to those observed for the first two terms of the oligophosphate series. For n = 3, the geometry is that of a central PO_4 tetrahedron sharing three of its oxygen atoms with three CrO_4 tetrahedra. For n = 4, the geometry is very similar to that observed for the Si_4O_{16} group of the zunyite-like silicates, in the present case a central PO_4 tetrahedron sharing its four oxygen atoms with four CrO_4 tetrahedra. Figure 5.6.7a and 5.6.7b represent these various heteropolyanions.

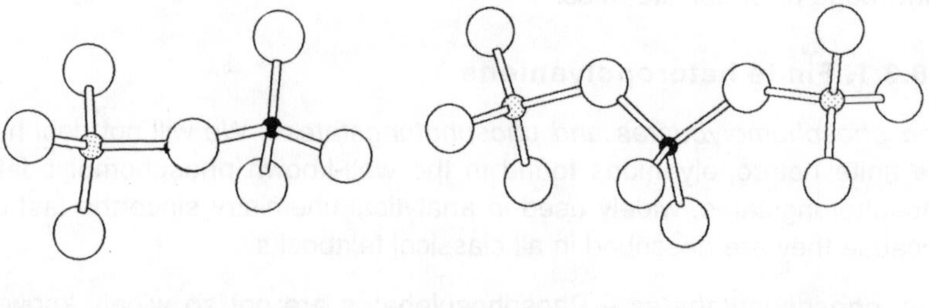

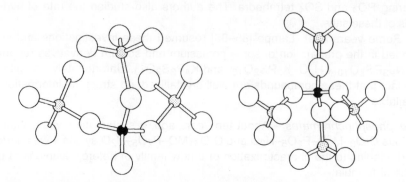

Figure 5.6.7a. The condensed anions observed in phosphochromates. The PCr_2O_7 group in $CuK_2H_2(PCrO_7)_2$ is represented on the left-hand side, the Cr_2PO_{10} group in $K_2HCr_2PO_{10}$ on the right-hand side. The black circles represent the phosphorus atoms, the spotted circles the chromium atoms and the large open circles the oxygen atoms.

Figure 5.6.7b. The condensed anions observed in phosphochromates. The PCr_3O_{13} group in $Na_3PCr_3O_{13}.3H_2O$ is represented on the left-hand side, the PCr_4O_{16} group in $(NH_4)_3PCr_4O_{16}$ on the right-hand side. The drawing conventions are identical to those of Figure 5.6.7a.

It is worth mentioning that one of these compounds, $(NH_4)_3PCr_4O_{16}$ was first prepared as early as 1894 by Friedheim and Moskin [96].

The chemical procedures of synthesis for these salts are very similar to those used in preparing condensed chromates. The condensation is always performed in a highly acidic medium. We have reported as an example of chemical preparation the procedure used for the synthesis of $K_3PCr_4O_{16}$ [94]. A concentrated solution of tripotassium monophosphate K_3PO_4 and chromic acid with a

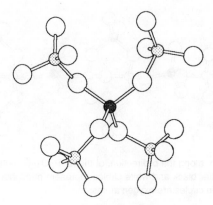

Figure 5.6.8. The zunyite-like AlP_4O_{14} group observed in $Ca_2Na_5AlP_4O_{16}$. The black circle represents the Al atom, the spotted circles are the P atoms and the larger open circles the O atoms.

ratio Cr: P = 4 is kept boiling for some minutes and then left at room temperature. A first precipitation of $K_2Cr_2O_7$ occurs some hours later. After elimination of this precipitate, one observes the formation of orange-red pseudo-rhombohedral crystals of $K_3PCr_4O_{16}$, in the remaining solution after some days.

In many cases, the corresponding arsenatochromate exists.

The phosphoaluminates – $Ca_2Na_5AlP_4O_{16}$, recently described by Alkemper *et al.* [97], provides another new example of finite heteropolyanion that we represent in Figure 5.6.8. This heteropolyanion built by a central AlO_4 tetrahedron sharing its four corners with four PO_4 tetrahedra is also very similar to a zunyite group and comparable to the PCr_4O_{16} heteropolyanion.

5.6.3.2. Infinite linear heteropolyanions

The phosphoborates – A typical example of infinite linear heteropolyanion is provided by the recent structural investigation of $Na_5B_2P_3O_{13}$ by Hauf *et al.* [98]. This compound is monoclinic, $P2_1$, with $Z = 2$ and the following unit-cell dimensions:

$$a = 6.71(2), \quad b = 11.62(2), \quad c = 7.69(2) \text{ Å}, \quad \beta = 115.17(2)°$$

The unit cell is crossed by an infinite ribbon built by a linkage of BO_4 and PO_4 tetrahedra represented in Figure 5.6.9. In this arrangement, the main buiding unit is a central chain, $-PO_4-BO_4-BO_4-PO_4-BO_4-BO_4-$, of corner-sharing PO_4 and BO_4 tetrahedra. Along this arrangement, each pair of BO_4 share its four external oxygen atoms with two PO_4 groups located on both sides of the central chain.

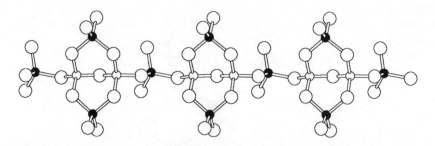

Figure 5.6.9. Projection, along the **b** direction, of the $[B_2P_3O_{13}]^{5-}$ heteropolyanion observed in $Na_5B_2P_3O_{13}$. The smaller black and white circles represent phosphorus and boron atoms respectively. The larger open circles are oxygen atoms.

The phosphosilicates – Among many examples of phosphosilicates, we have selected the synthetic compound, $Cd_2SiP_4O_{14}$, to illustrate this family. This salt, prepared by Trojan et al. [99] is monoclinic, *C2/c*, with $Z = 4$ and the unit cell:

$$a = 17.191(3), \quad b = 5.136(1), \quad c = 12.486(2) \text{ Å}, \quad \beta = 103.39(1)°$$

In this compound, the heteropolyanion is also, as in the latter example, a ribbon, but this time built by an association of P_2O_7 groups and SiO_4 tetrahedra. Inside the ribbon, each SiO_4 shares its four corners with four adjacent P_2O_7 groups. Such a configuration generates a succession of six-member rings interconnected through the SiO_4 groups. A representation of this heteropolyanion is given in Figure 5.6.10.

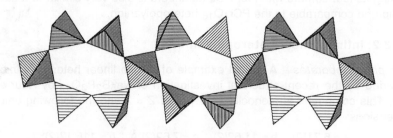

Figure 5.6.10. The $[SiP_4O_{14}]^{4-}$ heteropolyanion observed in $Cd_2SiP_4O_{14}$. The tetrahedra sharing their four corners correspond to the SiO_4 groups.

5.6.3.3. Infinite two-dimensional heteropolyanions

Multiple arrangements are given as containing the so-called infinite di- or tridimensional heteropolyanions. For the reasons that we have discussed at the beginning of this section, it is difficult to verify these assumptions in many cases and for several reasons. Thus, we have restricted the illustration of this type of heteropolyanions to only one example which is unambiguous, in our opinion. It is found in $FeAlPO_5$, recently investigated by Hesse and Cemic [100].

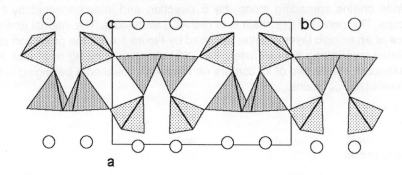

Figure 5.6.11. Projection, along the **c** direction, of the atomic arrangement in $FeAlPO_5$. The open circles represent the Fe atoms, the dark grey tetrahedra the AlO_4 entities, and the pale grey the PO_4 tetrahedra.

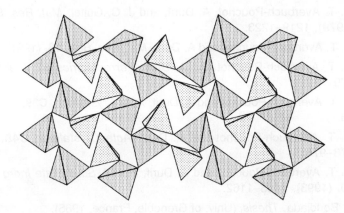

Figure 5.6.12. Projection of an $AlPO_5$ isolated layer approximately along the [20$\overline{1}$] direction. The light grey tetrahedra are the PO_4 groups, the dark ones the AlO_4 tetrahedra.

This compound is monoclinic, $P2_1/c$, with the parameters:

$$a = 7.141, \quad b = 10.549, \quad c = 5.493 \text{ Å}, \quad \beta = 98.19° \text{ and } Z = 4$$

This salt may be described as an arrangement built by two kinds of layers which alternate parallelly to the (b, c) plane. The first ones are layers of Fe (II) atoms located around the planes $x = 0$ and 1.0. They are separated by a thick anionic layer centered by the plane $x = 1/2$. In Figure 5.6.11 a projection along the **c** direction shows clearly the arrangement of these two types of layers. The infinite anionic entity may be described as being built by corrugated $(AlO_3)_n$ infinite chains spreading along the **c** direction and interconnected by PO_4 groups. The period of this chain is of two AlO_4 tetrahedra. The internal arrangement of an anionic layer is better analyzed by Figure 5.6.12, the projection of an isolated layer along approximately the $[20\overline{1}]$ direction showing how each PO_4 tetrahedron shares two of its corners with two AlO_4 tetrahedra belonging to two different $(AlO_3)_n$ chains.

References

1. R. F. Weinland and H. Prause, *Z. anorg. allg. Chem.*, **28**, (1901), 45–70.

2. A. Durif, M. T. Averbuch-Pouchot, and J. C. Guitel, *Acta Crystallogr.*, **B35**, (1979), 1444–1447.

3. M. T. Averbuch-Pouchot, *Acta Crystallogr.*, **B36**, (1980), 2405–2406.

4. M. T. Averbuch-Pouchot, A. Durif, and J. C. Guitel, *Mat. Res. Bull.*, **14**, (1979), 1219–1223.

5. M. T. Averbuch-Pouchot and A. Durif, *Mat. Res. Bull.*, **16**, (1981), 71–76.

6. M. T. Averbuch-Pouchot and A. Durif, *Ferroelectrics*, **52**, (1984), 271–279.

7. M. T. Averbuch-Pouchot and A. Durif, *Acta Crystallogr.*, **C39**, (1983), 27–28.

8. M. T. Averbuch-Pouchot and A. Durif, *Acta Crystallogr.*, **C48**, (1992), 973–975.

9. M. T. Averbuch-Pouchot and A. Durif, *Eur. J. Solid State Inorg. Chem.*, **30**, (1993), 1153–1162.

10. N. Boudjada, *Thesis*, (Univ. of Grenoble, France, 1985).

11. M. T. Averbuch-Pouchot and A. Durif, *Acta Crystallogr.*, **C43**, (1987), 1653–1655.

12. M. T. Averbuch-Pouchot, *Acta Crystallogr.*, **C44**, (1988), 1166–1168.

13. N. Boudjada, M. T. Averbuch-Pouchot, and A. Durif, *Acta Crystallogr.*, **B37**, (1981), 645–647.

14.   N. Boudjada, M. T. Averbuch-Pouchot, and A. Durif, *Acta Crystallogr.*, **B37**, (1981), 647–649.

15.   N. Boudjada and A. Durif, *Acta Crystallogr.*, **B38**, (1982), 595–597.

16.   N. Boudjada, A. Boudjada, and J. C. Guitel, *Acta Crystallogr.*, **C39**, (1983), 656–658.

17.   A. Durif, M. T. Averbuch-Pouchot, and J. C. Guitel, *J. Solid State Chem.*, **41**, (1982), 153–159.

18.   M. T. Averbuch-Pouchot and A. Durif, *Acta Crystallogr.*, **C43**, (1987), 1245–1247.

19.   M. T. Averbuch-Pouchot and A. Durif, *C. R. Acad. Sci.*, Sér. 2, **304**, (1987), 269–271.

20.   M. T. Averbuch-Pouchot and A. Durif, *Acta Crystallogr.*, **C46**, (1990), 2236–2238.

21.   J. E. Salmon and H. Terrey, *J. Chem. Soc.*, (1950), 2813–2824.

22.   J. Komrska and V. Satava, *Silikati*, **13**, (1969), 135–141.

23.   R. Herak, B. Prelesnik and M. Curic, *Z. Kristallogr.*, **164**, (1983), 25–30.

24.   A. Boudjada, A. Durif, and J. C. Guitel, *Acta Crystallogr.*, **B36**, (1980), 133–135.

25.   J. C. Elliot, *Structure and Chemistry of the Apatites and Other Calcium Orthophosphates*, (Elsevier, Amsterdam, 1994).

26.   A. Modaressi, A. Courtois, R. Gerardin, B. Malamann, and C. Gleitzer, *J. Solid State Chem.*, **47**, (1983), 245–255.

27.   M. Bouchdoug, A. Courtois, R. Gerardin, J. Steinmetz, and C. Gleitzer, *J. Solid State Chem.*, **42**, (1982), 149–157.

28.   M. Ijjali, M. Malaman, C. Gleitzer, and M. Pichavant, *Eur. J. Solid State Inorg. Chem.*, **26**, (1989), 73–89.

29.   L. Merker and H. Wondratschek, *Z. anorg. allg. Chem.*, **306**, (1960), 25–29.

30.   L. H. Brixner and C. M. Foris, *J. Solid State Chem.*, **7**, (1973), 149–154.

31.   I. V. Tananaev, E. V. Maksimchuk, Yu. G. Busshuev, and S. A. Shestov, *Izv. Akad. Nauk SSSR, Neorg. Mater.*, **14**, (1978), 719–722.

32.   J. J. Serra, *Thesis*, (Univ. of Perpignan, France, 1977).

33.   J. J. Serra, J. Coutures, A. Rouanel, H. Dexpert, and G. Garon, *Rev. Int. Hautes Tempér. Réfract. Fr.*, **15**, (1978), 287–313.

34.   A. Rouanet, J. J. Serra, K. Allaf, and V. P. Orlovskii, *Izv. Akad. Nauk SSSR, Neorg. Mater.*, **17**, (1981), 104–109.

35. J. R. Rea and E. Kostiner, *Acta Crystallogr.*, **B28**, (1972), 2525–2529.

36. J. R. Rea and E. Kostiner, *Acta Crystallogr.*, **B28**, (1972), 2505–2509.

37. J. R. Rea and E. Kostiner, *Acta Crystallogr.*, **B28**, (1972), 3461–3464.

38. E. N. Matvienko, O. V. Yakubovich, M. A. Simonov, and N. V. Belov, *Dokl. Akad. Nauk SSSR*, **246**, (1979), 875–878.

39. O. V. Yakubovich, M. A. Simonov, E. N. Matvienko, and N. V. Belov, *Dokl. Akad. Nauk SSSR*, **238**, (1978), 576–579.

40. M. Greenblatt, E. Banks, and B. Post, *Acta Crystallogr.*, 23, (1967), 166–171.

41. O. V. Yakubovich, O. A. Evdokimova, and O. K. Mel'nikov, *Dokl. Akad. Nauk SSSR*, **282**, (1985), 625–630.

42. O. V. Yakubovich and O. K. Mel'nikov, *Kristallografiya*, **36**, (1991), 334–341.

43. Yu. A. Ivanov, M. A. Simonov, M. Sirota, and N. V. Belov, *Dokl. Akad. Nauk SSSR*, **231**, (1976), 856–859.

44. M. T. Averbuch-Pouchot and A. Durif, *Eur. J. Solid State Inorg. Chem.*, **30**, (1993), 447–459.

45. K. K. Palkina, S. I. Maksimova, N. T. Chibiskova, B. F. Dzhurinskii, and L. Z. Gokhman, *Izv. Akad. Nauk SSSR, Neorg. Mater.*, **20**, (1984), 1063–1067.

46. J. C. Harris in *Phosphorus and its Compounds*, Vol. 2, 1816–1817, ed. Van Wazer (Interscience Publishers, New York).

47. T. A. Kurova, N. G. Shumyatskaya, A. A. Voronkov, and Yu. A. Pyatenko, *Dokl. Akad. Nauk SSSR*, **251**, (1980), 605–607.

48. E. V. Sokolova and A. P. Khomyakov, *Dokl. Akad. Nauk SSSR*, **322**, (1992), 531–535.

49. W. Frazier and J. R. Lehr, *J. Agr. Food Chem.*, **16**, (1968), 388–390.

50. A. P. Belopolskii, M. T. Serebrennikova, and S. Y. Shpunt, *J. Appl. Chem. (USSR)*, **10**, (1937), 1523–1529.

51. A. P. Belopolskii, M. N. Shulgina, M. T. Serebrennikova, and S. Y. Shpunt, *J. Appl. Chem. (USSR)*, **10**, (1937), 403–413.

52. R. Flatt, J. Wilhelm, G. Brunisholz, and G. Fell, *Helv. Chim. Acta*, **37**, (1954), 607–619.

53. R. Flatt, G. Brunisholz, and G. Denereaz, *Helv. Chim. Acta*, **39**, (1956), 473–483.

54. A. Durif, I. Tordjman, R. Masse, and J. C. Guitel, *J. Solid State Chem.*, **24**, (1978), 101–105.

55 U. Schülke, private communication.

56. M. T. Averbuch–Pouchot and A. Durif, *Eur. J. Solid State Inorg. Chem.*, **29**, (1992), 1161–1172.

57. M. T. Averbuch–Pouchot and A. Durif, *Acta Crystallogr.*, **C48**, (1992), 1173–1176.

58. L. Fanfani and P. F. Zanassi, *Z. Kristallogr.*, **126**, (1968), 433–443.

59. R. A. Bideaux, M. C. Nichols, and S. A. Williams, *Amer. Mineral.*, **51**, (1966), 258–259.

60. A. Durif, *C. R. Acad. Sci.*, **244**, (1957), 2815–2817.

61. A. Durif, *C. R. Acad. Sci.*, **245**, (1957), 1151–1152.

62. V. M. Krutik, D. Yu. Pushcharovskii, A. P. Khomyakov, E. A. Pobedimskaya, and N. V. Belov, *Sov. Phys. Crystallogr.*, **26**, (1981), 679–682.

63. M. T. Averbuch-Pouchot and A. Durif, *Acta Crystallogr.*, **C46**, (1990), 965–968.

64. D. M. Wiench and M. Jansen, *Acta Crystallogr.*, **C39**, (1983), 1613–1615.

65. A. Larbot, J. Durand, S. Vilminot, and A. Norbert, *Acta Crystallogr.*, **B37**, (1981), 1023–1027.

66. A. Larbot, A. Norbert, and J. Durand, *Z. anorg. allg. Chem.*, **486**, (1982), 200–206.

67. M. De la Rochère, A. Kahn, F. d'Yvoire, and E. Bretey, *Mater. Res. Bull.*, **20**, (1985), 27–34.

68. B. Kojic-Prodic, M. Sljukic, and Z. Ruzic-Toros, *Acta Crystallogr.*, **B38**, (1982), 67–71.

69. K. K. Palkina and S. I. Maksimova, *Dokl. Akad. Nauk SSSR*, **250**, (1980), 1130–1134.

70. J. J. Chen, K. H. Lii, and S. L. Lang, *J. Solid State Chem.*, **76**, (1988), 204–209.

71. A. Leclaire, J. C. Monier, and B. Raveau, *J. Solid State Chem.*, **48**, (1983), 147–153.

72. A. Leclaire, M. M. Borel, A. Grandin, and B. Raveau, *Z. Kristallogr.*, **188**, (1989), 77–83.

73. A. Benmoussa, A. Leclaire, A. Grandin, and B. Raveau, *Acta Crystallogr.*, **C45**, (1989), 1277–1279.

74. G. G. Sadikov, V. P. Nikolaev, A. V. Lavrov, and M. A. Porai-Koshits,

Dokl. Akad. Nauk SSSR, **264**, (1982), 862–867.

75. G. G. Sadikov, V. P. Nikolaev, A. V. Lavrov, and M. A. Porai-Koshits, *Dokl. Akad. Nauk SSSR*, **266**, (1982), 354–358.

76. Yu. A. Ivanov, M. A. Simonov, and N. V. Belov, *Dokl. Akad. Nauk SSSR*, **242**, (1978), 599–602.

77. M. T. Averbuch-Pouchot, *Z. anorg. allg. Chem.*, **545**, (1987), 118–124.

78. V. P. Nikolaev, G. G. Sadikov, A. V. Lavrov, and M. A. Porai-Koshits, *Izv. Akad. Nauk SSSR, Neorg. Mater.*, **19**, (1983), 448–451.

79. I. Grunze, K. K. Palkina, S. I. Maksimova, and N. T. Chibiskova, *Dokl. Akad. Nauk SSSR*, **275**, (1984), 879–883.

80. M. T. Averbuch-Pouchot, *Z. anorg. allg. Chem.*, **529**, (1985), 143–150.

81. M. T. Averbuch-Pouchot and A. Durif, *C. R. Acad. Sci.*, Sér. 2, **317**, (1993), 1179–1184.

82. A. Gharbi, A. Jouini, M. T. Averbuch-Pouchot, and A. Durif, *J. Solid State Chem.*, **111**, (1994), 330–337.

83. M. Evain, R. Brec, G. Ouvrard, and J. Rouxel, *Mat. Res. Bull.*, **19**, (1984), 41–48.

84 E. Thilo and G. Blumenthal, *Z. anorg. allg. Chem.*, **348**, (1966), 77–88.

85 F. von Lampe, *Z. anorg. allg. Chem.*, **368**, (1969), 93–105.

86 F. von Lampe, *Z. anorg. allg. Chem.*, **367**, (1969), 170–188.

87. J. Coing–Boyat, A. Durif, and J. C. Guitel, *J. Solid State Chem.*, **30**, (1979), 329–334.

88. M. T. Averbuch–Pouchot, A. Durif, and J. C. Guitel, *Acta Crystallogr.*, **B34**, (1978), 3725–3727.

89. M. T. Averbuch–Pouchot, A. Durif, and J. C. Guitel, *Acta Crystallogr.*, **B33**, (1977), 1431–1435.

90. M. T. Averbuch–Pouchot, A. Durif, and J. C. Guitel, *J. Solid State Chem.*, **33**, (1980), 325–333.

91. M. T. Averbuch–Pouchot, *Acta Crystallogr.*, **B34**, (1978), 3350–3351.

92. Yu. N. Sal'yanov, R. I. Bochkova, E. A. Kuzmin, and N. V. Belov, *Dokl. Akad. Nauk SSSR*, **257**, (1981), 619–621.

93. M. T. Averbuch–Pouchot, A. Durif, and J. C. Guitel, *J. Solid State Chem.*, **36**, (1981), 381–384.

94. M. T. Averbuch–Pouchot, A. Durif, and J. C. Guitel, *J. Solid State Chem.*, **38**, (1981), 253–258.

95.   M. T. Averbuch–Pouchot, *Z. Kristallogr.*, **155**, (1981), 315–317.

96.   C. Friedheim and J. Moskin, *Z. anorg. Chem.*, **6**, (1894), 273–280.

97.   J. Alkemper, H. Paulus, and H. Fuess, *Z. Kristallogr.*, **209**, (1995), 76.

98.   C. Hauf, T. Friedrich, and R. Kniep, *Z. Kristallogr.*, **210**, (1995), 446.

99.   M. Trojan, D. Brandova, J. Fabry, J. Hybler, K. Jurek, and V. Petricek, *Acta Crystallogr.*, **C43**, (1987), 2038–2040.

100. K. F. Hesse and L. Cemic, *Z. Kristallogr.*, **209**, (1994), 346–347.

CHAPTER 6

The Networks of Acidic Phosphate Anions

6.1. INTRODUCTION

It is a common observation that in any kind of oxosalts containing acidic anions these entities have a strong tendency to assemble by strong hydrogen bonds to build infinite anionic networks in most of the cases. Generally, the H–bonds connecting the anions correspond to O–O distances shorter than the O–O distances in the constituting anionic entities. In a basic investigation of H–bond geometries, Brown [1] estimated to approximately 2.70 Å the boundary between strong and weak H–bonds. In most of the networks that we will describe here, the O–O distances are much shorter than 2.70 Å, thus explaining the stability of these arrangements.

A variety of geometries of such networks could be achieved. Finite entities, infinite chains or ribbons or planes are commonly observed while three-dimensional arrangements are less common. Generally, very little attention is paid to the geometry of these networks in structural reports. The dimensionality of the H–bond network is usually reported and discussed, but only a few words to the examination of the geometry of the anionic network itself.

When describing such hydrogen-bond networks one must be very careful especially when discussing their dimensionality. Some of the examples that we have selected are hydrated compounds. If one considers the hydrogen-bond schemes taking or not into account the water molecules their dimensionality can vary. If the water molecules are included, sets of isolated chains or ribbons of acidic anions can be interconnected by additional H–bonds provided by these water molecules and so the newly obtained H–bond network becomes two- or three-dimensional.

The presence of monophosphoric acid molecules in some compounds either isolated or assembled to build clusters seems to be in our opinion, a special case to be examined only when a greater number of accurate examples becomes available. To date, only few recent data is available.

In the next few pages, we have selected some examples in phosphate chemistry in order to illustrate the most common geometries observed in these networks. For each example or category of examples, we will present the main geometrical features of the H–bonds responsible for the network cohesion:

P–O–H, O–H, H···O, O–O, and O–H···O

6.2. THE ACIDIC ANIONS IN PHOSPHATE CHEMISTRY

Acidic monophosphates and acidic diphosphates are common compounds and along the survey of their properties and of the present state of their chemistry (Chapter 5), we have reported numerous examples of such salts.

Acidic higher oligophosphates are less common, but not rare.

For monovalent-cation acidic triphosphates, one can report four ammonium compounds, $(NH_4)_3H_2P_3O_{10}$ (I), $(NH_4)_3H_2P_3O_{10}$ (II), $(NH_4)_4HP_3O_{10}$, and $(NH_4)_9H(P_3O_{10})_2.2H_2O$, characterized by Farr *et al.* [2] during a study of the $NH_3–H_5P_3O_{10}–H_2O$ system and $K_3H_2P_3O_{10}.H_2O$ whose chemical preparation and crystal structure were described by Lyutsko and Johansson [3].

Two divalent-cation salts are presently known, $Zn_2HP_3O_{10}.6H_2O$ reported by Averbuch-Pouchot and Guitel [4] and $Pb_2HP_3O_{10}$ described by Worzala and Jost [5].

Several anhydrous trivalent-cation triphosphates, with general formula $M^{III}H_2P_3O_{10}$ (M^{III} = Al, V, Fe, Ga, Yb), and one monohydrate, $FeH_2P_3O_{10}.H_2O$, are also well characterized. We have listed them below:

– $AlH_2P_3O_{10}$ Lyutsko *et al.* [6]
– $VH_2P_3O_{10}$ Lyakhov *et al.* [7]
– $FeH_2P_3O_{10}$ Lyutsko and Johansson [3]
– $GaH_2P_3O_{10}$ Yakhov *et al.* [8]
– $YbH_2P_3O_{10}$ Palkina *et al.* [10]
– $FeH_2P_3O_{10}.H_2O$ Averbuch-Pouchot and Durif [9]

Ten trivalent-monovalent cation derivatives, reported below, are also known. Their common chemical formula is $M^{III}M^IHP_3O_{10}$ with M^{III} = Al, V, Fe, Cr, Bi and M^I = K, NH_4, Rb, Cs.

– $AlNH_4HP_3O_{10}$ Averbuch-Pouchot *et al.* [11]
– $AlKHP_3O_{10}$ Averbuch-Pouchot *et al.* [11]
– $FeNH_4HP_3O_{10}$ Krasnikov *et al.* [12]
– $VCsHP_3O_{10}$ Klinkert and Jansen [13]
– $VNH_4HP_3O_{10}$ Teterevkov and Mikhailovskaya [14]
– β-$CrNH_4HP_3O_{10}$ Vaivada *et al.* [15]
– α-$CrNH_4HP_3O_{10}$ Vaivada *et al.* [15]
– $CrRbHP_3O_{10}$ Teterevkov and Mikhailovskaya [14]
– $CrCsHP_3O_{10}$ Grunze and Chudinova [16]

– $BiNH_4HP_3O_{10}$ Averbuch-Pouchot and Bagieu-Beucher [17]

Only one acidic tetraphosphate, $(NH_4)_4H_2P_4O_{13}$, is mentioned in the literature. It was studied by Waerstrad and Mc Clellan [18] and Farr et al. [19], but no structural investigation was performed for this compound.

Acidic long-chain polyphosphates are not common and, in addition, among the five reported examples that we have listed below, only the first four were the subject of satisfactory structure determinations. Thus, in this chapter, we will report with more details the networks observed in this category of compounds.

– $CaH_2(PO_3)_4$ Averbuch-Pouchot [20]
– $ErH(PO_3)_4$ Palkina et al. [21]
– $BiH(PO_3)_4$ Palkina and Jost [22]
– $(UO_2)H(PO_3)_3$ Linde et al. [23] and Sarin et al. [24]
– $Ba_2H_3(PO_3)_7$ Averbuch-Pouchot et al. [25] and El-Horr et al. [26]

Acidic cyclophosphates are still rarer than acidic long-chain polyphosphates. Only three examples have been reported so far.

Griffith [27] described the chemical preparation of $Na_2HP_3O_9$ and characterized two crystalline forms for this *cyclotriphosphate*. Crystal structure of one of these forms was later determined by Averbuch-Pouchot and Durif [28] and remains the only structural evidence for an acidic cyclotriphosphate to date.

In the family of *cyclotetraphosphates* two examples have been reported.

The chemical preparation of a salt said to be $Na_2H_2P_4O_{12}$ was described by Griffith [29] by the reaction of H_3PO_4 on NaH_2PO_4 at 673 K:

$$2NaH_2PO_4 + 2H_3PO_4 \longrightarrow Na_2H_2P_4O_{12} + 4H_2O$$

According to this author, the melting or, more probably, the decomposition point of $Na_2H_2P_4O_{12}$ is close to 673 K. A more detailed procedure for preparation of this salt was later reported by Griffith [27]. This salt is almost insoluble in water and has a fibrous crystal habit. These two characteristics are rather surprising for an alkali cyclotetraphosphate. In order to explain these properties, Gryder et al. [30] performed a careful crystallographic investigation of this salt. They suspected the so-called $Na_2H_2P_4O_{12}$ to be an intimate intergrowth of two distinct crystalline forms, both monoclinic and, from various considerations, concluded that the anionic framework must be built up by chains of P_4O_{12} rings sharing an oxygen atom, leading to $Na_2P_4O_{11}$, which is the formula of an ultraphosphate. Jarchow [31] suggested a structural model based on the $Na_2H_2P_4O_{12}$ formula proposed by Griffith. Owing to the poor quality of the experimental data, this work does not bring a conclusive solution and needs revision.

Averbuch-Pouchot and Durif [32] prepared $Sr_3Cs_4H_2(P_4O_{12})_3$ and performed the structure determination of this salt, closely related to the atomic arrangement of $Al_4(P_4O_{12})_3$, but could not determine the location of the hydrogen atoms.

Chemistry of higher phosphoric rings does not provide any example of acidic anions. No acidic ultraphosphate anion has been reported to date.

6.2.1. The Acidic Anion Networks in Monophosphates

6.2.1.1. Finite clusters

Finite clusters of acidic anions are relatively rare in phosphate chemistry but some examples are found in the family of monophosphates.

$MgHPO_4.3H_2O$ – In the well-known mineral newberyite, two $[HPO_4]^{2-}$ anions assemble to build a centrosymmetric group that we have represented in Figure 6.1. This figure was drawn from the data reported in the investigation of Abbona et al. [33].

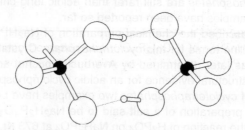

Figure 6.1. The centrosymmetric cluster $[H_2P_2O_8]^{4-}$ observed in newberyite [33]. The black circles represent the P atoms, the larger open circles the O atoms and the smaller the H atoms.

$ZnHPO_4.H_2O$ – This compound [34], which we have discussed in Chapter 5, provides another rare example for a finite arrangement of acidic monophos-

Table 6.1
Main geometrical features of the H–bond geometry in the two selected examples.

Formula	P–O⋯H (°)	O–H (Å)	H⋯O (Å)	O–O (Å)	O–H⋯O (°)	Ref.
$MgHPO_4.3H_2O$	111	0.98	1.67	2.615	162	[33]
$ZnHPO_4.H_2O$	148	1.07	1.74	2.576	131	[34]
	125	0.76	1.89	2.630	163	
	122	1.03	1.51	2.538	173	

phate anions. In this monophosphate, the HPO_4 groups assemble to build a centrosymmetric hexameric cluster, $(HPO_4)_6$. The distribution of these clusters inside the atomic arrangement of $ZnHPO_4.H_2O$ is shown in Figure 6.2.

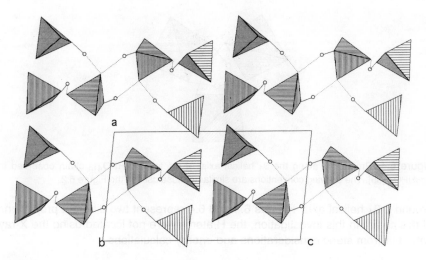

Figure 6.2. The arrangement of the hexameric clusters $(HPO_4)_6$ in $ZnHPO_4 \cdot H_2O$ [34]. The PO_4 tetrahedra are shown by polyhedral representation and the small open circles are H atoms.

In Table 6.1, we gather the main geometrical values of the hydrogen-bond schemes observed in these two examples of finite arrangements.

6.2.1.2. Chains

Chains arrangements are very common among the infinite networks formed by the acidic monophosphate anions. They can adopt various configurations. We have reported some examples below.

Ag₂HPO₄ – The atomic arrangement of this salt, determined by Tordjman *et al.* [35], provides one of the rare example of a chain of HPO_4 groups spiraling

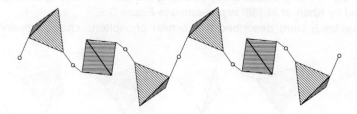

Figure 6.3. Projection, along the **a** axis, of the infinite $(HPO_4)_n$ chain observed in Ag_2HPO_4 [35]. The drawing conventions are identical to those used in the Figure 6.2.

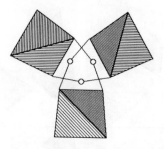

Figure 6.4. Projection, along the 3_1 helical axis, of the infinite $(HPO_4)_n$ chain observed in Ag_2HPO_4 [35]. The drawing conventions are similar to those used in the Figure 6.2.

around a 3_1 helical axis. Figures 6.3 and 6.4 represent two different projections of this chain. In this investigation, the H atoms were not located using the X-ray data, but from steric considerations and energy calculations.

Table 6.2
Numerical data for hydrogen bonds in a selection of $(HPO_4)_n$ chains.

Formula	P–O⋯H (°)	O–H (Å)	H⋯O (Å)	O–O (Å)	O–H⋯O (°)	Ref.
$(NH_4)_2HPO_4$	118	0.91	1.72	2.615	168	[36]
Ag_2HPO_4	126	1.22	1.22	2.420	162	[35]
$(eda)HPO_4$	117	0.71	1.90	2.578	160	[*]
$(sar)H_2PO_4$	113	0.95	1.61	2.566	180	[**]

[*] = Ref. [175] in the Chapter 5
[**] = Ref. [176] in the Chapter 5
eda = ethylenediammonium or $[NH_3–(CH_2)_2–NH_3]^{2+}$
sar = sarcosinium or $[CH_3–NH_2–CH_2–COOH]^+$

$(NH_4)_2HPO_4$ – A very simple chain geometry found in this atomic arrangement determined by Khan *et al.* [36] is presented in Figure 6.5.

In Chapter 5 were described two similar phosphoric chains observed in

Figure 6.5. The infinite $(HPO_4)_n$ chain as observed in $(NH_4)_2HPO_4$ [36]. The drawing conventions are identical to those used in the Figure 6.2.

organic-cation monophosphates. We include, in Table 6.2, the main geometrical features observed for their hydrogen bonds.

6.2.1.3. Two-dimensional networks

Glycinium dihydrogenmonophosphate – A bidimensional network of $[H_2PO_4]^-$ entities was recently described by Averbuch-Pouchot *et al.* [37] in the atomic arrangement of glycinium dihydrogenmonophosphate, $[C_2H_6NO_2]^+$ $[H_2PO_4]^-$. Figure 6.6 reports the internal arrangement of this network.

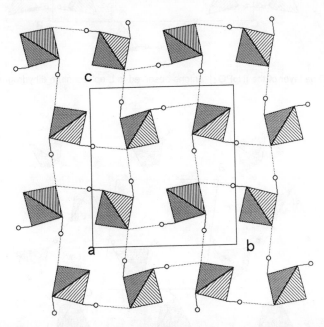

Figure 6.6. The layer of $[H_2PO_4]^-$ anions observed in glycinium dihydrogenmonophosphate.

L-ephedrinium dihydrogenmonophosphate – The crystal structure of this salt, performed by Hearn and Bugg [38], provides a similar example of network that we have reported in Figure 6.7.

Cytosinium dihydrogenmonophosphate – We present a third example reported by Bagieu-Beucher [39]: $[C_4H_6N_3O]^+$ $[H_2PO_4]^-$. In this last example, two of the three hydrogen atoms observed in the arrangement are statistically distributed around inversion centers. This layer is shown in Figure 6.8.

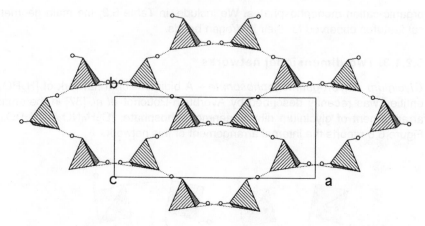

Figure 6.7. The layer of the [H$_2$PO$_4$]$^-$ anions observed in L-ephedrinium dihydrogenmonophosphate.

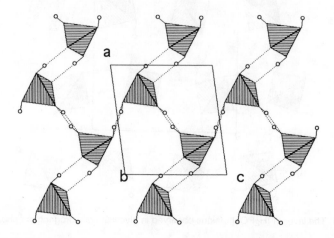

Figure 6.8. The layer of [H$_2$PO$_4$]$^-$ anions as observed in cytosinium dihydrogenmonophosphate.

L-histidinium dihydrogenmonophosphate-H$_3$PO$_4$ adduct – In some cases like in [C$_6$H$_{10}$N$_3$O$_2$]$^+$ [H$_2$PO$_4$]$^-$ H$_3$PO$_4$, a L-histidinium derivative investigated by Blessing [40], the two-dimensional network is built by both the [H$_2$PO$_4$]$^-$ groups and the H$_3$PO$_4$ molecules. Such a network is shown in Figure 6.9.

Table 6.3 report the main geometrical features of the hydrogen bonds in the selected examples.

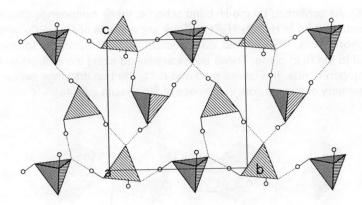

Figure 6.9. The layer of [H₂PO₄]⁻ groups and H₃PO₄ molecules in $[C_6H_{10}N_3O_2]^+$ $[H_2PO_4]^-$ H_3PO_4 [40].

Table 6.3

Some numerical data for hydrogen bonds in $(H_2PO_4)_n$ layers.

Formula	P–O···H (°)	O–H (Å)	H···O (Å)	O–O (Å)	O–H···O (°)	Ref.
(gly)H₂PO₄	115	0.87	1.70	2.568	175	[37]
	113	0.73	1.87	2.596	174	
(eph)H₂PO₄	108	1.06	1.60	2.632	164	[38]
	114	0.63	1.93	2.554	175	
(cyt)H₂PO₄	115	0.67	1.84	2.501	172	[39]
	114	0.73	1.76	2.476	167	
	113	0.73	1.89	2.611	178	
(L-hist)H₅P₂O₈ (*)	109	0.87	1.72	2.569	167	[40]
	127	0.92	1.54	2.460	173	
	114	0.72	1.85	2.554	168	
	112	0.90	1.68	2.573	176	
	119	0.75	1.87	2.583	157	

gly = glycinium or $[C_2H_6NO_2]^+$
eph = ephedrinium or $[C_{10}H_{16}NO]^+$ or $[C_6H_5CH[CH(NH_2CH_3)CH_3]OH]^+$
cyt = cytosinium or $[C_4H_6N_3O]^+$
hist = hystidinium or $[C_6H_{10}N_3O_2]^+$
(*) = $[C_6H_{10}N_3O_2]^+[H_2PO_4]^- H_3PO_4$

6.2.1.4. Three-dimensional networks

The $M^I H_5(PO_4)_2$ compounds – For M^I = Na, K, Rb, Cs, Tl, and NH₄, this family of salts provide good examples for three-dimensional hydrogen-bond networks

[41–46]. As confirmed by the H–bond scheme, these compounds should in fact be formulated as $M^IH_2PO_4.H_3PO_4$. They are adducts of $M^IH_2PO_4$ with mono-phosphoric acid. The atomic arrangement is built by thick anionic layers parallel to the (b,c) plane. These layers are linked along the **a** direction by a set of hydrogen bonds. In Figures 6.10 and 6.11, the two drawings explain clearly the geometry of this network in the case of the cesium salt [46].

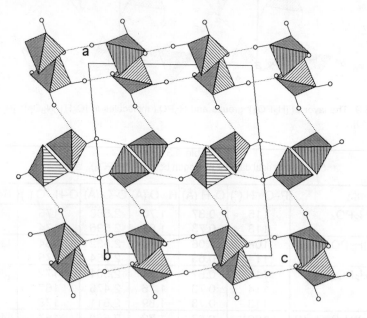

Figure 6.10. Projection along the **b** direction of the atomic arrangement in $CsH_2PO_4.H_3PO_4$.

[NH₂–(CH₂)₂–NH₂].3H₃PO₄ – A more complex three-dimensional anionic net-work was characterized by Bagieu-Beucher *et al.* [47] in an organic-cation derivative whose global formula is $[NH_2-(CH_2)_2-NH_2].3H_3PO_4$. The structural investigation showed that it must be formulated as $[NH_3-(CH_2)_2-NH_3]^{2+}$ $[H_5P_2O_8]^-$ $[H_2PO_4]^-$. The network can be described as built by two different enti-ties, a normal H_2PO_4 group and a centrosymmetric cluster made of two H_3PO_4 groups interconnected by an hydrogen atom statistically distributed around the inversion center leading to a formula $[H_5P_2O_8]^-$. Figure 6.12 reports a pers-pective view of this cluster. A very similar geometry has already been observed in a complex of antipyrine-H_3PO_4. But in this case, the hydrogen bonds con-necting the two H_3PO_4 entities are not performed through a statistical H atom and the cluster formula is $H_6P_2O_8$.

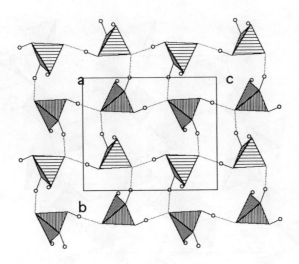

Figure 6.11. Projection, along the **a** direction, of one of the anionic layer in $CsH_2PO_4.H_3PO_4$.

Table 6.4

Main features of the H–bond geometry in the selected examples of three-dimensional networks.

Formula	P–O⋯H (°)	O–H (Å)	H⋯O (Å)	O–O (Å)	O–H⋯O (°)	Ref.
$CsH_5(PO_4)_2$	105	0.87	1.87	2.589	140	[46]
	109	1.01	1.57	2.538	159	
	115	0.64	1.97	2.534	146	
	112	0.89	1.66	2.544	173	
	125	1.02	1.43	2.426	162	
$(eda)(H_3PO_4)_3$	115	0.82	1.72	2.546	177	[47]
	113	0.77	1.68	2.432	167	
	112	0.59	1.96	2.542	173	
	114	0.73	1.80	2.533	166	

eda = ethylenediamine or $[NH_2–(CH_2)_2–NH_2]$

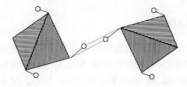

Figure 6.12. A perspective view of the centrosymmetrical $H_5P_2O_8$ cluster observed in $[NH_3–(CH_2)_2–NH_3]$ $[H_5P_2O_8]$ $[H_2PO_4]$.

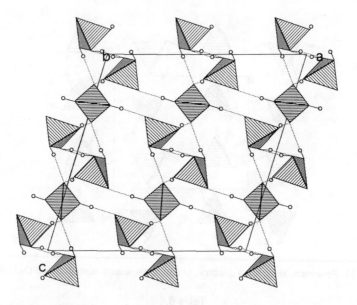

Figure 6.13. The anion network in $[NH_3-(CH_2)_2-NH_3]^{2+}$ $[H_5P_2O_8]^-$ $[H_2PO_4]^-$ seen along **b**.

The anionic part of $[NH_3-(CH_2)_2-NH_3][H_5P_2O_8][H_2PO_4]$ is shown by the Figure 6.13 in projection along the **b** axis.

The main geometrical values observed in these examples of three-dimensional network are reported in Table 6.4.

6.2.2. Acidic Anion Networks in Diphosphates

Up to now, four types of geometrical configurations are known:
- infinite chains
- infinite ribbons (generally double chains)
- infinite bidimensional layers
- infinite tridimensional network

6.2.2.1. Infinite chains

In this case, each diphosphate group is bonded to its two adjacent neighbors. In fact, one can distinguish two types of chains. In the first type, the H–bonds are established so that the P–O–P bond of the P_2O_7 group is parallel or quasi-parallel to the chain axis. An example of such an association is found in the atomic arrangement of *MnHP₂O₇* [48] that we have reported in Figure

6.14. In the second type of chain arrangement, the P–O–P bond is perpendicular or quasi-perpendicular to the chain direction. The chain arrangement is the most common for the diphosphates. Figures 6.15 and 6.16 report two exam-

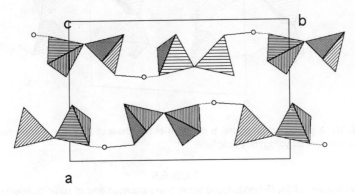

Figure 6.14. A projection, along the **c** direction, of the two chains of HP_2O_7 groups crossing the unit cell of $MnHP_2O_7$.

ples of this type of chains as observed in $K_2H_2P_2O_7$ [49] and in ethanolammonium dihydrogendiphosphate, $[NH_3–C_2H_4–OH][H_2P_2O_7]$ [50]. The various interatomic distances and bond angles observed for the hydogen bonds in the selected examples of chain networks are given in Table 6.5.

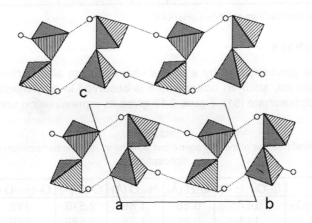

Figure 6.15. A projection, along the **a** direction, of the chain of $H_2P_2O_7$ groups crossing the unit cell of $K_2H_2P_2O_7$.

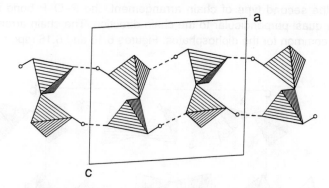

Figure 6.16. A projection along the **b** direction of the chain of $H_2P_2O_7$ groups observed in ethanolammonium dihydrogendiphosphate.

Table 6.5

Main features of the H–bonds found in the chain arrangements of some diphosphates

Formula	P–O···H (°)	O–H (Å)	H···O (Å)	O–O (Å)	O–H···O (°)	Ref.
$MnHP_2O_7$	121	0.85	1.75	2.577	163	[48]
$K_2H_2P_2O_7$	110	0.68	1.91	2.581	168	[49]
	118	0.69	1.89	2.578	178	
$(eta)H_2P_2O_7$	109	0.82	1.78	2.598	177	[50]
	112	0.87	1.70	2.553	165	

eta = ethanolammonium or $[NH_3–C_2H_4–OH]^+$

6.2.2.2. Ribbons

They are generally built by a double chain and are also frequently observed. For instance, such an arrangement is observed in ethylenediammonium dihydrogendiphosphate [51]. Figure 6.17 gives its representation and Table 6.6

Table 6.6

Main geometrical features of the hydrogen-bond scheme in ethylenediammonium dihydrogen-diphosphate.

Formula	P–O···H (°)	O–H (Å)	H···O (Å)	O–O (Å)	O–H···O (°)	Ref.
$(eda)H_2P_2O_7$	117	0.86	1.66	2.513	172	[51]
	113	0.85	1.75	2.588	169	

eda = ethylenediammonium or $[NH_3–(CH_2)_2–NH_3]^{2+}$

reports the main geometrical features of the hydrogen-bond scheme. In this type of association, each diphosphate group is connected to three adjacent groups.

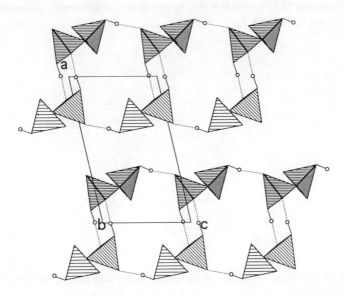

Figure 6.17. The ribbon of $H_2P_2O_7$ groups observed in ethylenediammonium dihydrogen-diphosphate.

Table 6.7

Main geometrical features of the H–bond schemes in the two reported examples of infinite layers.

Formula	P–O···H (°)	O–H (Å)	H···O (Å)	O–O (Å)	O–H···O (°)	Ref.
$Ag_2H_2P_2O_7$	120	1.226	1.226	2.442	169	[54]
	115	1.227	1.227	2.453	177	
$Rb_2H_2P_2O_7.1/2H_2O$	104	0.94	1.65	2.508	149	[52]
	110	1.12	1.41	2.519	169	
$Cs_2H_2P_2O_7$	114	1.05	1.49	2.534	170	[53]

6.2.2.3. Infinite bidimensional layers

They were observed in $Rb_2H_2P_2O_7.1/2H_2O$ [52] and in $Cs_2H_2P_2O_7$ [53], for instance. These two arrangements are represented in Figures 6.18 and 6.19. One can notice that each P_2O_7 group is connected to three other ones in the first example, whereas in the second example it is connected to four. Table 6.7 includes the numerical values observed in their hydrogen-bond schemes.

In the atomic arrangement of $Ag_2H_2P_2O_7$ [54] that we have described in Chapter 5, a two-dimensional network of $H_2P_2O_7$ groups also exists. The two H–bonds building this network are symmetrical. Their main geometrical features are given in Table 6.7.

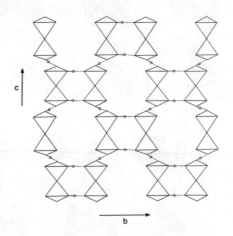

Figure 6.18. Schematic projection of the infinite bidimensional layer of $H_2P_2O_7$ groups in $Rb_2H_2P_2O_7.1/2H_2O$.

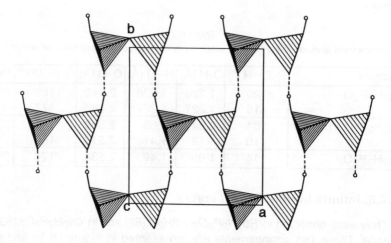

Figure 6.19. Projection, along the **c** direction, of the anionic layer in $Cs_2H_2P_2O_7$.

6.2.2.4. Infinite three-dimensional networks

This type of network is rare. The only known example was reported very recently in $Cs_2H_2P_2O_7.Te(OH)_6$ [55]. This very intricate arrangement is built by strongly corrugated layers containing all the components of the structure. These layers spread perpendicular to the **b** axis, separated from each other by a distance of about 4.19 Å. This short distance explains the fact that they are interconnected by H–bonds. Figure 6.20 shows the projection along the **b** direction of the H–bond network connecting the phosphoric entities. This arrangement is difficult to describe because of the special situation of some hydrogen atoms located close to symmetry elements. One hydrogen atom occupies a general position and establishes a normal H–bond responsible for the connection between the layers. The second hydrogen atom is statistically split on two general positions. On these general positions, the two fractional atoms remain close to symmetry elements ($\bar{1}$ and 2) so that the H–bond close to the inversion center is nearly symmetrical (Table 6.8). In this example, each diphosphate group is connected to four of its neighbors.

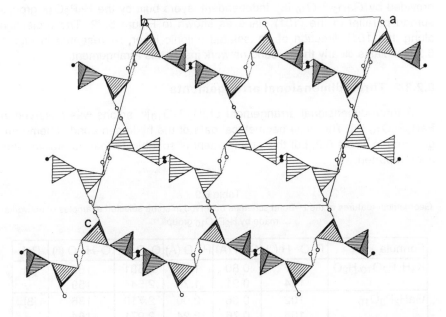

Figure 6.20. Projection, along the **b** direction, of the network built by the $H_2P_2O_7$ groups in $Cs_2H_2P_2O_7.Te(OH)_6$. The black H atoms are those that are establishing the connection between the layers along the **b** direction.

Table 6.8

Main geometrical features of the hydrogen-bond scheme in $Cs_2H_2P_2O_7.Te(OH)_6$.

Formula	P–O···H (°)	O–H (Å)	H···O (Å)	O–O (Å)	O–H···O (°)	Ref.
$Cs_2H_2P_2O_7.Te(OH)_6$	115	0.99	1.59	2.538	160	[55]
	110	1.11	1.36	2.403	153	
	124	0.80	1.73	2.517	171	

6.2.3. Acidic Anion Networks in Triphosphates

6.2.3.1. Infinite bidimensional layers

In $K_3H_2P_3O_{10}.H_2O$ [3], the $[H_2P_3O_{10}]^{3-}$ anionic network forms an infinite bidimensional layer. This compound is a layered arrangement built by sheets of potassium atoms alternating with anionic layers including the water molecules. Figure 6.21 represents one of this anionic layer in projection along the **b** axis.

Another example of a two-dimensional network of triphosphate groups is provided by $GaH_2P_3O_{10}$ [8]. Independent layers built by the $H_2P_3O_{10}$ groups spread parallel to the (101) plane as shown in Figure 6.22. The projection along the [$\bar{1}$01] direction of an isolated anionic layer, represented in Figure 6.23, explains clearly the H–bond network inside this arrangement.

6.2.3.2. Three-dimensional arrangements

A three-dimensional arrangement of $[H_2P_3O_{10}]^{3-}$ anions was observed in $FeH_2P_3O_{10}$ [3]. The main geometrical data of the hydrogen-bond scheme are gathered in Table 6.9, but this arrangement is so intricate that no drawing has been reported.

Table 6.9

Geometrical features of the H–bond schemes observed in some selected examples of networks made by $H_2P_3O_{10}$ groups.

Formula	P–O···H (°)	O–H (Å)	H···O (Å)	O–O (Å)	O–H···O (°)	Ref.
$K_3H_2P_3O_{10}.H_2O$	117	0.80	1.79	2.581	160	[3]
	124	0.91	1.70	2.544	159	
$GaH_2P_3O_{10}$	102	0.86	2.00	2.710	136	[8]
	106	0.76	2.24	2.971	164	
$FeH_2P_3O_{10}$	112	0.77	1.98	2.716	157	[3]
	129	0.85	2.24	2.991	148	

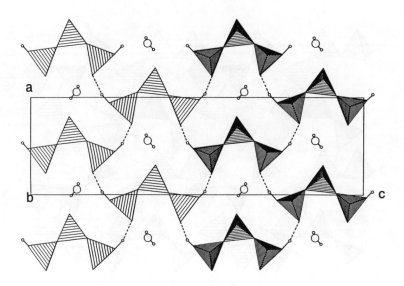

Figure 6.21. Projection, along the **b** axis (0 < x < 0.5), of a $[H_2P_3O_{10}]^{3-}$ layer in $K_3H_2P_3O_{10}.H_2O$ [3]. The water molecules included in this layer are shown.

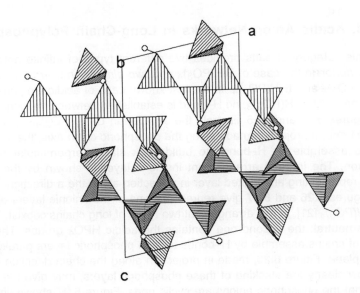

Figure 6.22. Projection, along the **b** axis, showing the layered arrangement of the anionic components in $GaH_2P_3O_{10}$.

Figure 6.23. Projection along the [$\bar{1}01$] direction of an isolated anionic layer built by the $H_2P_3O_{10}$ groups in $GaH_2P_3O_{10}$.

6.2.4. Acidic Anion Networks in Long-Chain Polyphosphates

This category of salts provides two different types of infinite networks, a *broad ribbon* in the case of $BiH(PO_3)_4$ and *two-dimensional arrangements* for $CaH_2(PO_3)_4$ and $ErH(PO_3)_4$. In the last known case of acidic long-chain polyphosphate, $(UO_2)H(PO_3)_3$, no H–bond is established between the chains.

Figures 6.24 and 6.25 illustrate the situation in *$CaH_2(PO_3)_4$*. As shown by the first one, a projection made along the phosphoric chain axis, the long-chain anions assemble by H–bonds to build thick slabs perpendicular to the **a** direction. The internal arrangement inside a layer is shown by the second figure representing an isolated layer in projection along the **a** direction.

Figures 6.26 and 6.27 give representations of the anionic layers observed in *$ErH(PO_3)_4$* [21]. In this arrangement two types of long chains coexist. The first one is neutral, the second one contains the acidic HPO_4 groups. These two kinds of chains assemble by H–bonds to form phosphoric layers parallel to the (101) plane. Figure 6.26, made in projection along the chain direction in order to show clearly the stacking of these phosphoric layers, may give the impression that the phosphoric anions are cyclic ones. Figure 6.27 shows, in projection along the **a** direction, the internal arrangement inside an isolated layer.

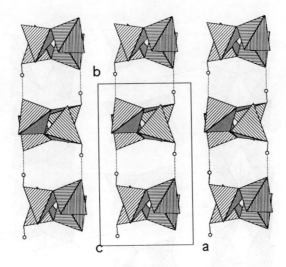

Figure 6.24. The anionic network in $CaH_2(PO_3)_4$ [20] in projection along the **c** axis. The small open circles are the hydrogen atoms and the H–bonds are represented by dotted and full lines.

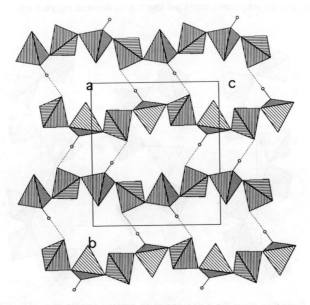

Figure 6.25. An isolated anionic network in $CaH_2(PO_3)_4$ [20] in projection along the **a** direction. The drawing conventions are identical to those of Figure 6.24.

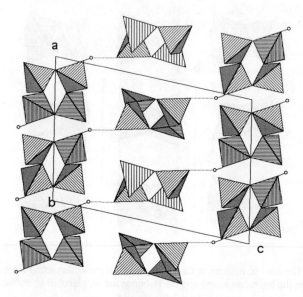

Figure 6.26. The arrangement of the anionic layers in ErH(PO$_3$)$_4$ [21] seen in projection along the **b** direction. The drawing conventions are identical to those of Figure 6.24.

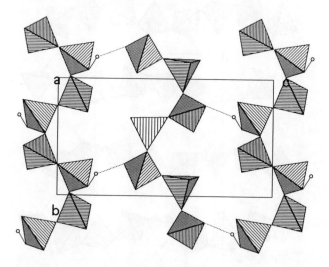

Figure 6.27. Projection along the **a** direction of an isolated anionic layer of ErH(PO$_3$)$_4$ [21]. The drawing conventions are identical to those of Figure 6.24.

In the case of the bismuth derivative, $BiH(PO_3)_4$ [22], the infinite chains assemble by pairs building infinite ribbons parallel to the **c** axis. Figure 6.28, a projection made along the axis of these ribbons, shows their arrangement inside the structure. Figure 6.29 is the perspective view of an isolated ribbon.

Table 6.10 gathers the main geometrical features of the hydrogen bonds in the three atomic arrangements discussed in this section.

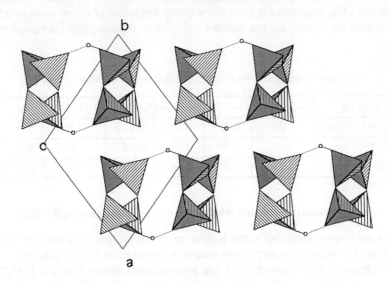

Figure 6.28. Repartition of the anionic ribbons in $BiH(PO_3)_4$ [22] shown by a projection along their axis. The drawing conventions are identical to those of Figure 6.24.

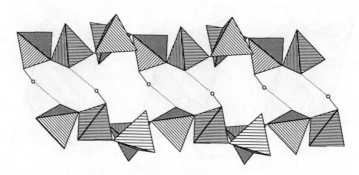

Figure 6.29. Perspective view of an isolated anionic ribbon in $BiH(PO_3)_4$ [22]. The drawing conventions are identical to those of Figure 6.24.

One can notice that in the three well investigated types of acidic long-chain polyphosphates the associated cation polyhedra are isolated. Palkina *et al.* [56] tried to explain this behavior which is rather unusual in condensed phosphate arrangements. According to these authors, the linkage of the anionic chains by H–bonds reduces their flexibility and hinders the possibility of forming aggregates of associated cation polyhedra, whereas in simple polyphosphates where the flexibility of chains is not violated by the presence of H–bonds the association of the MO_n polyhedra is possible with the formation of either finite or infinite aggregates depending on the cation sizes and on their oxygen coordinations.

Table 6.10
Hydrogen-bond geometries in the three acidic long-chain anion networks.

Formula	P–O···H (°)	O–H (Å)	H···O (Å)	O–O (Å)	O–H···O (°)	Ref.
$CaH_2(PO_3)_4$	124	0.83	1.76	2.587	176	[20]
$ErH(PO_3)_4$	89	0.96	1.82	2.650	143	[21]
$BiH(PO_3)_4$	117	0.97	1.95	2.84	151	[22]

6.2.5. Acidic Anion Networks in Cyclophosphates

As we have mentioned at the beginning of this chapter, the only convincing example of a network of acidic ring anions is provided by $Na_2HP_3O_9$ [24].

In Figure 6.30 a projection of this arrangement shows how the $[HP_3O_9]^{2-}$ groups assemble to build a ribbon spreading along the **c** direction. In this uni-

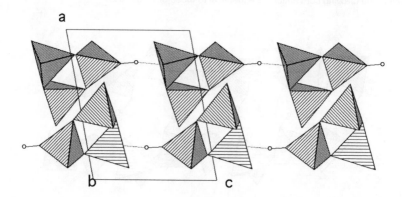

Figure 6.30. The two infinite ribbons of $[HP_3O_9]^{2-}$ groups crossing the cell of $Na_2HP_3O_9$ [24].

que example of ring-anion network, the main geometrical features of the H–bond are:

$P–O–H = 128°, \quad O–H = 0.86$ Å, $\quad H\cdots O = 1.59$ Å, $\quad O–O = 2.441$ Å, $\quad O–H\cdots O = 173°$

References

1. I. D. Brown, *Acta Crystallogr.*, **A32**, (1976), 24–31.

2. T. D. Farr, J. D. Fleming, and J. D. Hatfield, *J. Chem. Eng. Data*, **12**, (1967), 141–142.

3. V. Lyutsko and G. Johansson, *Acta Chem. Scand.*, **A38**, (1988), 663–669.

4. M. T. Averbuch-Pouchot and J. C. Guitel, *Acta Crystallogr.*, **B32**, (1976), 1670–1673.

5. H. Worzala and K. H. Jost, *Z. anorg. allg. Chem.*, **445**, (1978), 36–46.

6. V. A. Lyutsko, A. S. Lyakhov, G. K. Tuchkovskii, and K. K. Palkina, *Russ. J. Inorg. Chem.*, **36**, (1981), 662–664.

7. A. S. Lyakhov, K. K. Palkina, V. A. Lyutsko, S. I. Maksimova, and N. T. Chibiskova, *Izv. Akad. Nauk SSSR, Neorg. Mater.*, **26**, (1990), 1064–1068.

8. A. S. Yakhov, V. A. Lyutsko, G. K. Tuchkovskii, and K. K. Palkina, *Russ. J. Inorg. Chem.*, **36**, (1991), 803–805.

9. M. T. Averbuch-Pouchot and J. C. Guitel, *Acta Crystallogr.*, **B33**, (1977), 1613–1615.

10. K. K. Palkina, S. I. Maksimova, and V. G. Kuznetsov, *Izv. Akad. Nauk SSSR, Neorg. Mater.*, **15**, (1979), 2168–2170.

11. M. T. Averbuch-Pouchot, A. Durif, and J. C. Guitel, *Acta Crystallogr.*, **B33**, (1977), 1436–1438.

12. V. V. Krasnikov, Z. A. Konstant, and V. S. Fundamenskii, *Izv. Akad. Nauk SSSR, Neorg. Mater.*, **19**, (1983), 1373–1378.

13. B. Klinkert and M. Jansen, *Z. anorg. allg. Chem.*, **567**, (1988), 77–86.

14. A. I. Teterevkov and G. K. Mikhailovskaya, *Russ. J. Inorg. Chem.*, **25**, (1980), 781–782.

15. M. A. Vaivada, Z. A. Konstant, and V. V. Krasnikov, *Izv. Akad. Nauk SSSR, Neorg. Mater.*, **21**, (1985), 1555–1559.

16. I. Grunze and N. N. Chudinova, *Izv. Akad. Nauk SSSR, Neorg. Mater.*, **24**, (1988), 988–993.

17. M. T. Averbuch-Pouchot and M. Bagieu-Beucher, *Z. anorg. allg. Chem.* **552**, (1987), 171–180.

18. K. R. Waerstad and G. H. McClellan, *J. Appl. Crystallogr.*, **7**, (1974), 404–405.

19. T. D. Farr, J. W. Williard, and J. D. Hatfield, *J. Chem. Eng. Data.*, **17**, (1972), 313–317.

20. M. T. Averbuch-Pouchot, *Z. anorg. allg. Chem.*, **621**, (1995), 506–509.

21. K. K. Palkina, N. N. Chudinova, G. M. Balagina, S. I. Maksimova, and N. T. Chibiskova, *Izv. Akad. Nauk SSSR, Neorg. Mater.*, **18**, (1982), 1561–1566.

22. K. K. Palkina and K. H. Jost, *Acta Crystallogr.*, **B31**, (1975), 2285–2290.

23. S. A. Linde, Yu. E. Gorbunova, A. V. Lavrov, and V. G. Kuznetsov, *Dokl. Akad. Nauk SSSR*, **230**, (1976), 1376–1379.

24. V. A. Sarin, S. A. Linde, L. E. Fykin, V. Ya. Dudarev, and Yu. E. Gorbunova, *Russ. J. Inorg. Chem.*, **28**, (1983), 866–868.

25. M. T. Averbuch-Pouchot, J. L. Prisset, and A. Durif, *Eur. J. Solid State Inorg. Chem.*, **30**, (1993), 71–76.

26. N. El Horr, M. Bagieu, J. C. Guitel, and I. Tordjman, *Z. Kristallogr.*, **169**, (1984), 73–82.

27. E. J. Griffith, *J. Am. Chem. Soc.*, **78**, (1956), 3867–3870.

28. M. T. Averbuch-Pouchot and A. Durif, *Acta Crystallogr.*, **C39**, (1983), 809–810.

29. E. J. Griffith, *J. Am. Chem. Soc.*, **76**, (1954), 5862.

30. J. W. Gryder, G. Donnay, and H. M. Ondik, *Acta Crystallogr.*, **11**, (1958), 38–40.

31. O. H. Jarchow, *Acta Crystallogr.*, **17**, (1964), 1253–1262.

32. M. T. Averbuch-Pouchot and A. Durif, *Acta Crystallogr.*, **C41**, (1983), 809–810.

33. F. Abbona, R. Boistelle, and R. Haser, *Acta Crystallogr.*, **B35**, (1979), 2514–2518.

34. A. Riou, Y. Cudennec, and Y. Gerault, *Acta Crystallogr.*, **C43**, (1987), 194–197.

35.   I. Torjdman, A. Boudjada, J. C. Guitel, and R. Masse, *Acta Crystallogr.*, **B34**, (1978), 3723–3725.

36.   A. A. Khan, J. P. Roux, and W. J. James, *Acta Crystallogr.*, **B28**, (1972), 2065–2069.

37.   M. T. Averbuch-Pouchot, A. Durif, and J. C. Guitel, *Acta Crystallogr.*, **C44**, (1988), 99–102.

38.   R. A. Hearn and C. E. Bugg, *Acta Crystallogr.*, **B28**, (1972), 3662–3667.

39.   M. Bagieu-Beucher, *Acta Crystallogr.*, **C46**, (1990), 238–240.

40.   R. H. Blessing, *Acta Crystallogr.*, **B42**, (1986), 613–621.

41.   E. Philippot, P. Richard, R. Roudault and M. Maurin, *Rev. Chim. Minér.*, **9**, (1972), 825–835..

42.   A. Norbert and D. André, *C. R. Acad. Sci.*, Sér. C, **270**, (1970), 1718–1720.

43.   Y. Oddon, J. R. Vignalou, A. Tranquard, and G. Pépe, *Acta Crystallogr.*, **B34**, (1978), 3510–3514.

44.   E. Phillipot and M. Maurin, *C. R. Acad. Sci.*, Sér. C, **274**, (1972), 518–520.

45.   V. A. Efremov, E. N. Gudinitsa, I. Matsichek, and A. A. Fakeev, *Russ. J. Inorg. Chem.*, **28**, (1983), 973–976.

46.   V. A. Efremov, V. K. Trunov, I. Matsichek, E. N. Gudinitsa, and A. A. Fakeev, *Russ. J. Inorg. Chem.*, **26**, (1981), 1721–1723.

47.   M. Bagieu-Beucher, A. Durif, and J. C. Guitel, *Acta Crystallogr.*, **C45**, (1989), 421–423.

48.   A. Durif and M. T. Averbuch-Pouchot, *Acta Crystallogr.*, **B38**, (1982), 2883–2885.

49.   M. T. Averbuch-Pouchot and A. Durif, *Eur. J. Solid State Inorg. Chem.*, **29**, (1992), 191–198.

50.   A. Larbot, J. Durand, A. Norbert, and L. Cot, *Acta Crystallogr.*, **C39**, (1983), 6–8.

51.   M. T. Averbuch-Pouchot and A. Durif, *C. R. Acad. Sci.*, Sér. 2, **316**, (1993), 187–192.

52.   M. T. Averbuch-Pouchot and A. Durif, *C. R. Acad. Sci.*, Sér. 2, **316**, (1993), 469–476.

53.   M. T. Averbuch-Pouchot and A. Durif, *C. R. Acad. Sci.*, Sér. 2, **316**, (1993), 41–46.

54. M. T. Averbuch-Pouchot and A. Durif, *Eur. J. Solid State Inorg. Chem.*, **29**, (1992), 993–999.

55. M. T. Averbuch-Pouchot and A. Durif, *Eur. J. Solid State Inorg. Chem.*, **30**, (1993), 1153–1162.

56. K. K. Palkina, S. I. Maksimova, and N. T. Chibiskova, *Russ. J. Inorg. Chem.*, **38**, (1993), 750–779.

CHAPTER 7

On the Nomenclature and Classification of Phosphates

7.1. INTRODUCTION

The nomenclature used in one of the various families of phosphates — that of condensed phosphates — remained for a long time very confusing. This very complex part of phosphate chemistry developed very slowly over more than one and half century. During our survey of phosphate chemistry (Chapter 5) we have mentioned some of the reasons for this slow development.

Today, the classification and the nomenclature used for condensed phosphates is satisfactory if one excludes the domain of ultraphosphates. The clarification in this particular area is the fruit of the huge development of X-ray structural analysis during the last thirty years. Nevertheless, in spite of that, the various appellations used today for condensed phosphates in crystallographic or chemical literature are in many cases as confusing as in the last century. The word "metaphosphate" is still used indifferently for both cyclo- and polyphosphates. The word "polymetaphosphate" is also used sometimes. Some authors who are more prudent, simply write "phosphate of...", but in a good number of cases their manner of writing the chemical formula of the title compound adds a new enigma to the article. For instance, one can often read that "the four-member ring anion found in copper polyphosphate, $Cu(PO_3)_2$ is..." for discussing copper cyclotetraphosphate, $Cu_2P_4O_{12}$. Fortunately, in most cases, a good quality structural determination is included and the reader finally finds his way.

We mentioned that the clarification in the area of condensed phosphates was the fruit of X-ray structural analysis but this same technique coupled with a renewed interest for phosphate compounds lead to the discovery of a number of new kinds of phosphates which received more or less empirical and fanciful denominations through the years and are often as confusing as those found in the former condensed-phosphate literature.

In this chapter, we will review some areas of confusion and ambiguity in the domain of phosphate nomenclature and suggest a few ideas to clarify it, but fully aware that these suggestions are probably only temporary improvements in view of the tremendous development of phosphate chemistry.

A general survey of the present state of the crystal chemistry of phosphates suggests clearly that the phosphates can be classified into four distinct families as we have done in Chapter 5:

- the monophosphates, substituted or not
- the condensed phosphates, substituted or not
- the adducts
- the heteropolyphosphates

In the present nomenclature or classification, the main ambiguities arise from the following kinds of salts:

- those in which some oxygen atoms of the basic phosphoric anion, condensed or not, are substituted by other atoms
- the ultraphosphates
- those classified as phosphate adducts
- those presently described as mixed-anion phosphates
- the heteropolyphosphates

7.2. THE CASE OF SUBSTITUTED ANIONS

At the beginning of this book, when trying to find a definition to the word "phosphate" we have discussed the substitution of some oxygen atoms by various other atoms as F, S, H in a phosphoric anion and said that the corresponding salts are generally called substituted phosphates. In this category of compounds, frequent confusing appellations appear in chemical or crystallographic literature. We report some examples involving well-known compounds:

i) $Mg_3(PO_4)_2.MgF_2$, also commonly written as $MgPO_4F$, is clearly an adduct between magnesium monophosphate and magnesium fluoride as demonstrated by its atomic arrangement and suggested by its chemical formula. Nevertheless, this compound is very often said to be a *fluorophosphate* suggesting that it could contain a fluor-substituted anion. For this compound, the proper name is *magnesium monophosphate fluoride*.

ii) K_2PO_3F, well-known to be built by a packing of $[PO_3F]^{2-}$ substituted tetrahedral anion and potassium atoms, is also said, but this time almost correctly, to be a fluorophosphate. In fact, it is a *potassium monofluoromonophosphate*.

iii) sometimes fluorinated compounds as $LiKPO_3F.H_2O$, β-Na_2PO_3F, and $NaK_3(PO_3F)_2$ are simply said to be *"Composés Oxyfluorés du P^V"* (oxyfluorinated compounds of pentavalent phosphorus) without any informative term about the geometrical nature of the basic phosphoric anion and its degree of substitution. They are also *monofluoromonophosphates*.

iv) in the original structural paper, $Fe_2K_2(P_2O_7)F_2$ is simply said to be a *pyrophosphate*, but then its crystal structure description shows clearly that it is an adduct and should be denominated an *iron-potassium diphosphate fluoride*.

In our opinion, the proper appellation and classification for phosphates including substituted anions must obey these three rules:

i) they must keep a name reflecting clearly the geometry of the basic non-substituted anion

ii) an indication of the degree of substitution must be attached to this name

iii) they must be classified in the category of compounds corresponding to that of the basic non-substituted derivatives.

Following these simple rules, phosphates including PO_3F groups should be called *monofluoromonophosphates* and those containing PO_2F_2 groups *difluoromonophosphates*.

When applied to sulfur-substituted compounds, these rules lead naturally to *monothiomonophosphates* for phosphates including PO_3S groups.

Such a coherent system of nomenclature is already used for sulfur-substituted cyclic phosphoric anions, *trithiocyclotriphosphates* for salts including $[P_3O_6S_3]^{3-}$ ring anions and *tetrathiocyclotetraphosphates* for salts including $[P_4O_8S_4]^{4-}$ ring anions.

The case of hydrogen-substituted anions as $[PO_3H]^{2-}$, corresponding to the well-known class of *phosphites*, is peculiar. If the above suggested nomenclature is applied, it leads to the term *monohydrogenomonophosphates* and so there is a confusion with the classical appellation of monoacidic non-substituted monophosphates. Therefore in this case, we suggest that we keep the well established term of *phosphites*. One must, nevertheless, insist on the use of a proper writing for the corresponding acid. In order to show clearly the nature of the substitution in the anionic entity, one must write $H_2(PO_3H)$ and not $P(OH)_3$. Moreover, phosphites must be classified into the family of monophosphates. This last example shows the difficulty of suggesting a unified and coherent nomenclature.

7.3. THE CASE OF ULTRAPHOSPHATES

We mentioned that no coherent nomenclature exists for these salts and that the empirical denominations based on the number of phosphorus atoms in the formula unit introduced some years ago in chemical and crystallographic literatures, have induced rather important confusions. A group of compounds with a general formula LnP_5O_{14}, thoroughly investigated by crystallographers and physicists for their interesting optical properties, are most of time known as *pentaphosphates*, a term used for oligophosphates containing a linear

$[P_5O_{16}]^{6-}$ anion.

The confusion is sometimes increased by the English translations. For instance, a recent Russian paper describing some new ultraphosphates appeared after its translation under the title "Binary super-phosphates of...".

When explaining the classification of condensed phosphates (Chapter 4), we have mentioned that for the presently well characterized ultraphosphates the general formula of the anion may be written:

$$[P_{(n+2)}O_{(3n+5)}]^{n-} \quad [1]$$

The formula [1] as given above is not very easy to memorize. By using various variable changes, one can transform it to:

$$[P_nO_{3n-1}]^{(n-2)-} \quad [2]$$

or

$$[P_{n+1}O_{3n+2}]^{(n-1)-} \quad [3]$$

The formulæ [2] and [3] are not any easier to memorize or to handle and, in addition, have the disadvantage of introducing for n = 1 in [2] and n = 0 in [3] a foreign term, PO_2. So, we suggest that we write [1] in the form:

$$[(PO_3)_nP_2O_5]^{n-} \quad [4]$$

and to use for the ultraphosphates corresponding to the successive anions the following appellations:

n = 1 $[P_3O_8]^-$ mono-ultraphosphates
n = 2 $[P_4O_{11}]^{2-}$ di-ultraphosphates
n = 3 $[P_5O_{14}]^{3-}$ tri-ultraphosphates

and so on...

7.4. THE CASE OF ADDUCTS

In our survey of phosphate chemistry (Chapter 5), we met a great variety of compounds that we have classified as adducts. In this category, we have suggested that we include the former class of oxyphosphates. To justify this last point, a parenthesis is necessary in explaining what is the present accepted definition for oxyphosphates.

For oxyphosphates, for a time erroneously named "basic phosphates", two definitions are usually found in the chemical literature. The first one describes them as compounds with some oxygen atoms not belonging to the anionic entity, the second one as salts with a global formula corresponding to a ratio O/P > 4 in the anhydrous state. The first definition is acceptable, but the second one, based on the fact that in all previous examples the anionic entity was an isolated PO_4 tetrahedron, is not strictly acceptable. There is no reason for the nonexistence of oxyphosphates containing condensed phosphoric anions.

Thus, one can imagine an hypothetical oxyphosphate containing a diphosphate anion: $M_2^{II}P_2O_7.M^{II}O$. In this case, the O/P ratio is 4. One can also imagine the possible existence of a cyclotetraphosphate anion in an oxyphosphate, for instance $M_2^{II}P_4O_{12}.2M^{II}O$, then the O/P ratio decreases to 3.5.

Another consequence of this discussion is to demonstrate once more, that no reasonable classification is possible if it is not based on the knowledge of the atomic arrangements.

Let us come back to the point by reporting some examples which justify an attempt of clarification. The part of Chapter 5 devoted to the survey of phosphate adducts shows which kinds of salts are to be included in this category and which terminology is suggested for them. So, we simply report some errors currently encountered in the chemical literature and emphasize the terminology that we suggest we use.

For instance, $Fe_2K_2(P_2O_7)F_2$ is simply called a *pyrophosphate* in the original paper. It is evidently impossible to know if such an addition compound is to be formulated as $Fe_2P_2O_7.2KF$ or $FeK_2P_2O_7.FeF_2$. One can even write it as the authors did, $Fe_2K_2(P_2O_7)F_2$, but add that one matters with a potassium–iron *diphosphate fluoride* as it is clearly demonstrated by the crystal structure.

Another example is given by $Cd_3(NH_4)Na_2(P_2O_7)_2Cl$ for which the name attributed by the authors, *Na, Cd ammonium diorthophosphate*, is not the most appropriate to avoid confusion. This compound is unambiguously an cadmium–ammonium–sodium *diphosphate chloride.*

As a consequence of the definition that we have given in Chapter 5, it is evident that phosphates containing two or more different phosphoric anions are also to be classified among adducts and denominated in a manner illustrating clearly the geometrical nature of the various anions. For instance, a salt including both PO_4 and P_2O_7 groups should be called a *monophosphate diphosphate*.

7.5. THE CASE OF HETEROPOLYPHOSPHATES

A strict definition of an heteropolyanion is relatively simple to express by saying that :

i) it is a condensed anion

ii) X and Y being two different elements, this anion includes X–O–Y and possibly X–O–X and Y–O–Y bonds

It is rather evident that in the case of *finite heteropolyanions* as in those that we have met when describing "phosphochromates" (p. 357–359) or for the well-known phosphomolybdates or phosphotungstates no major problem appears. Nevertheless a simple but very academic question remains: should we consider phosphochromates, for instance, as belonging to phosphate chemistry or chromate chemistry?

In this domain, as in all the other categories of phosphates, the terms used by the authors for the appellation of new compounds are often confusing. Let us take the example of a recent and accurate investigation that we have discussed in Chapter 5. A ribbon-like heteropolyanion, $[B_2P_3O_{13}]^{5-}$, exists in $Na_5B_2P_3O_{13}$, but this last derivative is said to be a pentasodium catena-(diboratotriphosphate). The word *catena* reflects well the chain configuration of this heteropolyanion, *diborato* depicts the existence of B_2O_7 groups inside the arrangement, but the word *triphosphate* is confusing since it corresponds to the second term of the oligophosphate family. According to us, such a compound must simply be denominated as a *boratophosphate*.

Some problems occur when heteropolyanions constitute *two- or three-dimensional networks*. We have discussed in Chapter 5 (p. 171–172) the ambiguity in the case of $Be_2NH_4P_3O_{10}$. Let us take another similar example to illustrate this type of ambiguities. The atomic arrangement in $Zn_3Rb_2(P_2O_7)_2$ is mainly characterized by a three-dimensional network built by ZnO_4 and P_2O_7 groups which creates large voids accommodating the rubidium atoms. Does such a compound include a $[Zn_3P_4O_{14}]^{2-}$ heteropolyanion or is it simply a zinc–rubidium diphosphate?

These two examples illustrate the difficulties, already evoked in Chapter 5, to establish clearly the bondaries of such anions.

* * *

In no way have we extrapolated these ideas to design possible but not yet existing compounds. It is certainly a pleasant game for rainy days to play with the infinite possibilities of tetrahedron-linkage geometry, but it remains a game. For instance, it is rather evident that branched poly- or cyclophosphates will one day be characterized but it is the future.

We understand perfectly well that such ideas may become the objects of controversies, so let us say clearly that we do not pretend to bring the most appropriate solution or to edict rules for such a problem, but to open a field of discussions among the scientists involved in the more and more complex problems of phosphate chemistry.

Index

T

1 Month